Investigations into Biology

Investigations into BIOLOGY

Second Edition

ROBERT W. KORN and ELLEN J. KORN

JOHN WILEY & SONS, INC. New York / London / Sydney / Toronto

Library of Congress Catalogue Card Number: 72-126228

ISBN 0-471-50380-0

Printed in the United States of America

10 9 8 7 6 5 4 3 2

Preface

Unlike the first edition of *Investigations into Biology,* which was designed for laboratories operating in conjunction with any one of several biology texts on the market, the second edition has been patterned to complement a specific text, *Contemporary Perspectives in Biology.* The patterning of the laboratory manual after our textbook required a reorganization of material and several new experiments.

Most of the conceptual material of the text is complemented by laboratory exercises. As always, however, it is impossible (and, in some cases, undesirable) to represent every chapter in the text by laboratory work. In other cases, it has been necessary to divide some text chapters into two laboratory exercises. For example, there are no laboratory exercises corresponding to the text chapters on "The Theory of Evolution," "Biosystematics," and "The Origin of Man," but there are two laboratory exercises for the first chapter of the text, which introduces the concepts of measurement and data analysis and levels of biological organization, and the text chapter on "The Genetic Control of the Cell," which includes the concepts of Mendelian Inheritance and Central Dogma.

In keeping with our original intention for the manual, we have tried to retain and add exercises that are experimental in nature. In phylogeny discussions, descriptive approaches are necessary but, wherever possible, these have been treated from a life-cycle point of view. To comply with suggestions that the first edition contained too much demonstration, we have incorporated more student work. One of the effects is that many laboratory exercises cannot be completed in a single laboratory period. Although this may be a disadvantage in the sense that a complete unit is not accomplished in a single week, it is a distinct advantage in that the student derives a more honest view of biological research, which is certainly not a two- or three-hour-a-week process. Individual schools have the option to: (1) prepare all materials for student laboratories and use each laboratory period as a unit of study, or (2) have the students prepare some or all materials for their own laboratories. For example, in Chapter 10 "The Theory of Particulate Heredity," the students will be etherizing and counting *Drosophila melanogaster.* During this week, the flies of the P_1 cross for Chapter 14, "Sources of Variation," must be prepared. This may be done either by the students or the instructor. Two weeks later, the students may etherize and collect virgin females in addition to their work on the Growth experiment. There is sufficient flexibility to run the laboratories with varying degrees of instructor or student preparation.

Another feature incorporated into this edition, is an alternate means of using the manual. In Appendix II there is a list of exercises proposed for a one-semester, experimentally oriented laboratory. The second semester can then be devoted to group-research projects. Our experience in overseeing this type of program convinces us that student-group projects are most profitable when they are either done on a single aspect of an organism, or when a single approach is used to assay several organisms, and when the central theme of the class is performed in areas of research in which the instructor is both interested and competent. And, since we have no way of knowing the interests or competency of individual instructors, we have not included any suggestions for research problems in the Appendix. At the end of a research problem, instructors generally request a research report. Therefore, we have included the outline to be followed in the preparation of such a report in Appendix II.

In the five years that have elapsed since the first edition, many schools have obtained more sophisticated equipment for use in their freshman laboratories. Hence we are not as concerned in this edition with designing exercises that can be run with simple equipment, and have included work on the Spectronic "20," clinical and Serval centrifuges, and a simple Warburg apparatus. Another change has been in the approach to some principles. Model building is an important aspect of biological research, and should be represented as a valid means of discovering phenomena. These are paper exercises, but their objectives, and the understanding of phenomena that can be gained from them, more than justify their inclusion.

We express our appreciation to William Bryden, Biology Editor of Wiley, for his aid and encouragement during the preparation of both the text and this manual, and to our children for putting up with us during the long, arduous, but always rewarding months required to prepare these books.

Robert W. Korn
Ellen J. Korn

Bellarmine College
Louisville, Ky.

To the Student

You have been our major concern in writing both the text and this laboratory manual. You have been the object of every question we have asked ourselves during the preparation of these books: Will *you* understand the material? Will *you* be stimulated intellectually by the approach? Will *you* find our presentation of the field of biology a rewarding experience? We hope that the answer to all of these questions is, "Yes."

However, you cannot relinquish the responsibility of working. We have written a scientific text and laboratory manual, not a workbook for demonstrations that allows you to sit back and passively be entertained. Reading scientific material is not like reading a novel. Scientific writing, its very nature, is intense, and no thought may be passed over lightly. Many students who underline the important passages in their humanities textbooks find themselves underlining entire pages in their scientific texts. Although we tried to make the explanations of biological concepts and experimental techniques as clear as possible, your comprehension of the material depends on the effort you are willing to expend. The following suggestions will help you to gain the greatest understanding and intellectual and academic satisfaction from your course in biology.

1. Read each exercise before the laboratory period. The introductions to the chapters are not to be construed as substitutes for text or lecture material, but serve to orient you for the laboratory work.
2. Understand each exercise before you enter the laboratory. This may involve reading the introduction and instructions several times. Notice that in several laboratories there are experiments in which long incubation times for materials are necessary. By making yourself thoroughly familiar with the entire program for the day, you can set up these materials for incubation so that they will be ready when you reach that part of the exercise.
3. Correlate the lecture and text material with your laboratory work in your mind. Be aware of the fact that the lecture and laboratory work complement each other and that both are aspects of the same body of information.

4. After performing the required experiments and observations, answer the questions at the end of each part. Sufficient information will be available either in the introduction or from your own observations to answer some of these questions. Others may require ingenuity or further reading on your part. These questions are intended to increase your working knowledge of the subject, and you will find them helpful in fixing the information in your mind.
5. As you finish each section of material, review all of the exercises to integrate all of the information. Try to envision the organizational pattern that has been developed throughout the section, and relate the specific information to this pattern.

Many aspects of this approach are not limited to the study of biology. They may be useful in almost every subject. We hope that, with an attempt for a high level of understanding and with a basic natural curiosity, you find this course in biology both stimulating and rewarding.

Robert W. Korn
Ellen Korn

Contents

Part I

CHEMICAL ENERGY AND BIOLOGICAL SYSTEMS

1

Measurement, Data Organization, and Analysis

MATERIALS

CLASS STOCK

radish seeds
analytical balances (in mg)
prepared slides early starfish development
branches of evergreen shrub Juniper

STUDENT STATION

weighing paper
microscope
microscope lamp
lens paper
millimeter ruler
razor blade

INTRODUCTION

The natural sciences are distinguished from the social studies and humanities by the translation of experience into numerical or other equally quantitative and abstract expressions. This translation involves (1) the art of measurement, (2) the organization of measurements as data, and (3) the changing of this data into a meaningful form. Measurement, data organization, and analysis can bring phenomena into focus, sorting them from other factors; the ultimate test of a science is that the conclusions drawn from the measurements, data organization, and analysis suggest a further quantitative approach to subsequent problems.

Chemistry and physics invariably fulfill the requisites of a science. In biology, measurement, data organization, and analysis may involve only the first phases of a scientific inquiry, and other, more descriptive means can then be employed to comprehend the phenomena. However, modern biological experimentation, has tended more and more toward complete fulfillment of those requisite for a science.

Your laboratory course will serve as an introduction to data gathering and manipulation. You will be asked to measure, and analyze data in such a way that meaningful conclusions may be drawn. Since metric system units are used for measurement in the natural sciences, all weight measures will be made in gram units (g = gram, mg = milligram, etc.), linear measures in meter units (m = meter, cm = centimeter, mm = millimeter), area measurements in units of $meters^2$, and volume measures in units of $meters^3$ or liters (l = liter, ml = milliliter, etc.).

MAKING ESTIMATES

In laboratory work the prediction of estimates can often appreciably reduce the amount of subsequent labor. For example, an investigator might estimate from his personal experience and experiments of others that an event he is to study occurs once in 1000 individuals. He also knows that he can reach a more accurate conclusion if, in his study, he observes several individuals who have experienced the event; this means that he should use at least 10,000 organisms in his study. But he must also take into account that estimates are often correct to only one order of magnitude (± 10 times), so that at least 100,000 organisms must be used.

$$\text{Estimation of frequency of event} = 1 \text{ in } 1000 = \frac{1}{1000} = \frac{1}{10^3}$$

$$\text{Analysis of 100,000 organisms} = \frac{100{,}000}{1000} = \frac{10^5}{10^3} = 10^2 \text{ or } 100$$

(To divide two exponential values that have the same base, subtract the exponents.)

If the event does indeed occur at the rate of one in 1000, the use of 100,000 organisms would result in the detection of about 100 affected individuals. If the estimate was too high, and the event occurs at a frequency of one in 10,000, ten affected individuals would still be detected.

$$\frac{10^5}{10^4} = 10^1 \text{ or } 10$$

Estimates are not random guesses. They are and must be based on past experience and valid information.

1. Weigh 1/2 g (500 mg) of radish seeds on the balance. Your instructor will demonstrate the operation of the balance. <u>Be sure you understand how to operate it before you proceed</u>.

   Initially, did you over- or underestimate the number of radish seeds that would weigh 500 mg?

2. How was the number of radish seeds that you initially placed on the weighing paper an estimate on your part, and how was it more of a wild guess?

3. Count out ten seeds, and weigh them on the weighing paper. Enter the weight on Chart 1.1.

CHART 1.1

Total Sample	Weight of Ten Seeds	Weight of One Seed	No. Seeds in Total Sample	Range and Possible No. of Seeds
500 mg				

4. Calculate the weight of one seed. For example:
10 seeds weight approximately 8 mg
$$\frac{10 \text{ seeds}}{8 \text{ mg}} = 1.25 \text{ mg per seed}$$
Enter the weight of a single seed on Chart 1.1.

5. Calculate the number of seeds in the 500 mg sample by multiplying the weight of one seed by 500. Enter your result on Chart 1.1.

6. Each balance has a limit of accuracy. Most are accurate only for measurements above 1 mg, and some are only accurate for measurements above 10 mg. Assuming a 50 percent error in the calculations of the weight of 10 seeds, what is the range and number of seeds in the total sample? For example: A 50 percent error in the data given would be 8 mg ± 50 percent or a range from 4 mg to 12 mg. Enter your results on Chart 1.1.

7. You plan to use radish seeds in studies on germination. Each experiment will have ten duplicate trials with 100 seeds per trial. Fifty experiments are planned. How many grams of seeds should you buy so that you are sure you will have enough?

8. Using the following information, estimate the number of gas molecules of air in the laboratory room within one magnitude of order.
   (a) 1 cm^3 = 1 ml.
   (b) 1000 cm^3 = 1000 ml = 1 l.
   (c) There are approximately 6.023×10^{23} molecules of gaseous substances in 22.4 l of volume.
   (d) Room height (in cm) X room length (in cm) X room width (in cm) = room volume (cm^3 or liters).

SURFACE AREA AND VOLUME

In accordance with principles of geometry cells, spores, and organisms have evolved to assume forms and shapes that balance the surface area/volume ratio with a multitude of other factors involved in organismal activities. Some insight can be achieved into the coupling of form and function by an examination of various surface area/volume relationships.

1. Draw the following objects.
   (a) A cube each side of which is 4.7 cm.
   (b) A rectangular solid with the following dimensions: length (l) = 20 cm; height (h) = 3 cm; depth (d) = 1.7 cm.
   (c) A sphere with a radius of 2.92 cm.

2. Find the volume of each of the three dimensional figures. (The volume of a sphere $= 4/3\ \pi r^3$; use the value 3.14 for π). Enter your data on Chart 1.2.

CHART 1.2

Shape	Volume	Surface Area	Surface Area/Volume Ratio
Cube			
Rectangular solid			
Sphere			

3. Find the surface area of the figures. (Surface area of a sphere $= 2\ \pi r^2$.) Enter your answers on Chart 1.2.

4. Arrange your values according to a surface area/volume ratio, and give this ratio as a fraction with the volume equal to one (1.0). For example: surface area = 10,000 cm^2, volume = 2000 cm^3. Surface area/volume ratio = 10,000/2000 or, setting 2000 equal to 1.0, 5/1. Enter the surface area/volume ratio of the figures on Chart 1.2.

5. Which of the figures has the highest surface area/volume ratio. Which has the smallest?

6. The surface area is indicative of the degree to which the contents of containers shaped like these figures would be exposed to the environment. Which shape would provide the least exposure to outside influences, and hence would afford the greatest protection for its contents.

7. Is your answer borne out by the fact that some cells, spores, seeds and many aquatic organisms are approximately spherical?

8. Chemical reactions take place at the surfaces of biological systems. What shape would you expect for a cell that functions primarily in contributing products of chemical reactions to other parts of the organism?

9. Can you think of other ways that would increase the surface area without an appreciable increase in volume other than overall shape modifications?

Many examples of variation in surface area/volume relationship will be studied in the exercises dealing with biological material. For the present, a simple example will suffice. However, for this example a microscope is necessary.

CARE AND PREPARATION OF THE MICROSCOPE

The microscope you have been assigned is a precision optical instrument, and should be handled with proper use and care. Using Fig. 1.1, acquaint yourself with its parts.

Carrying the Microscope

A microscope must always be carried with two hands, one holding the arm of the scope, and the other supporting the base. If you must also carry a microscope lamp to your seat, make another trip. This may seem wasteful in time and energy, but laboratories have successfully been based on the assumption that student energy is more expendable than expensive equipment. As you can see, the base is attached by a joint, which, if loose, could throw the body of the scope backward, and the ocular, which is loosely inserted at the top of the body tube, could fall out and be damaged. For this reason, you must always carry a microscope vertically with both hands.

Cleaning the Microscope

A microscope is only as good as the quality and condition of the lenses. Cleaning the lenses with a cloth or tissue leaves lint and tends to scratch the lenses. The only acceptable way to clean lenses is with a piece of silicon paper (lens paper), which is both lint-free and smooth enough to avoid scratching. The ocular, objective lenses, and the mirror should be cleaned prior to and after each laboratory period with lens paper only.

Preparing the Microscope for Use

As you will note in Fig. 1.1, the microscope has two lens systems for magnification. Either of the two objectives may be lined up with the ocular at a given time by turning the revolving nosepiece. The ocular usually has a magnification power of 10X, which means that it magnifies an image to ten times its size. The objectives on the revolving nosepiece are marked according to their magnification, usually 10X and 43X. When the low-power objective is lined up with the body tube, the total magnification is the product of the magnification power of the ocular and objective lenses, or 100X. The total

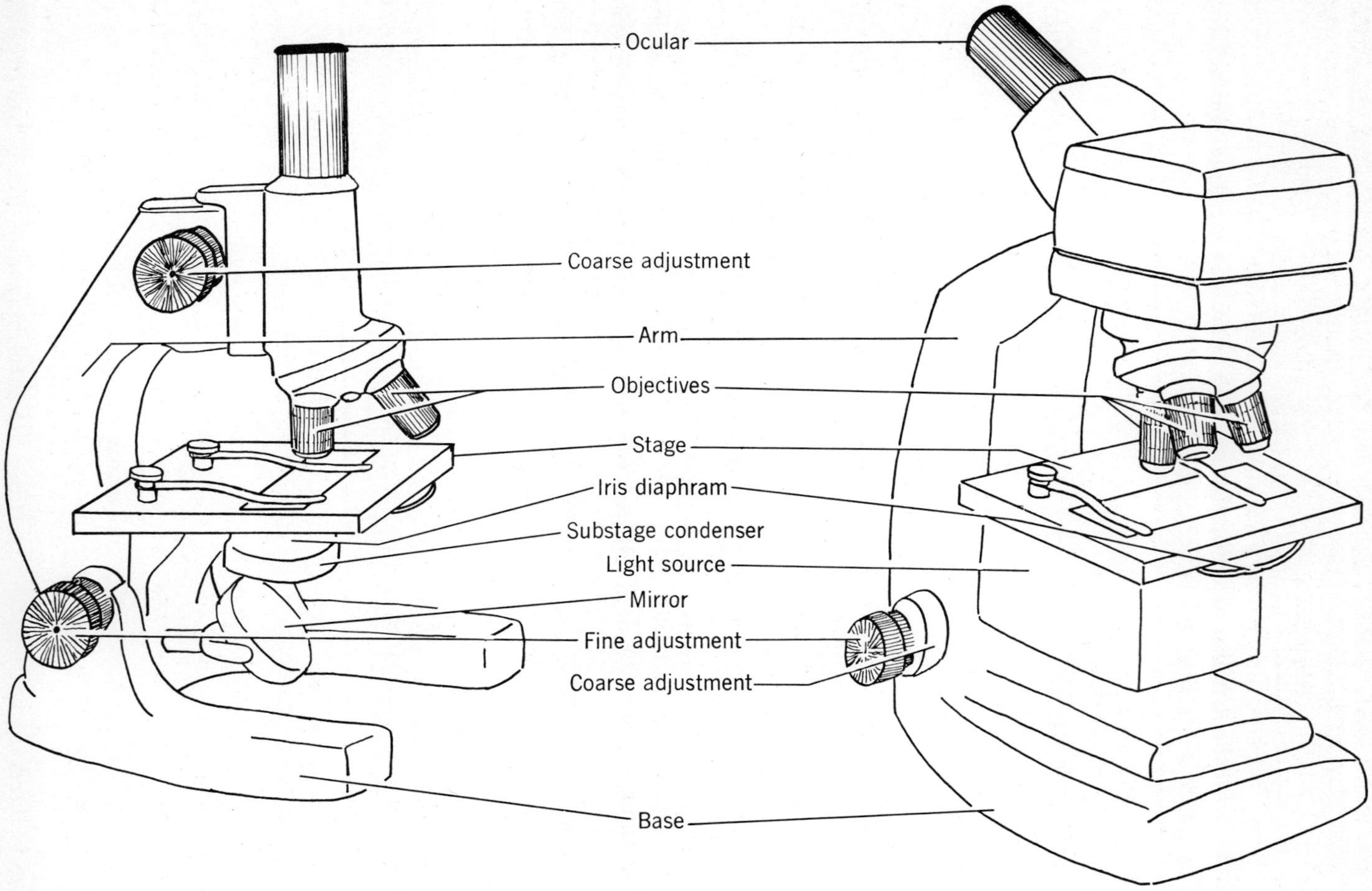

1. Always keep one hand on adjustment
2. Keep other hand on slide

Figure 1.1

magnification when the high-power objective (43X) is lined up with the body tube is 430X. What is the total magnification in each of the following lens systems?

Ocular	Objective	Magnification
10X	97X	
6X	10X	
2.5X	17.3X	

1. When you receive the microscope, the low-power objective should be in line with the ocular. By using the coarse adjustment, raise the body tube to the point where the low-power objective is *1 in.* from the stage. Turn the revolving nosepiece until the high-power objective is in line with the body tube. You can tell that they are completely lined up when the nosepiece clicks into place. This may require a slight pressure.
2. Now turn the nosepiece so that the low-power objective is firmly in position. Using the coarse adjustment, lower the body tube as far as it will go. Notice that the objective will not move closer than about 1/2 in. from the stage. The level at which the low power objective is in focus is always above this point. But this is not true of the high-power objective. The lens of the high-power objective could contact the condenser or a slide, causing serious damage to the lens. For this reason you should never lower the body tube with the high-power objective in position.

   With the low-power objective in line and as close to the stage as it will go, prepare the microscope for actual use.
3. Place the microscope lamp directly in front of the substage mirror. The mirror should have its flat side facing the lamp. Adjust the mirror so that the light is reflected into the condenser directly below the body tube.
4. Open the iris diaphragm of the condenser to admit the maximum amount of light.
5. Look through the ocular, keeping the other eye open to prevent eyestrain and fatigue.
6. Adjust the mirror again to make sure you have the maximum light coming through to your eye.
7. Adjust the iris diaphragm to reduce the glare. You are now ready to use the microscope.

Measurements Using the Microscope

1. Obtain a slide of early starfish development.
2. With the low-power objective still in place, raise the body tube by using the coarse adjustment to *1 in.* above the stage. If you attempt to put the slide in place with the low-power objective lowered, you can scratch the lens. Always raise the body tube before inserting or removing a slide.
3. Place the slide on the stage, securing it by the clips. Make sure the cover slip is directly above the condenser.
4. Looking directly at the slide from the side, lower the body tube as far as it will go. You may have made an error and be at high power instead of low, in which case the objective could hit the slide, and cause serious damage to the lens. Never lower the body tube while looking through the ocular.
5. Look through the ocular and slowly move the body tube upward by turning the coarse adjustment. Since you are starting from the lowest point, the plane brought into focus first will be the dust on the condenser (if there is one), then the dust on the bottom of the slide, and finally the specimen itself. If you see objects quickly flashing and out of focus, you have passed the plane of the specimen and must begin again. Lower the body tube, looking at it from the side. Move it up again more slowly.
6. Once you are satisfied that you are looking at the specimen, move the slide slowly with one hand, keeping the other hand on the fine adjustment to maintain focus. Notice that when you move the slide to your right, the objects move to the left. This is because the image is inverted by the lens.
7. There should be several starfish eggs on a slide. Select one that is intact, and center it under the microscope.
8. Take a millimeter ruler (one is provided on the inside back cover of the manual). Supporting the ruler with books, place it at the level of the virtual image, that is, from where the image appears to originate. This level is most easily found by looking through the microscope with both eyes open and moving your finger up and down the scope until it looks as if it is on the same plane as the specimen. The virtual image is usually found a little below the stage.
9. Keeping both eyes open, look at the specimen with one eye and the ruler with the other. Measure the diameter of the starfish egg directly.

CHART 1.3

Specimen	Diameter	Radius	Surface Area	Volume	Surface Area/Volume
Egg					
Four-celled stage					

10. You are examining the egg under low magnification (usually 100X), and determination of the actual diameter should take this magnification into consideration. Divide by the magnification in order to obtain the diameter and enter it on Chart 1.3.

11. The diameter = $2r$. Enter the value for the radius of the egg cell on Chart 1.3.

12. Calculate the surface area, volume, and surface area/volume ratio for the egg. Enter this data on Chart 1.3.

13. Find a four-celled stage of starfish development on the slide. In the course of starfish development, a cleavage occurs that divides the egg into two approximately equal blastomeres. The next cleavage divides the two-celled stage into four cells. The four blastomeres are approximately equal in size.

14. Measure the surface area, volume, and surface area/volume ratio of one of the blastomeres. Enter it on Chart 1.3.

15. How does the surface area/volume ratio of the blastomere compare with that of the egg?

16. What conclusions can you draw from these measurements? Could these conclusions be checked by purely descriptive means?

LINEAR MEASUREMENTS AND STATISTICAL ANALYSIS

Linear measurement, as well as surface area and volume measurements, can reveal certain phenomena. We shall exemplify this with a statistical analysis of the data. When a pattern is recognized through measurement and analysis of data, it may be nothing more than variations generated by pure chance. It is important to recognize the extent to which chance is operating, and the extent to which a fixed pattern is generated by biological processes.

1. You have been provided with a segment of a branch from an evergreen shrub. Carefully remove 20 needles (with a razor blade, don't pull them off) and measure each with a millimeter ruler. Enter the data on the first two columns of Chart 1.4.
2. Translate the charted data to graph form on the graph provided.

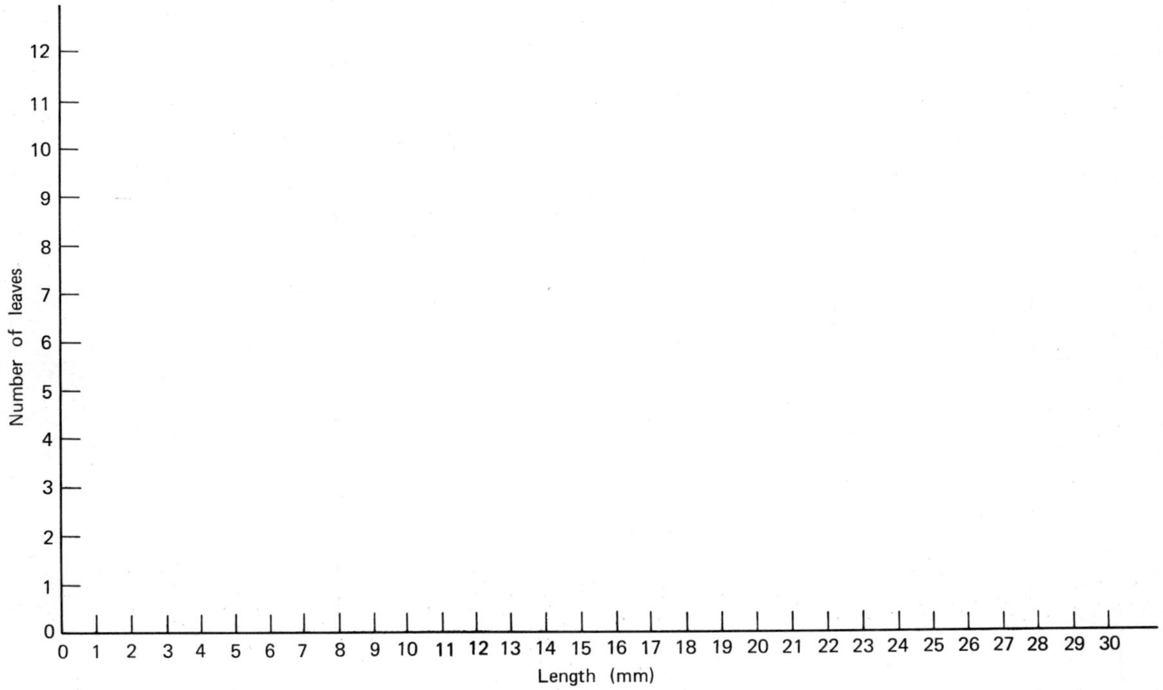

Graph 1.1

3. The values generate a bell-shaped curve on the graph. The fact that so many of the needles measured are similar in value would tend to indicate that there is a system within the pine tree that controls the length of the needle. However, the graphic pattern could be generated through pure chance. To test whether chance may be producing this pattern, the data must be submitted to statistical analysis.
4. There are many different types of statistical analyses. The one used here is for the calculation of <u>variance</u> (V) and <u>standard deviation</u> (SD). The first step in obtaining the variance and standard deviation is to obtain the <u>arithmetic mean</u> or average. This

CHART 1.4

Length (o)	Number (n)	Mean or Expected (e)	$o - e$	$(o - e)^2$	$n(o - e)^2$
	$\Sigma n =$				$\Sigma n(o - e)^2 =$

$$\frac{\Sigma\, n(o - e)^2}{\Sigma n - 1} = V$$

$$V = (SD)^2$$

$$\sqrt{V} = SD$$

is done by multiplying each measured value by the number of times it was found and dividing this total by the number of samples measured. For example, given the following data

length (o) x	No. times found (n)	=	($n\ o$)
4.2 cm	8		33.6
4.3 cm	5		21.5
4.4 cm	3		13.2
Total	16	Total	68.3

$$\frac{68.3}{16} = 4.26 = \text{arithmetic mean } (\overline{m})$$

Figure arithmetic mean of your data and enter it on Chart 1.4.

5. Then find the difference between the values obtained (o) and the arithmetic mean ($\overline{m}$), ($o - \overline{m}$). The arithmetic mean is standardly interpreted as the value expected (e) if there is a factor controlling needle length, so that $\overline{m} = e$, and the above difference may be given as $o - e$. Following Chart 1.4, enter the different lengths found and subtract the expected value (arithmetic mean) obtained from the previous calculation.

6. Next, square the $o - e$ value; For example, $(o - e)^2$ for the data given above would be:

o	n	e	$o - e$	$(o - e)^2$
4.2	8	4.26	−.06	.0036
4.3	5	4.26	.04	.0016
4.4	3	4.26	.14	.0190

7. For the next step, include the number of times each value was found by calculating $n(o - e)^2$. For example:

o	n	e	$o - e$	$(o - e)^2$	$n(o - e)^2$
4.2	8	4.26	−.06	.0036	.0282
4.3	5	4.26	.04	.0016	.0080
4.4	3	4.26	.14	.0190	.0570

Evaluate and enter on Chart 1.4.

8. Obtain the sum of all $n(o - e)^2$ values $\Sigma n(o - e)^2$. In the above example, .0282 + .0080 + .0570 = .0932 = $\Sigma\, n(o - e)^2$.

9. Divide the $\Sigma\, n(o - e)^2$ value by the number of times measurements were taken minus one ($\Sigma n - 1$). For example, in the above data, $\Sigma n = 8 + 5 + 3 = 16$; $\Sigma n - 1 = 16 - 1 = 15$. The operation in this step is then

$$\frac{\Sigma\, n(o - e)^2}{\Sigma n - 1} \quad \text{or} \quad \frac{.0932}{15}$$

And the variance of the above data is .0062. Find the variance for your data.

10. The standard deviation is the square root of the variance ($\sqrt{V}$) or in the above data $\sqrt{.0062}$ = approximately 0.0785 mm. Notice that the variance has no label, although the standard deviation is in measurement units.
11. The proper expression for the standard deviation is $\overline{m} \pm$ S D, or in the above data, 4.26 ± 0.0785 mm.
12. Express your data as a mean ± standard deviation.

13. If the mean is greater than the variance ($\overline{m} > V$), then the factors controlling the trait are the same for all individuals. Conversely, if the mean is less than the variance ($\overline{m} < V$) then different factors control the trait in different individuals. How does your obtained mean compare with your calculated variance?

14. What can you conclude about the factors controlling the trait of needle length?

15. How could this information be used to devise further means of study?

2

Levels of Biological Organization

MATERIALS

CLASS STOCK

- cultures
  - *Closterium*
  - *Nostoc*
  - *Spirogyra*
- dropper
- prepared slides
  - cross section of leaf
  - pancreatic tissue
- small plant with root, stem and leaf systems intact
- model of human systems

STUDENT STATION

- clean slides
- cover slips
- microscope
- microscope lamp
- lens paper

INTRODUCTION

Biological organization is characterized by various levels of complexity. The cumulative physical properties of molecules and atoms provide the chemical characteristics of the material. Cumulative chemical and physical (spacial) properties lend character to structural components, organelles and macromolecular complexes of a cell. Many cells containing the same type and approximate multiplicities of organelles and macromolecular complexes comprise a population if the cells themselves are organisms, or a tissue if the cells are part of a multicellular organism. Different tissues contribute to the functionality of an organ. And several organs with related or sequential functions, comprise a system. A group of organisms with the same structural and physiological features of their systems, comprise a population.

A listing of the biological hierarchy could be extended to include the catagories of classification. For example, the populations of organisms would be groups as species. Since the classification of organisms is a foundation for a later period in this course, only the aformentioned levels of biological organization will be discussed.

You should also be aware of the means by which the various levels are detected. The physical and chemical levels of biological organization are in the submicroscopic range, and by definition cannot be viewed directly and must be understood through experimental data. These levels will be the topic of the next four chapters. The organelle and macromolecular complex levels are in the ultramicroscopic range, and their internal structure can only be discerned by use of the electron microscope. Cell and tissue levels are in the microscopic range and can be studied from magnifications possible with the ordinary compound microscope. Organs and systems are in the macroscopic range and can be studied through direct observation. The means of observing and studying populations depends on the type of organisms in question. Virus populations are ultramicroscopic; bacterial, protozoan, and most algal populations are microscopic; and some fungi and most multicellular plants and animals are macroscopic.

GENERAL STUDY OF CELLS

Each of the levels of biological organization that have been mentioned can be subdivided further. Subdivision of the cellular level includes two basic types of cells: the procaryotic and the eucaryotic. The procaryotic cell type is characterized by a complex cytoplasm in which different regions are poorly defined. Bacteria, Rickettsia, and blue-gree algae are all procaryotic cell types. The eucaryotic level is characterized by a localization of different regions of the cell into organelles and macromolecular complexes. Red, green, brown, and golden algae, Protozoa, fungi, and all animals and plants are comprised of eucaryotic cells.

1. Prepare a wet mount of *Closterium* in the following manner:
   (a) Clean and dry a slide.
   (b) Take up some of the culture with a dropper and place two drops of the cell suspension in the center of the slide.
   (c) Place the edge of the cover slip at the edge of the drop of closterium.
   (d) Slowly lower the cover slip, letting it drop only when there is no more room for your fingers. Do *not* drop it from an upright position. Doing so will cause air bubbles to form under the cover slip (see Fig. 2.1).
2. Place the slide on the stage of the microscope and focus using the low-power objective.
3. If you wish to view the cells under high power, adjust the slide so that a cell is in the center of the field. This is important because the area of the field is reduced with higher magnification and many objects that were in view under low power will be out

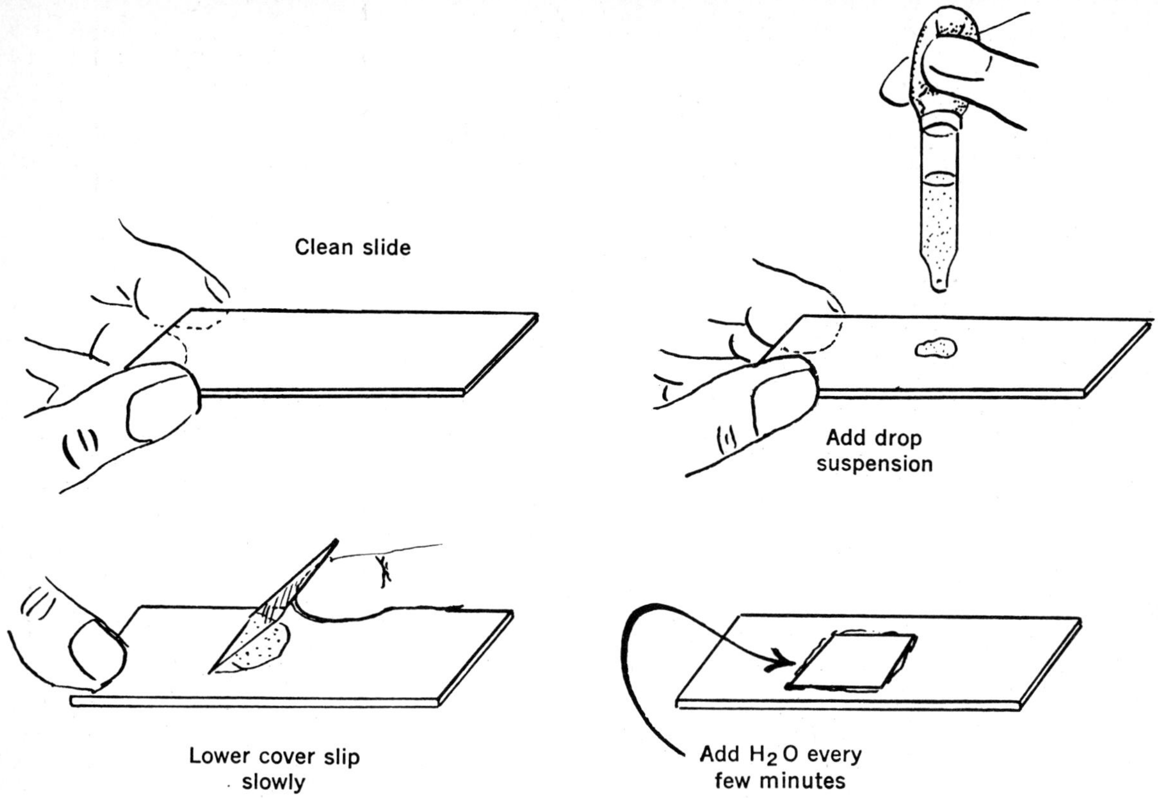

Figure 2.1

of the field under high power. Without changing the focus (do not touch either the fine or coarse adjustment), turn the revolving nosepiece until the high-power objective clicks into place. When the image is in focus under low power you will automatically be nearly in focus under high power. Although it may seem that the lens is too close to the slide, do not make any adjustments. When looking through the microscope, use the fine adjustment to bring the image into full clarity. Never use the coarse adjustment when the high-power objective is in place. You may have to increase the amount of light admitted through the condenser. Do this by adjusting the iris diaphragm.

4. Examine a single cell of *Closterium.* The green color of the cell is due to the presence of chlorophyll. Is the chlorophyll in *Closterium* diffuse or localized in organelles known as chloroplasts?

5. The nucleus, another organelle, is characteristic of eucaryotic cells. It is generally centrally located, is less dense than the surrounding cytoplasm, and is circular. Can you discern a nucleus in *Closterium?*

6. Is *Closterium* a procaryotic or eucaryotic cell type?

7. Draw the cell of *Closterium* in the space provided, following the rules for biological drawings:
   (a) A 3H or 4H pencil should be used. These are hard leads and do not smear.
   (b) Using the side of the pencil to shade areas is *not* acceptable. Shading may be done by stippling, crosshatching, or lining. Acceptable techniques of drawings are given in Fig. 2.2 to serve as guidelines.

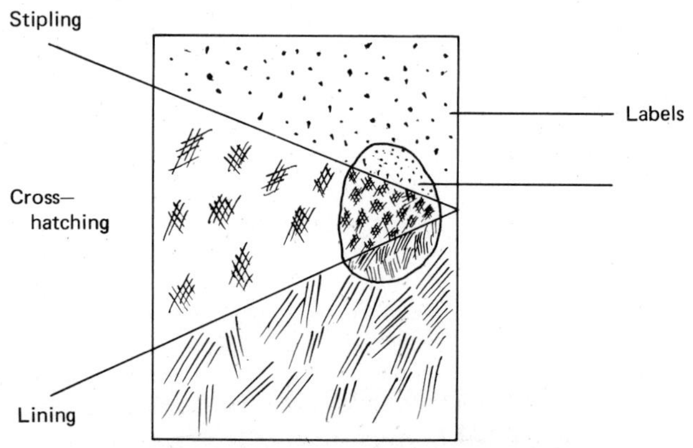

Figure 2.2 Biological drawing.

   (c) Study the cell to be drawn until you are sure of its general shape. Observe a number of cells and choose one that appears to be typical.
   (d) Draw the shape lightly on the page, making sure that the size of your drawing fits comfortably on the paper. There should be a 1½-inch margin around the drawing, with a 3-in margin on the right to provide room for labeling.

Closterium

(e) When you are satisfied that the shape is correct, darken the lines to complete your outline.

(f) Turn your attention to the interior of the cell. By focusing up and down with the fine adjustment, notice the conformations of organelles. When you are certain of the forms, relative sizes, and shapes of the structures within the cell, draw them lightly within your outline.

(g) When you are satisfied with the position, size, and shape, darken or shade according to the depth of color.

8. When your drawing is complete, wash and dry your slide and coverslip.

9. Prepare a wet mount of *Nostoc,* stirring the culture before taking the two drops to put on your slide.

10. Examine a single cell of *Nostoc. Nostoc* grows as filaments. Notice that the central region of the cell containing the metachromatic granules is lighter than the peripheral portion. Are the boundaries of this lighter central region as distinct as the nucleus of *Closterium?*

11. Is the chlorophyll in *Nostoc* diffuse or localized in chloroplasts?

12. Are there discrete organelles of any type in the *Nostoc* cell?

13. Is *Nostoc* a procaryotic or eucaryotic organism?

14. *Nostoc* is a blue-green alga; *Closterium* is a green alga. Do these facts fit your conclusions regarding the types of each cell?

15. Which has the greater surface area/volume ratio; *Closterium* or a *Nostoc* cell? Explain.

16. Draw a cell of *Nostoc* in the space provided.

Nostoc

Variations on a Cellular Theme

There are hundreds of different types of eucaryotic cells and the type difference lies in the form of the cell and multiplicities of the various organelles. Chloroplast structure and cell shape (the cell membrane is a macromolecular complex) can be used to exemplify some of the variations possible.

1. Prepare a wet mount of *Spirogyra.* The *Spirogyra* culture need not be mixed. Using a forceps take a few filaments. Too many filaments on a slide will make an individual filament difficult to see.
2. Examine the wet mount under low and high powers, and concentrate on a single cell.
3. How does the chloroplast in *Spirogyra* differ from those in *Closterium?*

4. A cell in a filament has two sides that are next to other cells and that, therefore, are not exposed directly to the environment. How does the elongated shape of the cell compensate for the lack of exposure on two sides?

5. How does the surface area/volume ratio of *Spirogyra* cell compare with *Closterium?*

6. What modifications of the chloroplastic structure distinguish the *Spirogyra* cell from a *Closterium* cell?

7. The cell membrane is a macromolecular complex. The chloroplast is an organelle consisting of several macromolecular complexes. How can different multiplicities of macromolecular complexes and organelles and different orientations of these factors to one another result in very different cells?

8. Draw several adjacent *Spirogyra* cells in the space provided.

Spyrogyra

FROM CELLS TO TISSUES TO ORGANS TO SYSTEMS

Just as the form and multiplicities of macromolecular complexes and organelles determine the morphology and physiology of a cell, the form and multiplicities of cells determine the morphology and physiology of tissues, and the form and multiplicities of tissues determine the features of organs, and organs for systems. Although systems and organs exist at the macroscopic level, their functions are traceable to the character of the cells of which they are composed.

Cells Through Systems in a Plant

1. Obtain a prepared slide of a cross section of a leaf and examine it under low and high power.
2. There are four layers of tissue in the typical leaf—a layer of epidermis, under which is a layer of palisade tissue composed of long columnar cells, under which is a layer of spongy tissue composed of irregularly shaped cells, and finally, another layer of epidermis. In the lower epidermis you may see openings bordered by pairs of cells. These cells are the guard cells that surround the stomatal openings through which a gas exchange occurs between the environment and the leaf tissues. Somewhere in the cross section you will see a cluster of large, red staining cells. These compact cells collectively deliniate the vein of the leaf.
3. What features of cell shape, cell organelles, and position of cells characterize the cells of the epidermis? The palisade tissue? The spongy tissue?

4. Do the epidermal cells have many chloroplasts? What would this condition indicate about their participation in the photosynthetic contribution to the plant? What would you deduce their function to be? How is the tissue distributed in the leaf to afford this function?

5. Which tissues appear to be primarily responsible for the photosynthetic function of the plant? On what do you base your conclusion?

6. How can you justify the statement that "a leaf is a plant organ?" What would the leaf system of a plant be?

7. Draw a cross section of the leaf in the space provided, labeling the different tissues (cell types).

cross section of leaf

8. Draw the small plant with which you were provided and label the following: root system, stem system, and leaf system.

small plant

Cells Through Systems in an Animal

1. Obtain a prepared slide of pancreatic tissue, and view it under low and high power. Pancreatic tissue is found in the pancreas, an organ associated with the digestive system.
2. There are different types of cells within the pancreatic tissue. The large cells that surround an opening comprise the acinar tissue, which produces pancreatic juice with its digestive enzymes. The acinar cells surround duct openings. Many of these ducts join to form the pancreatic duct which enters the digestive system below the stomach. The patches of small, compact cells are the islets of Langerhans, which produce and secrete insulin, a hormone that aids in the regulations of the level of blood sugar. The presence of the acinar cells which empty their secretions into a duct, and the islets of Langerhans, which secrete insulin into the blood vessels coursing through the pancreas, cause this organ to be considered both an exocrine (secretions through ducts) and an endocrine (ductless secretion) gland.

   The small irregularly shaped openings that contain shrunken cells in the center are the blood vessels running through the pancreas.
3. How does the regularity of cell shape in the pancreatic tissue differ from that of the plant leaf?
4. Does the somewhat spherical shape of the cells of the acinar tissue in the pancreas strike you as serving their function? If so, why? If not, why not?
5. On the basis of examining the tissue, would you expect the pancreas as an organ to be smooth or spongy, soft or hard?
6. What tissues comprise the pancreas?

7. Draw and label a section of pancreatic tissue.

pancreatic tissue

8. Examine a sample or a model of a pancreas. Does the form and texture of the organ correspond to what you envisioned when viewing the pancreatic tissue?

9. How do the different tissues of the pancreas contribute to the organ as a whole?

10. Draw the pancreas, labeling blood vessels and ducts that you see entering the organ.

the pancreas model

11. Examine a model of the human, or if available, the demonstration model of a dissected cat.

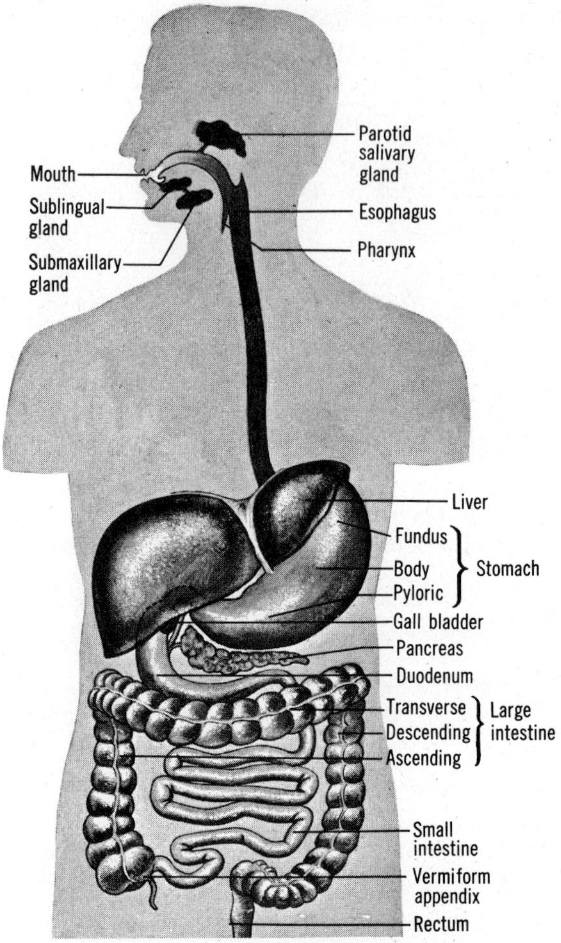

Figure 2.3 Model of the mammalian digestive system.

12. Trace and identify the organs that comprise the digestive system:
   (a) Mouth.
   (b) Salivary glands.
   (c) Esophagus—leading from mouth to stomach.
   (d) Stomach—J-shaped organ into which food is brought from the mouth.
   (e) Duodenum—small thickened region into which food passes from the stomach.
   (f) Pancreas—surrounding the duodenum and connected to it by a duct.
   (g) Liver—large, dark red, lobular organ overlapping the stomach and connected to the duodenum by a duct.

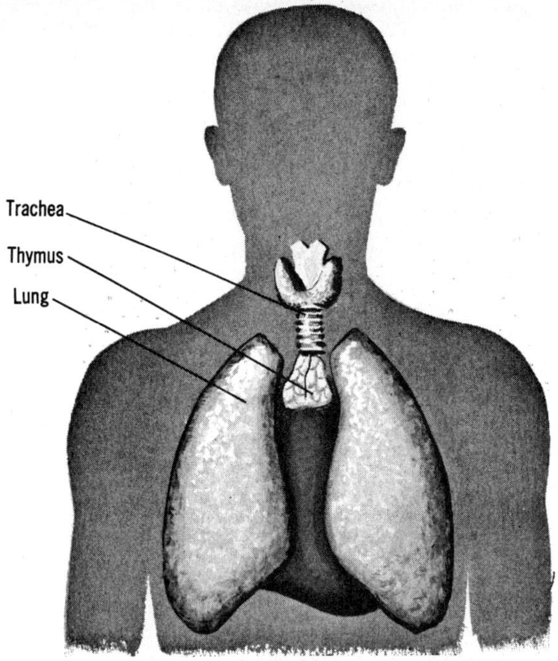

Figure 2.4 Model of Respiratory System.

(h) Small intestine–region (including the duodenum) into which food passes from the stomach.
(i) Large intestine–large tubular structure into which undigested food passes from the small intestine. The large intestine ascends, transverses, and descends within the abdominal region, bordering the small intestine, and terminates in the anus.
(j) Anus–opening through which fecal material (undigested food and certain metabolic products) pass.
(k) Rectum–muscle tissue surrounding the anus.
(l) Blood vessels–associated with all organs in the system.
(m) Nerve tracts–associated with all organs in the system.

14. Label the organs of the model provided in Fig. 2.3, and examine the models presented in Figs. 2.4 and 2.5.

General Considerations of a Cellular System

1. Drawing from your observations of systems, justify the statement that "the component organs of a system need not be directly connected to one another."

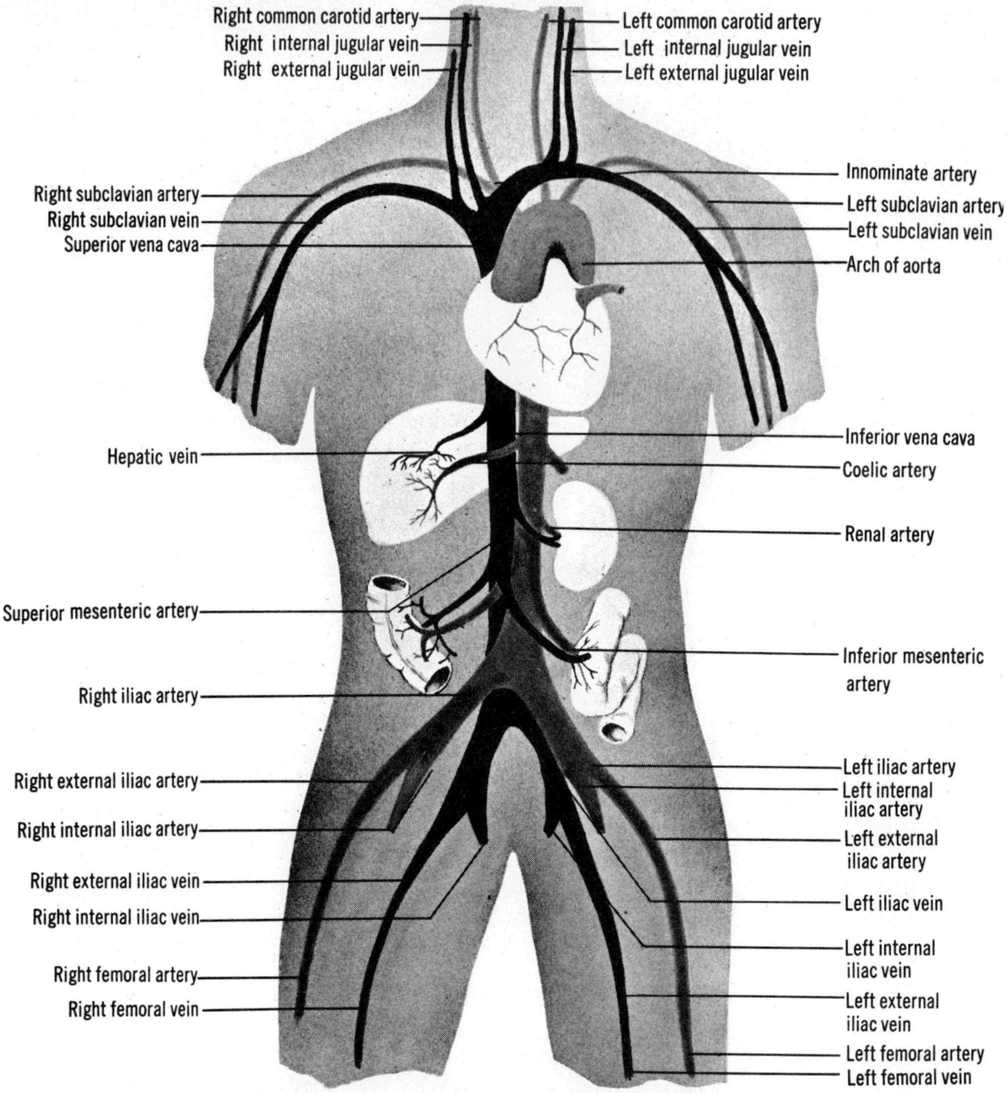

Figure 2.5 Circulatory system of man

The example chosen for the animal system happened to be one in which the organs were physically associated. Of the following animal systems, which one(s) has (have) organs that are not physically associated with one another? Justify your answer.

skeletal system
muscular system
circulatory system
nervous system
endocrine system
reproductive system

3. Are the systems of animals and plants isolated, self-sustaining entities, or do the systems overlap in contributing to common functions? Justify your conclusions on the basis of your observations of both plant and animal systems in this laboratory.

3

Chemical Background for Biology

MATERIALS

CLASS STOCK

chemicals
- KOH pellets
- NH_4Cl crystals
- 1% $CaHPO_4$ solution
- 0.1% K_2SO_4 solution
- dextrose powder
- soluble starch
- Benedict's solution
- IKI solution
- 1 N HCl
- 0.1 N KOH solution
- 0.1 N HCl solution
- H_2PO_4 + KH_2PO_4 buffer
- conc. NaOH solution
- 0.5% $CuSO_4$ solution
- yeast hydrolyzed (60°-10 min) solution
- Sudan IV dye powder
- Diché reagent

water baths
- 60°C; 70°C; boiling

red cabbage leaves

STUDENT STATION

2 100-ml beakers
millimeter ruler
wax pencil
0°-100°C thermometer
aluminum foil
4-16 test tubes
bunsen burner
500-ml beaker
pH paper of different ranges
Pasteur pipettes
spatula
forceps
glass stirring rod

CLASS STOCK (cont'd)

pH meter
onion slices
potato tubers
egg yolk solution
egg albumen solution
salad oil

INTRODUCTION

Biological material has mass and density, and therefore, is considered to be matter. It is composed of atoms that are physically associated into molecules, which possess specific chemical properties and act in accordance with chemical laws. The functions of the organelles, macromolecular complexes, cells, tissues, organs, and systems in accordance with chemical laws endows these biological entities with their unique biological properties of form and function. Hence, a meaningful perspective of form and function can only be achieved through a study of the chemical basis of biological material.

The key to biological structure is the conformation of molecules. The key to biological function is the chemical reaction.

The Chemical Reaction

Chemical reactions involve changes in the physical relationships between atoms, and therefore, follow the laws of physics as well as those of chemistry. Important physical directives in chemical reactions are the law regarding the conservation of matter and the first and second laws of thermodynamics. The conservation of matter law states that "matter can be neither created nor destroyed." For application to a chemical reaction this means that the same atomic particles present in the reactants (A,B) must be present in the products (B,C).

$A \longleftrightarrow B + C$ or

$A + B \longleftrightarrow C$ or

$A + B \longleftrightarrow B + C$

The reaction arrows point to both products and reactants in order to indicate that the reaction proceeds until a dynamic equilibrium is reached, that is, to a point where for every set of reactants converted to a product, there is a set of products converted to reactants.

The first law of thermodynamics is similar to the conservation of matter law, except that it applies to energy instead of matter. The first law simply states that "energy can be changed from one form to another, but never created or destroyed." The energy present in a chemical system at the beginning of a reaction equals the energy present in that chemical system when the reaction reaches equilibrium.

The direction in which the reaction will preferentially proceed is governed by the second law of thermodynamics, which simply states that "change tends to occur in the direction of a loss of useful energy (free energy), that is, energy that can be used for work." Thus, reactions tend to proceed from levels of high potential energy to levels of low potential energy, and in the course of the energy change free energy is lost.

The change in the energy relationships during a reaction can be mathematically represented:

$$\Delta G = \Delta H - T\Delta S$$

Where ΔG represents the change in the potential energy of the chemical system; $T\Delta S$ represents the loss of usable energy (entropy change); and ΔH represents the kinetic energy that is available for work released from the reaction. A verbal description of the equation would be:

Original potential energy in reactants minus final energy in product (ΔG) equals kinetic energy with work potential (ΔH). Minus kinetic energy with no work potential ($T\Delta S$).

When ΔG is negative, the system has moved from a state of relatively high potential energy to a state of low potential energy. Thus, reactions in which ΔG is negative occur spontaneously.

When ΔG is positive, the system has moved from a state of low potential energy to a state of higher potential energy. This can only happen when energy is added to the system. Therefore, reactions with positive ΔG values must have energy added to the system in order to proceed.

When ΔG is zero, the system is at equilibrium, and energy would have to be added to make it go in either direction.

Equipment, such as a bomb calorimeter, is used in chemical laboratories for measuring ΔG values. For student laboratories, general indications of positive or negative ΔG values are sufficient.

In the chemical world there is no such occurrence as a single isolated reaction, although for the sake of simplicity we have been discussing single reactions. Changes in the course of a chemical reaction always involve an alteration of two or more entities. For example, the dissolution of a KOH (potassium hydroxide) pellet in water involves at least two reactions. One is the change in the associations of potassium and hydroxyl ions in the crystal as dissolution occurs. The lattice of the KOH crystal contains energy, and as the ions dissolve, the energy is released. The second reaction involves the association of potassium and hydroxyl ions with water molecules, the hydrating process that requires energy. If the energy from the lattice of the crystal is sufficient to foster the hydrating process, the ΔG of the reaction is negative. The excess energy is released as heat, which is detected as a rise in temperature. If the energy is not sufficient, and energy from the system must be used, the ΔG value is positive. Since heat energy is drawn out of the system, the change is detected by a drop in temperature.

1. Using your millimeter ruler, measure 4 cm from the bottom of each of two 100-ml beakers, and mark the line with a wax pencil.
2. Fill to the line of each beaker with tap water.

3. Place a thermometer in one beaker, and record the temperature on Chart 3.1.
4. With the thermometer still in the beaker (holding it with one hand) place 10-15 pellets of KOH in the beaker. Use forceps or spatula for handling the KOH. Never touch it with your hands.
5. Does the temperature rise or fall after the addition of KOH pellets? Enter the data in Chart 3.1.
6. Once you note the direction of the temperature change, remove and wash the thermometer. If the thermometer remains in the beaker too long, the pressure from the column of mercury may break the glass.
7. What is the origin of the heat?

8. What can be said regarding the comparative energy inherent in the KOH pellet and the energy necessary for hydrating potassium and hydroxyl ions?

9. Using the notations $T\Delta S = a$, $\Delta H =$ either $>$ or $< a$, is the ΔG positive or negative for this reaction?

10. Place the washed thermometer in the other beaker of tap water and record the temperature on Chart 3.1. During the period in which you made your observations on the KOH solution, the temperature may have changed, so one reading will not suffice for both beakers of water.
11. Place several crystals of ammonium chloride (NH_4Cl) into the beaker.
12. Note the direction of the temperature change, and record it on Chart 3.1.
13. What can be said regarding the energy in the lattice of a crystal of NH_4Cl, and the energy required for hydration of the ammonium ions and chloride ions?

14. What can be said regarding the energy in the lattice of an NH_4Cl crystal compared to that of a KOH crystal (assuming the energy necessary for hydration of all ions is the same)?

15. Evidently the energy from the lattice of the crystal is insufficient for hydration. Where does the additional energy for this process come from?

16. Using the notations $T\Delta S = a$, $\Delta H = >$ or $< a$, is the ΔG of the NH_4Cl reaction with water positive or negative?

17. Reactions with a negative ΔG are usually exergonic and reactions with a positive ΔG are usually endergonic. Define exergonic and endergonic in terms of energy change, and label the two reactions studied as either exergonic or endergonic.

CHART 3.1

Reaction	Water Temperature	Temp Change + or −	ΔG + or −	Exergonic or Endergonic
$KOH + H_2O \longrightarrow K^+ + OH^- + H_2O$				
$NH_4Cl + H_2O \longrightarrow NH_4^+ + Cl^- + H_2O$				

Energy of Activation

Another factor that must be considered regarding chemical reactions is the stability of certain molecules at certain temperatures. This stability factor is evident in a consideration of the occurrence of the reaction itself, and in the product formed by the reaction. In accordance with the second law of thermodynamics, the product of a reaction has less potential energy than the reactants, but converting that potential energy to kinetic energy would require more energy than the system possesses. For example, the burning of wood (a chemical reaction) results in the evolution of carbon dioxide (CO_2), which still contains potential energy in bond form. But the CO_2 molecule is stable at the energy levels associated with burning wood and, therefore, the CO_2 is a product instead of carbon (C) and oxygen (O_2).

The occurrence of a chemical reaction depends on the instability of the reactants at the energy state associated with a given temperature. Some reactions, such as KOH + H_2O and NH_4Cl + H_2O, occur readily at the energy state associated with room temperature (25°C). For other reactants, the energy available in a room temperature solution is insufficient to create the instabilities necessary for the reaction, and energy, in the form of heat, must be added to the system. Once the instability is created, the reaction will proceed spontaneously. The energy that must be added to such a system is called the *energy of activation*. (see Fig. 3.1)

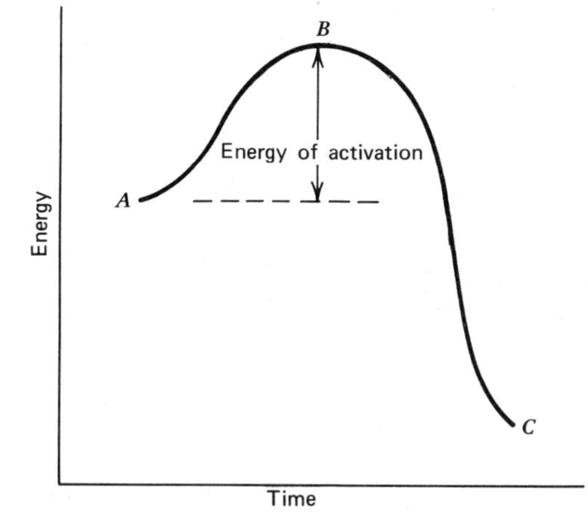

Figure 3.1 Energy of activation. Once the energy of activation has been added to a reaction, the reaction will proceed spontaneously.

1. Using your millimeter ruler, mark two test tubes at the 2 cm and 4 cm levels with a wax pencil.
2. Fill each of the test tubes to the first mark with 0.1% $CaHPO_4$ solution.
3. Fill each to the second mark with 0.1% K_2SO_4 solution.
4. Leave one test tube at your desk for observation, and place the other in a rack in a 60°C water bath for ten minutes.
5. It will take about five minutes for the temperature of the solution in the test tube to rise to 60°, so the solution will be at 60° for only five minutes. Remove to test tube from the water bath, place it next to the one kept at room temperature, and observe.
6. The reaction between $CaHPO_4$ and K_2SO_4 is as follows:

   $$CaHPO_4 + K_2SO_4 \longrightarrow CaSO_4 + K_2HPO_4$$

   in which $CaSO_4$ is a white precipitate. In which tube has the reaction taken place?

7. Is there sufficient energy at room temperature for the reaction to take place in the allotted time?

8. Chemical reactions depend on molecular collisions, which in turn may depend on either molecular speed or time. Would you expect to find a precipitate in the tube kept at room temperature if you left it for 10 days? Why?

9. Where did the energy come from to increase molecular collisions in the tube placed in the water bath?

10. How did the water bath provide the energy of activation for this reaction?

THE CHEMICAL BOND

There are two major types of chemical bonds – the ionic bond in which electrical forces keep ions in physical association, and covalent bonds, in which atoms share outer orbits. As a general rule, ionic bonds are not as rich in energy as covalent bonds and, hence, are not as strong as covalent bonds. However, there are many exceptions to this rule.

Ionic Bonds

The experiments you have done thus far have all involved ionic bonds, that is, an electrical association between ions.

Positively charged ions	*Negatively charged ions*
K^+	OH^-
NH_4^+	Cl^-
Ca^{++}	$SO_4^{=}$
K_2^{++}	$HPO_4^{=}$

1. At what temperature did the KOH reaction take place? The NH_4Cl reaction? The $CaHPO_4 + K_2SO_4$ reaction?

2. Temperature is a reflection of the heat energy in a system. The amount of heat energy necessary to bring about a chemical reaction is a direct reflection of bond energy. Using the scale on Chart 3.2, give the relative bond energies involved in the ionic associations of the molecules listed.

Covalent Bonds

Covalent bonds, in which atoms share orbits and, therefore, their electrons, have a great range of energy and hence, strength values. The carbon-carbon (C:C) bond is extremely rich in energy and is one of the strongest bonds in the chemical world. The ester linkages formed by hydrolytic reactions in which an oxygen is bonded by carbons on either side, are much weaker, and more easily broken.

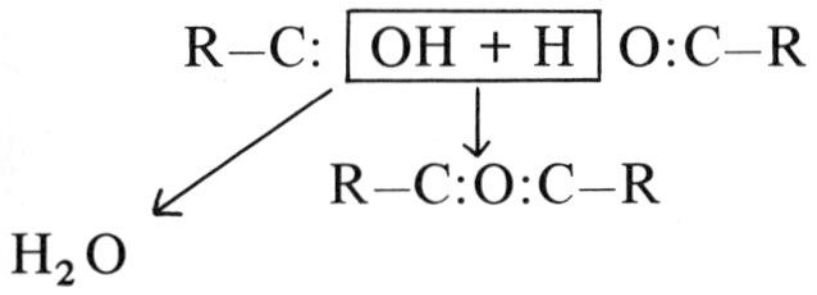

1. Place a spatula of dextrose powder (glucose) on a piece of aluminum foil, and spread it thinly over the surface of the foil.
2. Place the foil with the thin layer of dextrose in a 425°F oven (170°C) for 20 minutes, or on the surface of a hot plate for 5 min.
3. Next mark a test tube with a millimeter ruler 2 cm from the bottom and fill with tap water to the mark.
4. Place a spatula of dextrose powder in the test tube, and mix until dissolved.
5. Place the test tube in a 70° water bath for twenty minutes.
6. Dextrose (glucose) is an organic molecule with the following structure:

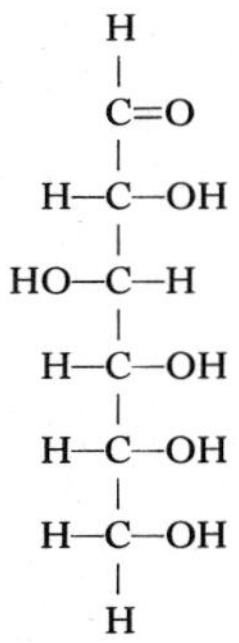

Breakage of the glucose bonds give carbon dioxide, water, and black carbon residue. What temperature is required for breakage of the carbon-carbon bonds of the glucose molecule? Record on Chart 3.2 using the bond energy according to the scale.

CHART 3.2

Molecule	Type of Bond	Temp Necessary to break bond[a]	Relative Bond Strength[b]
KOH	ionic		
NH_4Cl	ionic		
$CaHPO_4$	ionic		
Glucose	covalent (C:C)		
Starch	covalent (ester)		

[a]In terms of +, ++, +++, etc.
[b]Scale for chart, + = 25°C. For example, reaction occurs at 100°C = ++++.

7. Using a millimeter ruler, mark four test tubes at the 2 cm level and label the test tubes A, B, C, and D.
8. Fill each test tube to the mark with the soluble starch solution.
9. Keep test tubes A and C at your desk, and place tube B and D in the 70° water bath for 30 minutes.
10. Starch is a large molecule consisting of glucose units in ester linkages. The presence of glucose in either or both sets of test tubes would indicate a breakage of the ester linkages.
11. After removing the test tubes from the water bath, mark test tubes A and B at a level 4 cm from the bottom.
12. Add Benedict's solution up to this mark. Benedict's solution is specific for the presence of reducing sugars, that is, sugars that have a free aldehyde or ketone group. Glucose is a reducing sugar.
13. Heat test tubes A and B by either of the following methods:
    (a) Place test tubes in a boiling water bath for three minutes.
    (b) Carefully hold the test tube with a test tube holder and gently move it back and forth over a bunsen burner flame. Be sure the open side of the test tube is not in the direction of any other students, as the liquid may spurt. Bring to a boil. (Glass beads may be used to prevent sudden spurting.)
14. A color change towards a brick red indicates the presence of a reducing sugar, in this case, glucose. Yellow or green coloration would indicate a partial breakdown of the starch. What color reaction did the test tube with soluble starch kept in the water bath give? The one kept at room temperature? Why?

15. To show that the glucose came from a breakdown of the starch, place three drops of iodine solution into the other set of test tubes (B and D).
16. Iodine turns black in the presence of starch. How does the depth of color in test tube B compare with that in test tube D? How does this indicate the origin of the glucose?

ACIDS, BASES AND BUFFERS

Acidity and alkalinity play important roles in cellular environments, and therefore, are as pertinent to biology as to chemistry.

The acidity is determined chemically by the concentration of free hydrogen in the ionic state (H^+). Certain chemicals, when placed in water, have a tendency to dissociate into their ionic forms. For example hydrochloric acid (HCl) breaks down in solution to H^+ and Cl^-.

$$HCl \longrightarrow H^+ + Cl^-$$

The greater the concentration of H^+ the greater the acidity of the solution.

In the same way, sodium hydroxide (NaOH) dissociates or ionizes in solution:

$$NaOH \longrightarrow Na^+ + OH^-$$

The presence of OH^- or hydroxyl ions causes the solution to be alkaline, and the greater the concentration of the hydroxyl ions, the greater the alkalinity.

The point at which there are equal numbers of hydrogen and hydroxyl ions in the solution, is neutrality:

$$H^+ + OH^- \rightleftharpoons H_2O$$

Mathematically the acidity of a solution is measured in terms of pH units.

pH = negative logarithm of hydrogen ion concentration.

The lowest number of hydrogen ions that can exist in solution in a gram of water is 10^{-14}, or $1H^+/100{,}000{,}000{,}000{,}000$. The negative log of this number is 14. Therefore, pH or hydrogen ion concentration is expressed over the range of pH 0 to pH 14. The neutrality point is pH 7, when the hydroxyl ions are equal in number to the hydrogen ions.

pH 0 to pH 6.9 is the acidic range
pH 7.0 = neutral
pH 7.1 to pH 14 is the basic or alkaline range

The closer the pH is to zero, the stronger the acid. The closer the pH is to 14, the stronger the base.

1. Measure 275 ml of tap water with a graduated cylinder, and place the water in a 500-ml beaker.
2. Cut up three red cabbage leaves to a size that will fit into the beaker.
3. Boil for ten minutes on a ring stand over a bunsen burner. The red color coming into solution is the pigment anthocyanin.
4. After extraction is complete, pour 100 ml of the anthocyanin solution into a graduated cylinder.
5. Using a forceps, drop two KOH pellets into the cylinder.
6. Mark a test tube at the 5-cm level, using your millimeter ruler, and fill the test tube to this mark with 1 N HCl.
7. Pour the HCl from the test tube gently down the side of the cylinder so that it does not penetrate the anthocyanin solution too deeply.
8. With a glass stirring rod, gently swirl to mix. Do not shake up and down. Mixing is complete when a spectrum ranging from deep yellow to deep pink forms.
9. Using the cylinder in Fig. 3.1, draw the colors, indicating the acidic, basic, and neutral range.
10. With a Pasteur pipette and bulb, take a small amount of the solution from the top. Using a strip of pH paper that registers in the acid range, take a small drop from the Pasteur pipette and determine the pH. What is the pH of the solution on or near the top?

11. Wash the Pasteur pipette thoroughly, squeezing tap water in and out, and shaking it on a paper towel to dry it as much as possible.
12. Insert the pipette to the 75-ml level of the graduated cylinder and take up some of the solution.
13. When you remove the pipette, wipe it with a damp paper towel to eliminate the higher pH solutions it has passed through.
14. Place a drop on a pH paper in the medium acid range, and determine the pH. What is the pH at the 75 ml level?

15. Repeat this procedure, taking samples at the 50 ml and 0 ml level of the graduated cylinder. Wash the pipette between samples in fresh water, and wipe the pipette with a damp towel after each immersion in the cylinder. Use pH paper with appropriate

ranges for each sample. What is the pH at the 50-ml level? At the 0-ml level?

16. Record the pH of the different levels as labels on Fig. 3.1. What colors are associated with the pH readings?

Your instructor will take a variety of samples from one student's graduated cylinder, and demonstrate the use of the pH meter. A graduated cylinder will be drawn on the blackboard for instructor and students to enter the pH readings associated with the different spectral colors. You are now ready to use the remainder of the anthocyanin solution as a pH indicator.

17. Mark four test tubes at the 3-cm and 4-cm level with a wax pencil.
18. Fill each test tube with fresh anthocyanin solution to the 3-cm mark.
19. Divide the test tubes into two sets, each with two test tubes.
20. In one set, put H_2PO_4 + KH_2PO_4 buffer up to the 4-cm mark.
21. In the other set, add water up to the 4-cm mark.
22. Mark the two test tubes with buffer *A* and *B*; and the two without buffer *C* and *D*.
23. Add 0.1 N KOH drop by drop to test tube *A* until a color change occurs. How many drops did you have to add?

24. Add 0.1 N HCl drop by drop to test tube *B* until a color change occurs. How many drops did you have to add?

25. Add KOH drop by drop to test tube *C* until a color change occurs. How many drops did you have to add?

26. Add 0.1 N HCl to test tube *D* drop by drop until a color change occurs. How many drops did you have to add?

27. Was there a significant difference between the effect of acid and base addition to the water tubes and those containing the buffer?

28. H_3PO_4 and KH_2PO_4 undergo the following dissociations in solution:

$$H_3PO_4 \longleftrightarrow H^+ + H_2PO_4^- \longleftrightarrow H^+ + H^+ + HPO_4^{=}$$

$$\updownarrow$$

$$H^+ + H^+ + H^+ + PO_4^{\equiv}$$

$$K^+ + H^+ + H^+ + PO_4^{\equiv}$$

$$\updownarrow$$

$$KH_2PO_4 \longleftrightarrow K^+ + H_2PO_4^- \longleftrightarrow K^+ + H^+ + HPO_4^{=}$$

Indicate the direction of the shift in equilibrium when an acid is added to a system buffered with these compounds. How does the presence of the buffer offset the effect of additional H^+ ions?

29. The fact that you eventually obtained a change in the pH of the buffered system indicates that sufficient addition of an acid or a base can overcome the effectiveness of a buffer. In what state would the compounds of the buffer be when the system is overcome by the addition of acid? Of base?

30. Add the same number of drops of 0.1 N KOH to tube *C* that was added to tube *A*. To what pH does the color correspond?

31. Add the same number of drops of HCl to tube *D* that was added to tube *B*. To what pH does the color correspond?

32. What is the effective range of the $H_2PO_4 + KH_2PO_4$ buffer?

33. Judging from the natural color of the cabbage leaves, what is the pH of the cell's central vacuole?

MONOMERS, POLYMERS, AND HETEROCONJUGATES

Biologically active organic compounds have been classified according to chemical properties alone. However, in recent years awareness of the roles of molecules have lead biochemists to seek a more meaningful way of grouping molecules. Although not all molecules fit comfortably into the monomer, polymer, and heteroconjugate classification, this system does reflect the levels in the chemical reaction of the body (metabolism) in which the molecules operate.

Monomers are small carbon containing molecules, which are divided according to chemical type. Monosaccharide or sugar monomers include glucose, galactose, and fructose. Amino acid monomers include the 20 amino acids, each of which contains both an amino group and an acid group.

Some polymers are ester linkages of monomers. The polymeric form of sugar monomers may be starch, cellulose, or glycogen. The polymeric form of amino acids are the different proteins.

Heteroconjugates are combinations of two or more classes of monomers. For instance, a nucleotide heteroconjugate contains a purine or pyrimidine monomer, a ribose (sugar) monomer, and one or more phosphate groups. Polymerized heteroconjugates also exist. DNA is a polymerized form of nucleotide heteroconjugate.

Nutrient material enters the cell in monomer form. In some cases, the organism itself takes in monomers; in other cases, the organism takes in polymeric forms and the digestive processes break the ester linkages to form monomers. Sugar monomers are either broken down further with a concomitant extraction of energy, or linked by ester bonds to form starch (in plant cells), cellulose (in plant cells), or glycogen (in animal cells). Amino acid monomers are converted to sugars or linked in the specific arrangements to form native protein.

Polymers

Some tests for polymers are based on their chemical properties instead of their physical properties as large molecules.

Testing for Starch

1. Place a small slice of onion on a slide.
2. On the same slide place a small piece of potato tuber.

3. Add 2 drops of IKI solution to the slide.
4. Note the color change.
5. In what form does the potato store excess sugars?

6. How would the potato obtain the sugars when energy is needed?

7. Do all plants store excess sugars (glucose) as starch?

Testing for Protein

1. Mark two test tubes with a wax pencil at levels 3 cm and 5 cm from the bottom.
2. Put egg yolk solution into the first test tube up to the 3 cm mark; put albumen solution in the second test tube up to the 3 cm mark.
3. Add concentrated NaOH up to the second mark in each test tube. If handled carelessly, NaOH can burn your skin severely. If you accidently get some on either your skin or clothing, wash immediately at the nearest faucet and inform the instructor.
4. Add a few drops of 0.5% copper sulfate ($CuSO_4$) solution.
5. Note the color change.
6. In the presence of protein, the solution will turn violet; but in the presence of a polypeptide (units made up of several amino acids) the solution will turn pink. What can you conclude about the relative amounts of protein and polypeptide material in the egg yolk as compared with that in the albumen?

7. From which part of the egg would the developing chick draw most of its nutritional material? Why are the amino acids in this part more accessible?

Test for Nucleic Acids

1. Mark a test tube at 1 cm and 3 cm from the bottom with a wax pencil.
2. Add yeast hydrolyzed (60°-10 minutes) solution up to the 1 cm mark.
3. Add Diché reagent (diphenylamine) to the second mark.
4. Heat in a water bath, almost at boiling, for 20-40 minutes. A marble may be placed in the mouth of the tube to act as a condenser.

5. Let stand for 10 minutes and note the color change. A blue color indicates the presence of DNA, a green color is indicative of RNA. Which type of nucleic acid appears to be more prevalent in yeast?
6. Can you think of any type of cell that would give the opposite result? (Hint: the cell would have to have structures containing this type of nucleic acid as a major part of the cell.)

Monomers

The only class of monomers that can easily be tested for are the reducing sugars. You have already performed the Benedict's experiment earlier in the laboratory, so you should be familiar with the color changes noting the presence, partial presence, or absence of reducing sugars.

The reducing ability of a sugar is involved in its capacity to reduce $Cu^{++} \longrightarrow Cu^{+}$, thereby changing the color of the copper solution. This reduction of copper ions is brought about by the aldehyde or ketone group on a carbon-chain molecule.

$$2Cu(OH)_2 + R{-}C{:}O \longrightarrow Cu_2 + 2H_2O + R{-}COOH$$

By definition, those sugars known as aldoses and ketoses contain the necessary aldehyde or ketone group for reaction with Benedict's reagent. Benedict's test also distinguishes between monosaccharides and dissacharides. All monosaccharides give a color change with Benedict's reagent, but only some disaccharides do.

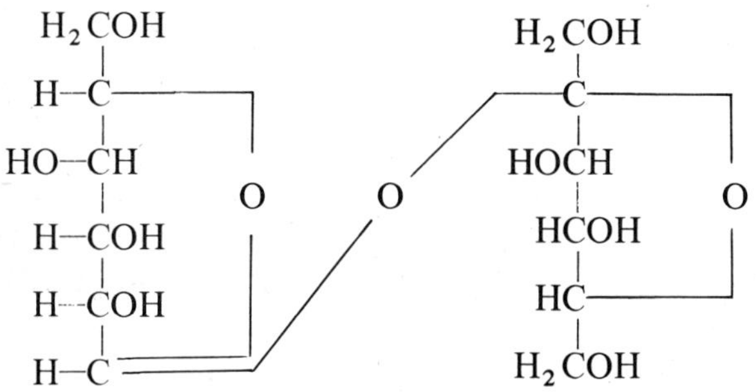

sucrose

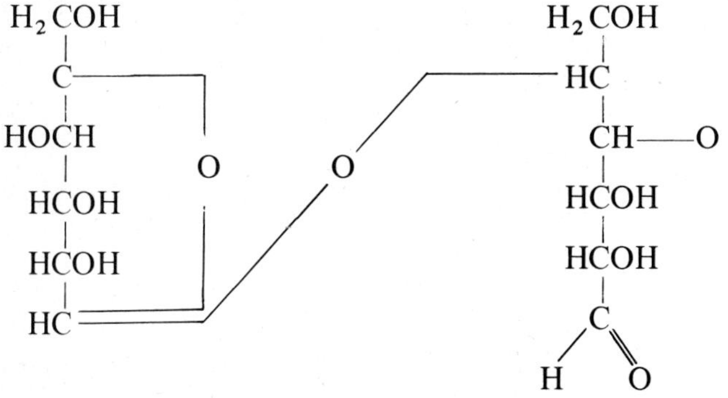

maltose

Which of the dissacharides above would give a positive Benedict's reaction (color change) and why?

Heteroconjugates

There is no biological material that contains a sufficient amount of heteroconjugate material for detection in a student laboratory. In addition to this factor, there are no easy tests for determining the presence of heteroconjugates. Thus they are merely mentioned in this exercise and will not be tested for.

Molecules Which Do Not Fit into the Above Classification

There are several types of molecules that do not fit comfortably into the classification of compounds according the polymers, monomers, and heteroconjugates. Disaccharides, for example, are made up of two sugars joined by an ester linkage. In this system of classification, the monosaccharides are the monomers; polymerized glucose comprise the polymers; and the dissacharides are difficult to classify. From a metabolic standpoint, they fit with the polymers because they are the storage form of sugar in plants. For instance, barley stores excess sugar as maltose; sugar cane reserves sucrose. However, from a chemical standpoint they are relatively small molecules compared to the typical polymers.

One major group of compounds that does not fit into the classification comfortably is the *lipids* or *fats*. Lipids are composed of one molecule of glycerol and three molecules of fatty acid chains. In one sense, since lipids are a combination of compounds, they could be considered heteroconjugates. But the fatty acid chains – about 20 carbons long – cannot truly be considered monomers. Hence, lipids are religated to a somewhat "orphan" category in this classification. Structure of a fat:

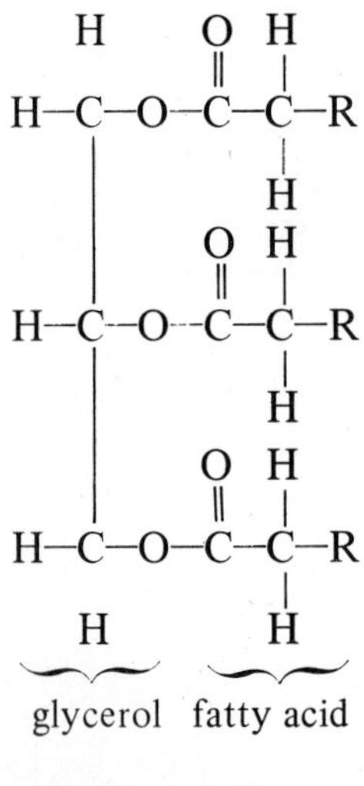

Tests for Lipids

1. Mark a clean test tube with a wax pencil 2 cm and 3 cm from the bottom of the tube.

2. Put distilled water up to the 2-cm mark.
3. Add a few grains of Sudan IV dye (the few grains that coat the tip of a toothpick are sufficient).
4. Mix by swirling the test tube. Does the dye dissolve in the water or remain in granular form?

5. Add salad oil to the 3-cm mark and mix again. Does the dye dissolve in the fat or remain in granular form?

6. How would this test indicate the presence of fat in an unknown sample?

4

Protein Chemistry and Enzymes

MATERIALS

CLASS STOCK

chemicals
- 0.5% albumen solution
- 1% soluble starch solution
- 1% dextrose solution
- 10% NaCl solution
- Benedict's reagent
- 0.1% mercuric chloride
- 1 N HCl
- 1 N KOH
- 0.1% catechol solution
- enzyme solution
- acetone
- 0.5% trypsin solution

dialysis sacs
string
water baths
- 37°C; 50°C; 70°C; boiling

"Spectronic 20" spectrophotometer
crushed ice
potato extract[a]
distilled water

STUDENT STATION

100-ml graduate cylinder
2 250-ml beakers
500-ml beaker
8-22 test tubes
2 colorimeter tubes
dropper
wax pencil

[a]keep in ice bucket throughout laboratory period

INTRODUCTION

In the final analysis , organismal characteristics are expressions of the ensemble of proteins that the organism is capable of producing (synthesizing). Protein specificity is an expression of the linear sequence of amino acids within the protein, and their chemical and/or physical properties that lead to the secondary, tertiary and quartenary conformations of the molecule.

This laboratory exercise provides a study of some of the salient features of proteins in order to reveal the different levels of bonding, and therefore to indicate how proteins function as enzymes.

SIZE OF A PROTEIN MOLECULE

A protein, being a polymer, is an extremely large molecule. It is possible to test molecular sizes by *dialysis,* a method that employs a porous membrane sac called a *dialysis sac.* The pores of the sac are sufficiently large for small molecules to pass through, but are too small for the passage of large molecules.

1. Pour each of the following solutions into one 250-ml beaker. Use a 100-ml graduate cylinder to measure the solutions and wash the cylinder after each use.
   (a) 50 ml of 0.5% albumen
   (b) 50 ml of 1% soluble starch solution
   (c) 50 ml of 1% dextrose solution
2. Tie a dialysis sac tightly at one end (the bottom) and pour the mixture of solutions into the sac until 1½ in. of the sac is filled.
3. Tie the top of the sac about 2 in. above the level of the mixture. Leave about 4 in. of string (see Fig. 4.1).

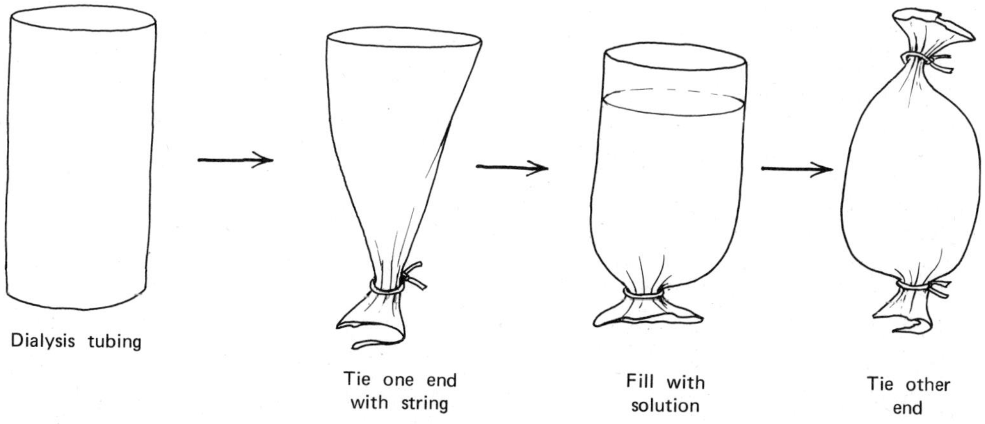

Figure 4.1

4. Place the sac in a 500-ml beaker containing 200 ml 10% NaCl solution for 30 min, agitating the sac about every 5 min.
5. Mark two test tubes with a wax pencil 2 cm and 3 cm from the bottom, and label *A* and *B* respectively. These tubes are for the Bendict's test.
6. Mark two test tubes with a wax pencil 3 cm and 5 cm from the bottom and label them *C* and *D* respectively. These tubes are for the biuret test, which will be performed later.
7. Mark two test tubes with a wax pencil 2 cm from the bottom and label them *E* and *F* respectively. These tubes are for the starch test.
8. After 30 min. have elapsed since step 4 was completed, lift the dialysis sac out of the beaker. Do not discard the contents of the beaker. Note changes in the firmness (turgidity) of the sac, and record.

9. Cut the dialysis sac near the top and empty the contents into a clean 250-ml beaker.
10. Pour the solution taken from the dialysis sac to the first mark in one test tube from each set prepared in steps 5, 6, and 7 (test tubes *A, C,* and *E*).
11. Pour the solution that surrounded the dialysis sac (originally 10% NaCl solution) to the first mark in the other set of tubes (test tubes *B, D,* and *F*).
12. Perform the Benedict's, biuret, and iodine tests, using the appropriate test tubes.
13. Record your results on Chart 4.1.

CHART 4.1

Solution	Benedict's	biuret	iodine
Inside sac			
Outside sac			

14. Molecules tend to pass from areas of higher concentrations to areas of lower concentrations (diffusion). In what directions would the glucose, starch, protein, and NaCl molecules tend to pass?

15. Which of the above molecules passed in this direction and why?

16. From experiments done in the preceding laboratory exercise, you know that starch is a polymerized form of glucose. Did the starch pass out of the dialysis sac?

17. Did the protein pass out of the dialysis sac?

18. What can you conclude regarding the relative size of the protein molecules of albumen?

PROTEIN STRUCTURE

From molecular behavior proteins are known to exist at four structural levels:

(1) the sequence of amino acids;
(2) the twisting of the chain to form a helix (because of the positions of the carboxyl and amino groups);
(3) the association of amino acid side groups to form a stable three-dimensional molecule;
(4) the association of two or more polypeptides to form the final, large protein.

The means that have been used to elucidate the structural levels of proteins are beyond the scope of a student laboratory. A workbook exercise can be used to illustrate the determination of the primary structure of a protein, and chemical tests can illustrate the third structure.

The Primary Structure of a Protein

The primary structure of a protein is determined by a jig-saw puzzle method. The first step in the procedure is a breakage of the peptide bonds holding the amino acids together. This is usually done by acid hydrolysis. The addition of large quantities of H^+ ion shifts the equilibrium stage of the protein in such a way that the C:N bonds are broken.

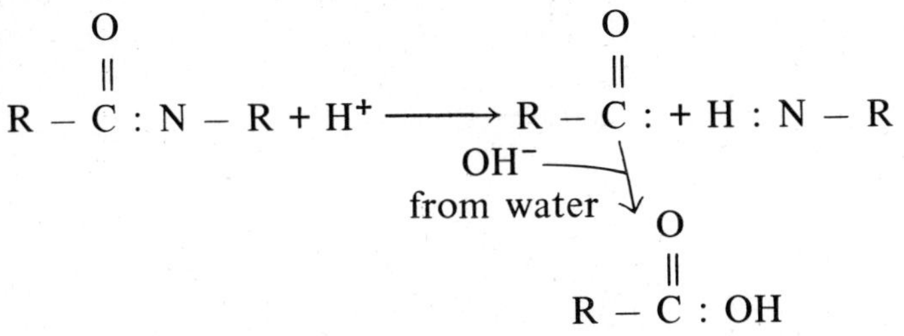

Separate amino acids are analyzed for type and quantity. These amino acids are from many identical protein or polypeptide molecules in the orginal solution, each of which is identical in its amino acid sequence. Thus the relative amounts of the individual amino acids reflect their relative numbers in a single protein.

Assume that you have hydrolyzed 7 mg of a polypeptide solution of a protein with a known molecular weight of 1265; you have also analyzed the type and quantity of amino acids. Your results are as follows:

	Amount (in mg)	Molecular Weights (Approximate)
alanine	.97	89
aspartic acid	2.19	133
cysteine	1.32	120
glutamine	1.30	119
serine	1.15	105

Find the ratios of the amino acids to one another according to the formula

$$\frac{\text{weight in mg}}{\text{molecular weight}}$$

2. What is the ratio of alanine to aspartic acid? Remember, the ratio must be in whole numbers.

3. What is the number of alanine molecules in the polypeptide chain?

The number and type of amino acids do not indicate their position or sequence in the polypeptide molecule. The next step in the analysis of the primary structure of a protein or polypeptide is to begin once again with the original solution, and with more gentle methods, break the protein into fragments. The polypeptides in solution will fragment in different places, so that there will be different size fragments, some of which might overlap. The fragmented solution is then separated into similar fragments, and these solutions are tested for amino acid content and sequence. A dye is placed in the solution that combines with the free amino group of the terminal amino acid. The solution is then treated with an enzyme that breaks the peptide bond between the amino acid with the terminal amino group and the next amino acid. The solution is then analyzed to find the amino acid with the dye molecule attached. The procedure is repeated in order to find the amino acid next in line, and so forth, through the three or four amino acid fragment. Then by fitting the fragments together, the primary structure of a polypeptide is reconstructed. For example the following three fragments

alanine-glutamine-phenylalanine-serine
tyrosine-serine-alanine-glutamine
phehylalanine-serine-cysteine-alanine
can only be lined as follows:

```
                   alanine–glutamine–phenylalanine–serine
                      |        |           |         |
tyrosine–serine–alanine–glutamine          |         |
                                   phenylalanine–serine–cysteine–alanine
```

Therefore, the primary structure is

tyrosine–serine–alanine–glutamine–phenylalanine–serine–cysteine–alanine

1. Assume that you have analyzed the amino acid content and arrived at the data given in the previous section of this exercise. What is the total number of amino acids in a polypeptide molecule?

2. How many of each of the following amino acids are there in the molecule?
   alanine
   aspartic acid
   cysteine
   glutamine
   serine

3. During fragmentation and analysis, you found the fragments to contain the following amino acid sequences:
   (a) aspartic acid
   (b) glutamine-cysteine-alanine-alanine
   (c) cysteine-aspartic acid-serine-aspartic acid
   (d) aspartic acid-glutamine-serine-glutamine
   (e) alanine-cysteine-aspartic acid
   (f) glutamine-serine-glutamine-cysteine
   Match the fragments by writing them down, positioning same sequences.

4. Does the number of amino acids in the polypeptide agree with your calculation made from earlier data. If not, then check the sequence carefully. What is the sequence of amino acids in the polypeptide?

As a point of interest, the insulin molecule, which was the first protein for which the primary structure was elucidated, has 103 amino acids.

Tertiary Structure of Proteins

A protein in solution bends and twists upon itself, and in doing so exposes the side chains of the amino acids to one another. When two cysteine molecules are in relative proximity, their terminal sulfur radicals (R–SH HS–R) tend to form a *disulfide bridge* (R–S–S–R). one of the strongest side group bonds in a protein. Side groups with terminal carboxyl groups (COO^-) tend to associate with side groups with alkaline properties ($NH_3{}^+$) by forming ionic bonds. Hydrophobic groups tend to associate because of their aversion to water. Van der Waals forces, an attraction between any two molecules regardless of their chemical nature tend to keep proximal groups in place at a distance equivalent to the sum of their atomic radii. These bondings generate the three dimensionality of the protein and stabilize it in a particular conformation.

Every molecule of a given protein has the same sequence of amino acids, and therefore, the same potential form for side-group associations among its amino acids. The amino acid side groups that do react are determined by the same factor that determines the direction and equilibrium point of all reactions—the second law of thermodynamics. In accordance with the second law of thermodynamics, a reaction proceeds to a state manifesting the lowest free energy. And, since the lowest free-energy state would be the same for all molecules of a given protein, all molecules of that protein will express the same three dimensional structure.

1. Using the amino acid sequence of the polypeptide chain you solved in previous section, and the following chemical structures, draw the sequence and show which side groups would tend to bond to form the tertiary structure by drawing a line between them.

<u>alanine</u>

$H_2N - CH - COOH$, with CH_3 on the CH

<u>aspartic acid</u>

$H_2N - CH - COOH$, with $CH_2 - COO^-$ on the CH

cysteine

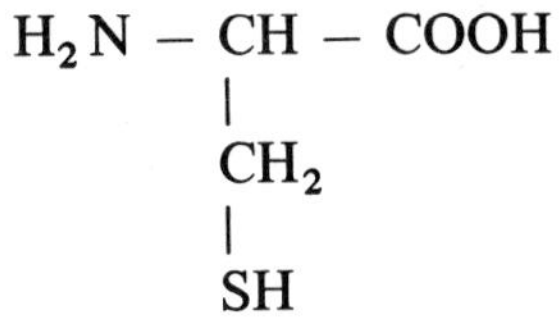

glutamine

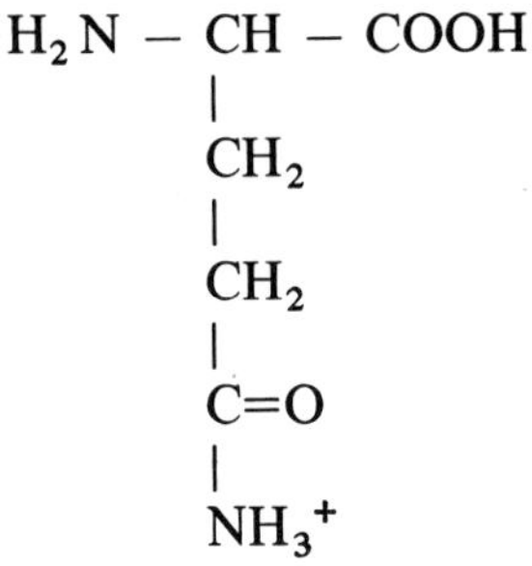

serine

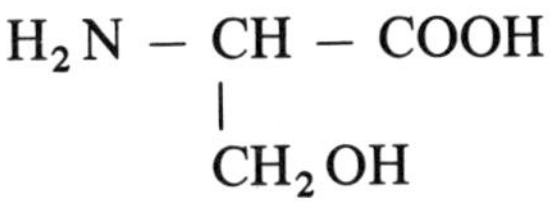

2. Draw the shape of your polypeptide, using the protein provided in Fig. 4.2 as a model.

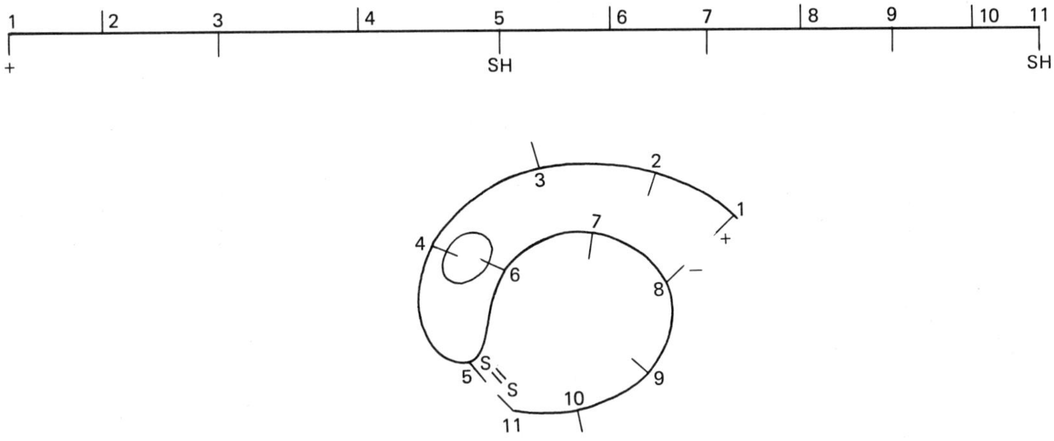

Figure 4.2

Indirect Detection of Tertiary Bonding

When a protein solution is heated, the secondary, tertiary, and quarternary bondings will dissociate in accordance with their bond energies. If the heat added to the system is insufficient to disturb the primary structure (by breaking the peptide bonds), some proteins will return to their folded state on cooling.

However, some proteins will undergo permanent denaturation when exposed to heat, a phenomenon that appears to be due to the exposure of the hydrophobic amino acids when the tertiary structure is altered. As nonpolar groups, the hydrophobic amino acids cannot form associations with water, and tend to bring the protein out of solution in the form of a precipitate. Those proteins that contain a preponderance of amino acids with hydrophobic radicals will not return to their folded state on cooling.

The pH of the solution has a striking effect on the tertiary structure of a protein. The side chains of a protein act as buffers, reacting with excess H^+ and OH^- ions in the solution. Under the acid conditions the –COO– radical will act as a hydrogen acceptor, and become COOH. When there is an excess of OH^- ion, the amide groups present on the side chains will tend to be in the form NH_2.

In biological systems, when the protein undergoes folding into its three dimensional structure, the pH is close to neutral (7). The structure of the protein is at its most thermodynamic state for that pH. When the pH of a protein solution is drastically changed by the addition of either an acid or a base, the ionic bonds will break, and the acid and base radicals will function in their roles as buffers. The unfolding of the protein resulting from the breakage of the ionic bonds is denaturation; and if, in the course of this denaturation many hydrophobic groups are exposed, the protein will precipitate.

Certain heavy metals, such as mercury, can also cause the denaturation and subsequent precipitation of proteins. These metals can combine preferentially with certain radicals, usually the sulfur of cysteine, forcing them out of their associations with other amino acids and leading to an unfolding of the protein structure.

1. Mark five test tubes 3 cm from the bottom, and label them A, B, C, D, and E.
2. Also mark C, D, and E 4 cm from the bottom.
3. Fill all five test tubes to the 3 cm mark with the 0.5% albumen solution.
4. Do the following with the test tubes:
   A. The control–leave in a rack at your desk.
   B. Place in a boiling water bath for 5 min.
   C. Fill to the second mark with 0.1% mercuric chloride.
   D. Fill to the second mark with 1 N HCl (use a dropper and exercise care in dispensing).
   E. Fill to the second mark with 1 N KOH (using a dropper and exercising care in dispensing).
5. Enter your results on Chart 4.2.

CHART 4.2

Treatment	Type of Precipitate (grainy, spongy, etc.)	Amount of Ppt. (+, ++, +++, etc.)
A Control		
B Heat		
C Heavy Metal		
D Acid		
E Base		

6. Heat totally disrupts tertiary structures by separating all tertiary bonds regardless of their chemical nature. Heavy metals affect cysteine bonds, and acids and bases affect ionic bonds. How are these effects manifested in your results?

7. Was the amount and type of precipitate the same when you used acid and base? Why would you expect it to be the same?

8. Each of the treatments not only separates the tertiary bondings at different sides, but also forces the molecule into new conformations. How is this demonstrated in the types of precipitates found in each case?

PROTEINS AS ENZYMES

The cell is dependent on the maintenance of structural and functional integrity for its existence. However, as a highly organized and complex system it is subject to natural destructive forces operating in accordance with the second law of thermodynamics. To counteract this tendency, cells must constantly replace or repair the components of its organization. Thus, cells are involved in constant building processes, undergoing the chemical reactions necessary to provide the needed molecules for maintaining integrity.

Within cells, energy derived from exergonic (energy yielding) reactions is used to drive endergonic (energy requiring) reactions.

$$A - B \longrightarrow A + B + \text{Energy}$$

$$\downarrow$$

$$C + D + \text{Energy} \longrightarrow C - D$$

Although all reactions within the cell would progress spontaneously under proper energy conditions, the rates of these reactions vary considerably. Since cells are wholly dependent on these chemical reactions for the maintenance of life, it is important that reactions occur at rates suitable to the requisites of the cell.

There are a number of ways by which reactions can be controlled. Temperature and pH are variable factors that affect the rate of reaction. Cells also contain a group of organic catalysts, called enzymes. Enzymes affect the rate of a chemical reaction and emerge unaltered from the reaction or set of reactions.

Without Enzyme

$$A + B \xrightarrow{\text{time} = a} A - B$$

With Enzyme

$$A + B + \text{enzyme} \longrightarrow A\text{-enzyme-}B \xrightarrow{a/x} A - B + \text{enzyme}$$

[substrate + enzyme ⟶ enzyme-substrate complex ⟶ product + enzyme]

The presence of enzymes permits the cell to utilize metabolites or foodstuffs with a minimum expenditure of energy to obtain their bond energies. However, at the same time, enzymes serve to limit the cells' capacities. The cell can only perform those reactions for which it has the enzymes. An organism may be able to utilize glucose as a food source because it has the enzyme (hexokinase) to break it down, whereas it may not be able to utilize maltose (two glucose molecules bonded to form a dissacharide) because it does not have the enzyme (maltase) that cleaves that bond to form two glucose molecules.

Each enzyme is highly specific, acting as a catalyst for only a single reaction. Most enzymes contain a protein portion, which confers specificity and is called the apo-enzyme, and a co-enzyme, which confers activity and is often a nucleic acid derivative. Co-enzymes are often vitamins.

In a student laboratory, you are limited to experiments that may be done safely and quickly. Many of the more significant enzymes involved in biological processes have difficult tests, requiring equipment not available in a student laboratory. Although you will be performing your tests on enzymes of perhaps secondary importance in cell metabolism, remember that the properties of enzymes, and the consequences of enzyme action are universal.

Enzyme Activity

Enzymes are usually classified according to their function. Oxidases, for example, catalyze the oxidation of certain organic compounds and result in the oxidized form of that compound and a release of energy. One such oxidase is catecholase, which catalyzes the oxidation of catechol.

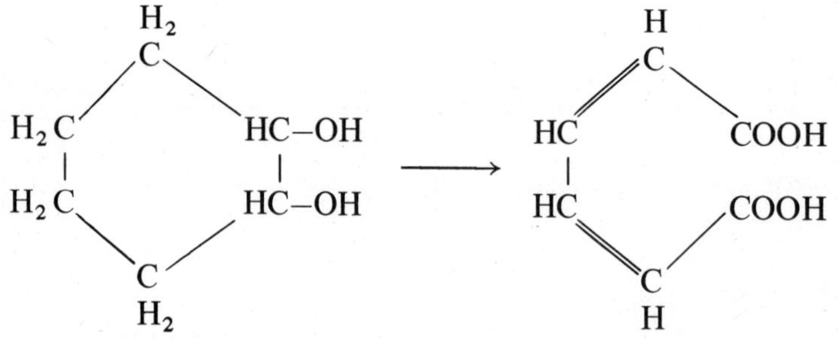

catechol   cis,cis muconic acid

Catechol is colorless in solution, while cis,cis muconic acid is black. The more catechol converted to cis,cis muconic acid, the darker the solution will be. Reactions of this type are responsible for black pigmentation in practically all forms of life. Note: all glassware must be scrupulously cleaned. Wash glassware carefully after each test.

Enzyme Action

1. Mark two clean test tubes at 2 cm and 3 cm from the bottom, and label them *A* and *B*, respectively, near the mouth of each tube.
2. Put 0.1% catechol solution in the tubes up to the first mark. In tube *A* add water to the second mark; in tube *B*, add potato extract (enzyme solution) to the second mark.
3. Shake the tubes for 5 min.
4. Turn on the "spectronic 20," and adjust to 0 percent transmission with the dial on the left, and a wavelength of 420 mu (see Fig. 4.3).

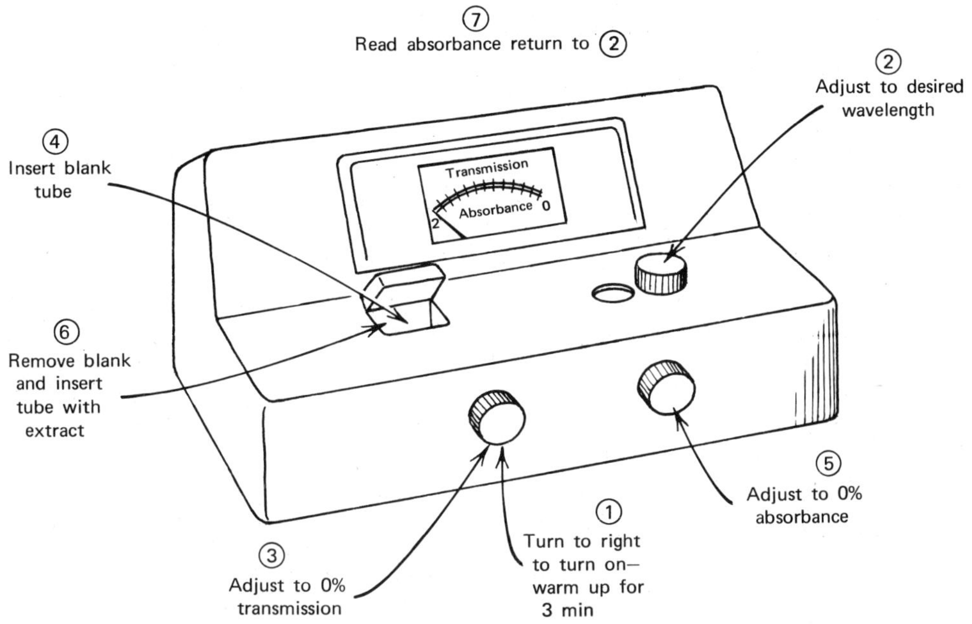

Figure 4.3 Operation of the "Spectronic 20."

5. Transfer mixture from tube *A* to a clean colorimeter tube.
6. Wipe the lower 2/3 of the tube and place it in the "Spectronic 20." Be sure to line up the ground glass mark on the tube with the line on the instrument.
7. With the dial on the left, set the meter at 0 percent absorption.
8. Remove the tube, and check the percent transmission.
9. Reinsert the tube and check to see if the meter is still at 0 percent absorption. If not, readjust.
10. Continue reinserting until you get two consecutive readings of 0 percent absorption with the tube in place. Do not handle the lower 2/3 of the tube while you are manipulating it.
11. Transfer mixture from tube *B* to a clean colorimeter tube and wipe the lower 2/3 of

the tube and place it in the "Spectronic 20." Read the percent absorption from the meter.

12. What is the relative absorption of the solution with enzyme compared to the one without enzyme?

13. What has happened to the catechol in the presence of the enzyme, and why is the occurrence of the event related to absorption properties of the solution?

14. Is the relationship between the enzymatic breakdown of catechol and the absorption of light by solutions a direct or indirect one? Why?

15. Why is absorption a relative value?

16. What would you use as a standard in order to arrive at an absolute, or comparatively absolute, absorption value?

17. Considering the properties of enzymes, would you expect there to be a formation of cis,cis muconic acid in tube *A* now? Later?

18. What would you have to do in order to determine whether a reaction has taken place in tube *A*? You may repeat your readings with an appropriate standard if you wish.

Effect of Time on Enzyme Activity

1. Wash the colorimeter tubes, and dry by rinsing with acetone.
2. Label the two regular test tubes *A* and *B*, and prepare as in the preceding section, marking the tubes and filling them with appropriate solutions.
3. Set tubes *A* and *B* in a rack or beaker at your desk. At 10-min intervals from the time you added your enzyme, shake each tube. Transfer the contents to colorimeter tubes, wipe the lower 2/3, and take an absorption reading on the "Spectronic 20," using distilled water as a standard.
4. Transfer solutions back into respective test tubes after readings.
5. Enter the percent absorption on Chart 4.3.

CHART 4.3

Tube	Minutes							
	10	20	30	40	50	60	70	80
A								
B								

6. Convert your data to graph form, using Fig. 4.4, connecting your points by adding straight lines. You should have two lines, or curves – one for the catechol with enzyme reaction and one for the catechol solution without enzyme.

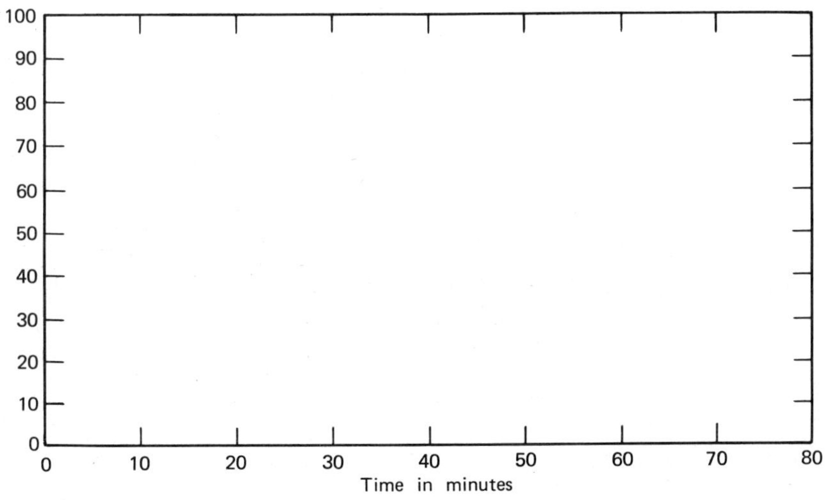

Figure 4.4

7. As time proceeds, what happens to the catechol with and without enzyme?

8. Does the solution without enzyme catch up to the one with enzyme? Would you expect it to?

9. Is the reaction taking place in the tube with enzyme limited by the catechol or the enzyme? Why?

10. If this reaction were constantly supplied with the limiting factor, would the reaction ever stop?

Effect of Temperature on Enzyme Activity

1. Label eight clean test tubes *A*1, *A*2, *B*1, *B*2, *C*1, *C*2, *D*1, and *D*2, and mark each with a wax pencil 2 cm and 3 cm from the bottom.
2. Add catechol to the 2 cm mark in each tube.
3. Add water to the second mark in test tubes *A*1, *B*1, *C*1, and *D*1.
4. Add enzyme solution to the second mark in test tubes *A*2, *B*2, *C*2, and *D*2.
5. Incubate each series at the following temperatures:
   *A*1, *A*2 at 0°C (ice water)
   *B*1, *B*2 at 37°C water bath or beaker of 37° water in incubator
   *C*1, *C*2 at 50°C (water bath or beaker of 50° water in incubator)
   *D*1, *D*2 at 70°C (water bath)
6. It will take the solutions about 5 min to reach the temperature of their environments. Wait a total of 15 min, then remove and shake the tubes.
7. Use distilled water as a standard. Pour the contents of tube *A*1 into a *clean* colorimeter tube and take an absorption reading. Enter it on Chart 4.4. Pour the solution back into the *A*1 test tube and clean the colorimeter tube thoroughly. Repeat the procedure with tubes *A*2, *B*1, *B*2, *C*1, *C*2, *D*1, and *D*2, entering the absorption readings on Chart 4.4.
8. Convert the data to graph form using Fig. 4.5.
9. Is there a direct or indirect relationship between temperature and the occurrence of the catechol to cis,cis muconic acid reaction without the enzyme?

CHART 4.4

	Absorbance		Absorbance
*A*1		*A*2	
*B*1		*B*2	
*C*1		*C*2	
*D*1		*D*2	

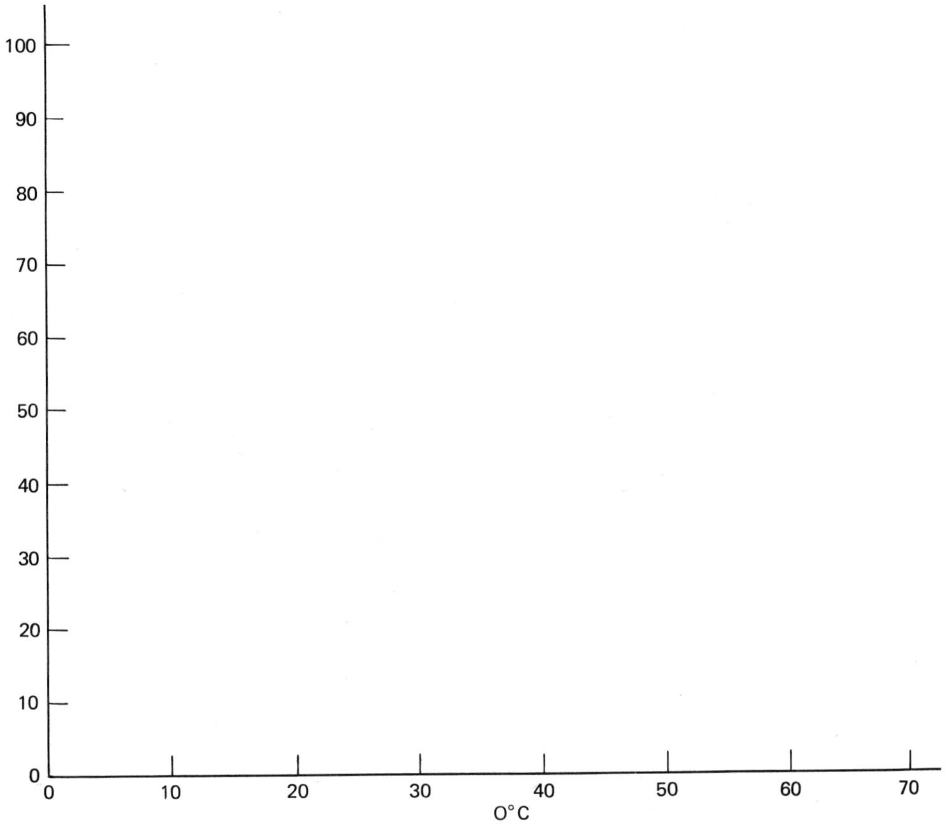

Figure 4.5

10. Can the same be said for the reaction taking place with the enzyme?

11. What happens to the *enzyme* mediated reaction after 50°C? Why? (Hint: The tertiary structure of proteins is affected at about 50°C.)

The Maxwell-Boltzman Distribution

Enzymes alter the rate of chemical reactions by combining with the substrate in such a manner that less energy is required for the completion of the reaction. In other words, enzymes reduce the activation energy of chemical reactions. But what determines which molecules will combine with the enzyme and subsequently be converted to product? The answer to this question is essentially the same as the answer to the question of why certain molecules undergo reaction at 20°C, while others do not. Let us first examine temperature effects on reactions.

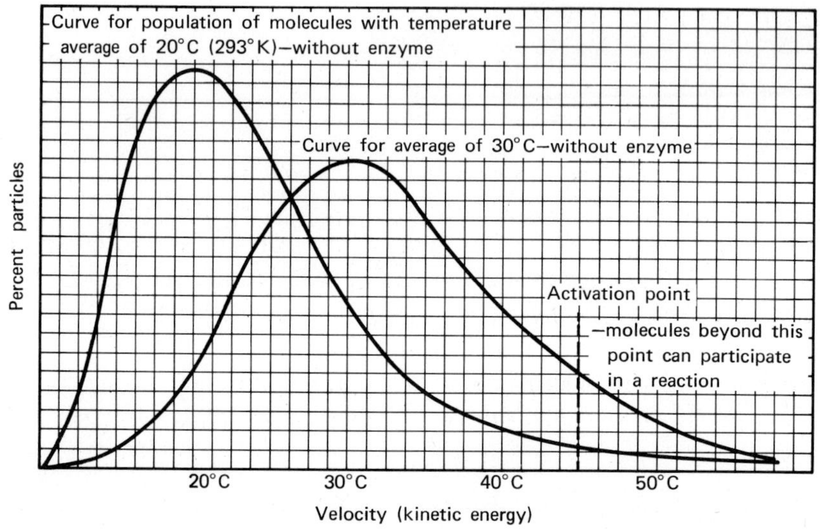

Figure 4.6 The Maxwell-Boltzmann distribution.

1. Notice that there are two curves on the temperature graph (Fig. 4.6); one for 20°C without enzyme, and the other for 30°C without enzyme. Find the area of the curve of 20°C without enzyme, bordered by the dark perpendicular line on the left parallel to the y-axis, by the x-axis, and by the slope of the line by adding the squares. This area value represents a relative number of molecules capable of undergoing a reaction in a solution at 20°C.
2. Enter the computed area on Chart 4.5.

CHART 4.5

	Parameters		
	20°C Without Enzyme	30°C Without Enzyme	20°C with Enzyme
Area			
Area Ratio	1[a]		

[a]Using the area of the 20°C without enzyme curve as 1.

3. Compute the area of the curve of 30°C without enzyme, bordered by the dark perpendicular line on the left parallel to the y-axis, by the x-axis, and by the slope of the line. This area value represents a relative number of molecules capable of undergoing reaction in a solution at 30°C.

4. Enter the computed area on Chart 4.5.
5. Calculate the area ratios, using the area of the 20°C without enzyme curve as 1. Enter your results on Chart 4.5.

At any given temperature, the molecules in a solution are moving at various speeds, including speeds of energy levels associated with other temperatures. For example, in a solution at 20°C, a small percentage of molecules are moving at speeds characteristic of molecules at 50°C or 60°C; other molecules are moving at speeds characteristic of the energy content of a solution at 0°C or 5°C. The actual temperature of the solution is an average of all molecular movement; it represents the energy content of most of the molecules in the solution. A graphed curve of molecular speeds, known as the Maxwell-Boltzmann distribution, would be an asymmetric bell-shaped curve, as illustrated in Fig. 4.6.

If the energy of activation necessary for a molecule to undergo reaction is the energy content associated with 45°C, and a molecule in a 20°C solution is manifesting this energy content, it will undergo the reaction. Thus, some molecules in a 20°C solution will take part in chemical reaction that requires a higher energy content than the average exhibited by most of the molecules in the solution. Therefore the area under the lines you have graphed indicate the relative numbers of molecules at each temperature that have the energy content sufficient to undergo reaction. That is, you have detected the small percentage pf molecules in the solution that are manafesting a higher energy content than the average of that solution, which corresponds to the bordered area on the Maxwell-Boltzmann distribution in Fig. 4.6.

6. Returning your attention to the graph in Fig. 4.6, what multiple of molecules with high enough energy content to undergo reaction are there at 30°C without enzyme compared to 20°C without enzyme?

7. To test your understanding of this relationship, construct a Maxwell-Boltzmann distribution of molecular speeds at 20°C to 30°C and explain how a 4 percent energy increase from 20°C to 30°C can result in a two- to three-fold increase in product formation.

Enzymes and the Maxwell Boltzmann Distribution

A reaction may have an activation energy associated with the energy content of a temperature of 45°C, but the combining of the enzymesand substrate for that reaction would give a considerably lower level of activation energy. Thus, enzymes would combine with substrates at, for example, an energy content associated with 30°C. In a solution 20°C there are many more molecules with energy contents associated with 30°C then there are molecules with energy contents associated with 45°C, and the former molecules will combine with the enzymes (see Fig. 4.7).

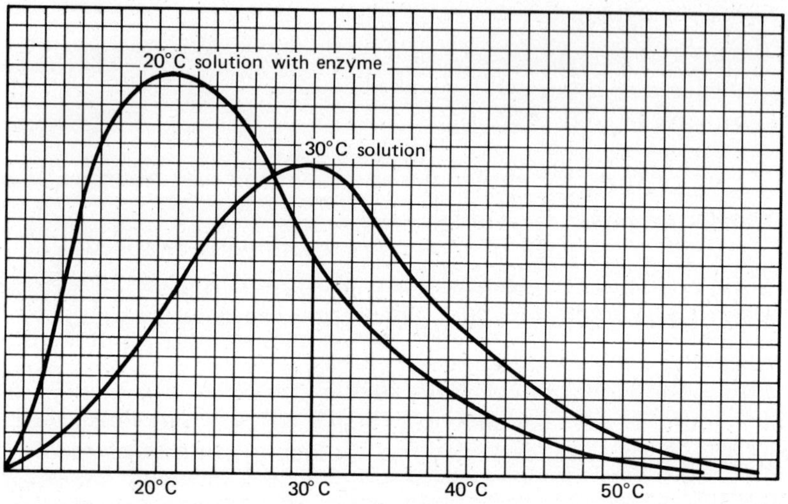

Figure 4.7

Because of the lowered level of activation energy in the enzyme-substrate complex, more reactions will take place at lower temperature. Again, however, only the molecules manifexting sufficient energy to combine with the substrate at lower temperature will do so.

8. Compute the area under the 20°C with enzyme curve and add it to Chart 4.5.

9. How does the area ratio of solution with enzyme at 20°C compare with that of the solution without enzyme at 20°C? At 30°C?

10. The optimal activity range for most enzymes is between 20°C and 40°C. Why?

The Chemical Nature of Enzymes

1. Label four test tubes *A*, *B*, *C*, and *D*. Mark each 2 cm from the bottom with a wax pencil.
2. Put catechol solution to the first mark and set the tubes aside.
3. Mark three other test tubes 1 cm and 2 cm from the bottom. Label these tubes 1, 2, and 3.
4. Put enzyme solution into test tubes 1, 2, and 3 to the first mark.
5. Add the following as indicated:

   Test tube 1 – add 5% trypsin to the second mark.
   Test tube 2 – place in boiling water bath for 10 min; Add water to 2-cm mark.
   Test tube 3 – place in 70° water bath for 10 min. Cool to room temperature and add water to 2-cm mark.
6. Add the treated enzyme solution from test tube 1 to test tube *A*; add the boiled enzyme solution from test tube 2 to test tube *B*; and add the heated solution from test tube 3 to test tube *C*. Shake all three tubes.
7. Add water to test tube *D* so that it contains the same amount of solution as test tubes *A*, *B*, and *C*.
8. Wait 5 min and read the solutions in the "Spectronic 20" using test tube *D* as a standard by pouring the contents of test tube *D* into a colorimeter tube for setting the spectrometer at zero percent absorption.
9. Wash and dry the colorimeter tube, and pour the contents of tube *A* into it. Take a reading, and enter it on Chart 4.6.

CHART 4.6

Treatment	Absorption Reading
Enzyme and trypsin	
Boiled enzyme	
Heated and cooled enzyme	

10. Do the same with test tube *B*.
11. Enter your results on Chart 4.6.
12. Trypsin is a <u>protease</u>, which causes a break in peptide bonds between arginine or lysine and another amino acid in the polypeptide chain. How has trypsin interfered with catechol activity, and what does this indicate concerning the chemical composition of the enzyme?

13. Is the effect of trypsin reversible?

14. Trypsin is also a protein. What explanation can you suggest for why one trypsin molecule does not destroy another?

15. What effect has boiling had on catechol? Is this consistent with what you would expect from a protein? Why?

16. Can the enzyme be restored to its active form after boiling?

17. What effect has heating to 70°C and cooling had on the enzyme?

18. What happens to a protein heated to 70°?

19. From your experiment, can you tell anything regarding the number of hydrophobic bonds in the apo-enzyme of catechol compared to albumen? Explain.

5

Intermediary Metabolism

MATERIALS

CLASS STOCK

water baths
  37° or Warburg bath, 50°C
wheat or barley seedlings
chemicals
  10% KOH solution
  Brodie solution
  Benedict's reagent
  Sudan IV solution
  IKI solution
  0.5% $CuSo_4$ solution
  conc. NaOH
  $(NH_4)_2SO_4$
  enzyme starch synthetase
  KOH and maleate buffer solution
  glucose-1-phosphate (30 mg/ml)
filter paper
potatoes
knives
potato peelers
blender
cheesecloth
Whatman #1 filter paper (10 cm)
clinical centrifuge

STUDENT STATION

Warburg flask or wide
  mouth bottles (2)
Pasteur pipette
wax pencil
2-hole stoppers (2)
syringe
plunger
stopcock
rubber tubing
forceps
vial
aspirator-Buchner funnel
beakers (10 and 125-ml)
15-ml conical centrifuge tubes
graduated cylinder
beam balance
spot plate
test tubes (25-ml)
test tube rack

INTRODUCTION

Enzymes operate in groups or enzyme systems. These systems activate a sequence of small changes in the molecular structure of a specific compound and ultimately result in the production of specific end products. Some enzyme systems are regarded as *exergonic* because the energy produced by one or more steps in the series of reactions is greater than the total amount of energy put into the system. Other enzyme systems, notably those responsible for *biosynthetic* or building processes of cells, are *endergonic* reactions.

The sum of the reactions of all enzyme systems is called metabolism. Generally, *intermediary metabolism* includes those pathways originating with foodstuffs and terminating in the biosynthesis of polymers or in the release of waste products.

Intermediary metabolism may be subdivided into distinct pathways with different names. The building processes of intermediary metabolism, such as protein synthesis, carbohydrate synthesis, and so forth, are grouped as biosynthetic pathways. The degradative processes through which organic compounds are broken down in a stepwise fashion for the release of energy is called *respiration.* In higher animals, respiration includes gas exchange with the environment, but in a metabolic perspective, respiration refers only to the chemical processes at the cellular level, common to all organisms.

The biosynthetic and degradative (or respiratory) processes of intermediary metabolism are closely related. Intermediate molecular forms that occur in the course of the stepwise degradative process are siphoned off as monomers or are converted to monomers for the biosynthesis of polymers. The links between biosynthetic and degradative pathways involve a controlled mass equilibrium among all products and reactants of all metabolic reactions.

The primary gain to the cell from respiration is the energy provided by the breakage of the bonds in organic molecules. This energy is directed toward the endergonic reactions of the cells for maintenance and growth. The chemical products resulting from the complete breakdown of organic materials are the waste products, which must be eliminated from the cell.

Respiration may be viewed as three integrated enzyme systems. The first, glycolysis or the Embden-Meyerhoff Scheme involves the conversion of simple sugars (hexoses) into two molecules of pyruvate (pyruvic acid), a three carbon molecule.

glucose (6-carbon molecule) $\xrightarrow{\text{(glycolysis)}}$ pyruvate (2 3-carbon molecules)

This process does not require oxygen and is referred to as anaerobic respiration. The pathways that follow after the formation of pyruvate depend on the enzymes present in the organism. All organisms have more than one pathway leading from pyruvate, but none have all the possible pathways known.

Several possible pathways

Pyruvate → ethyl alochol + CO_2
Pyruvate → lactic acid
Pyruvate → Kreb's cycle
Pyruvate → fatty acid

The formation of ethyl alcohol (ethanol) or lactic acid terminates the breakdown of the carbohydrate. Organisms in which respiration ends with the formation of these waste products directly from pyruvate are called anaerobic organisms because they survive in the absence of oxygen.

Both ethanol and lactic acid are organic molecules and contain bond energies. If these organized structures are released as waste products the energy within the bonds becomes unavailable to the cell. Most organisms contain the enzymes to complete the breakdown of a carbohydrate to the final products of CO_2 and H_2O, thus releasing the maximum amount of energy available.

If a carbohydrate is completely metabolized, the pyruvate undergoes reaction into a molecular form that enters the second major enzyme system, the Kreb's cycle. The reaction involves the breakage of a carbon-carbon bond of the pyruvate so that CO_2 and a two-carbon molecule are formed. The two-carbon molecule, acetyl CoA, joins a four-carbon molecule to form a six-carbon molecule. Fats, after a preliminary breakdown process, also enter the Kreb's cycle as the acetyl CoA fragment.

Within the cycle, the two carbons from pyruvate (or fatty acids) are eventually given off as CO_2 and the restored four-carbon molecule is in the form necessary to pick up another acetyl CoA from either pyruvate or fatty acids.(see Fig. 5.1).

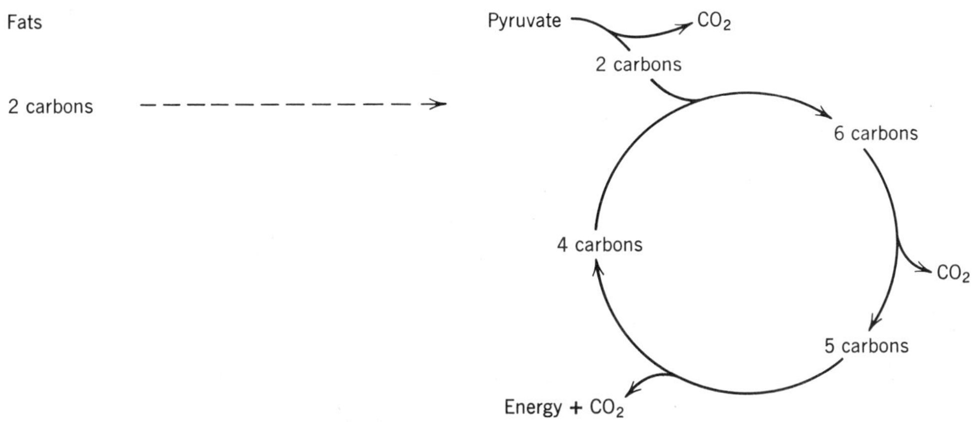

Figure 5.1

The CO_2 that has been eliminated does not contain the carbon and oxygen from the original carbohydrate or fatty acid. Although different carbon and oxygen atoms are given off for every acetyl CoA atom entering the Kreb cycle, there are two CO_2 groups formed during the cyclic process, balancing the atomic budget.

The hydrogen atoms that were part of the original carbohydrate are stripped from the molecule by dehydrogenases, enzymes that accept hydrogen atoms from molecules and incorporate them into their own structure. Three such dehydrogenases as NAD, NADP, and FAD.

molecule–H + NAD $\longrightarrow$ molecule + NADH + H^+

In the process of accepting a hydrogen atom from the molecule, NADH undergoes an ionic association with a proton of water.

The cell must rid itself of this excess hydrogen. The hydrogenated NADH + H^+ enters the third major process within the cell, called the electron transport system. The electrons from the hydrogens are passed through a series of molecules and finally join oxygen, which has been brought into the cell; water is formed. Transfer of the hydrogen electron results in energy production by a process called oxidative phosphorylation, which is closely linked to the electron transport system. Since oxygen is required by the respiration system as a final hydrogen acceptor, the process is called aerobic respiration.

The heat generated from exergonic reactions, such as the electron transport system series, is of little use to the cell. In moving the endergonic reactions in an orderly fashion, the cell must have a harnessed form of energy available for use. A group of nucleotides serve the cell as energy storage and carrier units. The principle carrier of the group contains adenine and ribose in a phosphorylated form, adenosine. Adenosine triphosphate (ATP)contains three phosphate groups, two of which are termed high-energy phosphates. In the high-energy phosphates there is a release of energy when a phosphate bond is broken. An energy gain to the cell refers to the number of ATP molecules formed from the bonding of a phosphate group to ADP (adenosine diphosphate) molecules.

ADP (adenosine diphosphate) + PO_4 + energy $\longrightarrow$ ATP (adenosine triphosphate)

The anaerobic respiratory process yields a gain of two ATP molecular per *original* glucose molecule. The complete aerobic process, because of the tremendous energy gain from the electron transport system, yields about *thirty-eight ATP molecules for each original* glucose molecule. This difference represents the relative efficiency of anaerobic and aerobic respiration.

RELATIVE EFFICIENCY OF FOOD UTILIZATION

Cells are capable of breaking down carbohydrates, fats, and proteins and of transferring the energy within these molecules to the form of high-energy phosphates. Despite the capacity of cells to utilize carbohydrates, fats, and proteins, the first two are the primary fuels used by organisms. The value of each type of fuel to the metabolizing organism may be seen in the net energy obtained by the breakdown of each of the above fuel types.

energy output − energy input = net energy gain

Direct measurement of the net energy change is impossible, so an indirect method must be used. Most of the energy from respiration is derived by the electron transport system. The number of hydrogens that pass through this system is directly proportional to the amount of energy that is derived from the system. It has been calculated that for every hydrogen molecule passing through the system from NADH + H^+, three ATP molecules are formed. The more hydrogen that is passed through the electron transport system, the more oxygen a cell must obtain in order to complete the conversion to water. Hence, the quantity of oxygen consumed is directly proportional to the amount of energy produced. Oxygen consumption can be used as a direct measurement of the relative efficiency of the breakdown of carbohydrates, fats, and proteins.

1. Hydrogen release indicates energy produced.
2. Oxygen intake indicates hydrogen release.
3. Oxygen intake indicates energy produced.

The calculation of the amount of oxygen used must be in terms of the original number of carbons present in the molecule, since the sizes of carbohydrates, proteins, and fats vary considerably. A more efficient method is to compare the oxygen intake with the CO_2 output, since each CO_2 molecule represents one carbon from the original molecule (see Fig. 5.1).

Calculation of the Respiratory Quotient

The respiratory quotient (RQ) is a value denoting the efficiency of respiration obtained by using oxygen and carbon dioxide as a reflection of the energy output of the system.

Carbohydrate

$$\begin{array}{ccc} \textit{Input} & & \textit{Output} \\ C_6H_{12}O_6 + \underline{6}\ O_2 & \longrightarrow & 6\ H_2O + \underline{6}\ CO_2 \end{array}$$

$$RQ = \frac{CO_2}{O_2} = \frac{6}{6} = 1$$

Carbohydrates characteristically have an RQ of 1.

Fats

$$\begin{array}{ccc} \textit{Input} & & \textit{Output} \\ 2C_{57}H_{110}O_6 + \underline{163}\ O_2 & \longrightarrow & 110\ H_2O + \underline{114}\ CO_2 \end{array}$$

$$RQ = \frac{CO_2}{O_2} = (2)\ \frac{57}{163} = \frac{114}{163} = 0.7$$

Fats are characterized by an RQ of 0.7.

Proteins

$$\begin{array}{ccc} \textit{Input} & & \textit{Output} \\ C_{492}H_{728}O_{59}N_{120} + \underline{615}\ O_2 & \longrightarrow & \underline{492}\ CO_2 + 364\ H_2O \end{array}$$

$$RQ = \frac{CO_2}{O_2} = \frac{492}{615} = 0.8$$

Proteins usually have an RQ of 0.8.

The RQ can be used to determine the primary material undergoing metabolism in an organism. The metabolic system you will examine is that of a seedling. The sprouts serve as a source of both fuel and enzyme. The growing plant embryo is nourished by material within the seed itself. You will determine the major type of fuel from which the organism is obtaining energy to grow by (1) by ascertaining the RQ, and (2) by chemically testing the dormant seed to find its main chemical constituent.

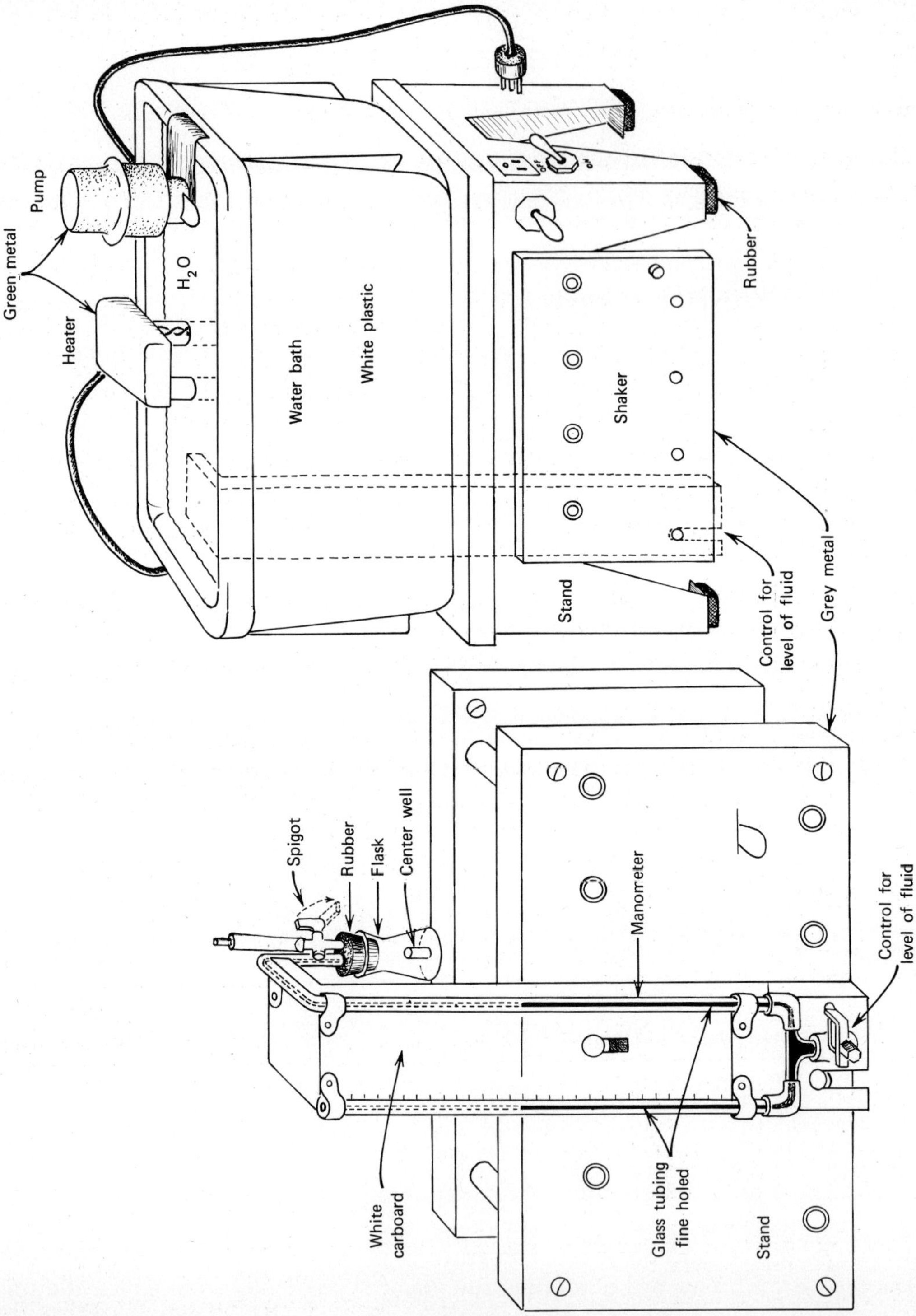
Green metal
Pump
Heater
H_2O
Water bath
White plastic
Shaker
Rubber
Stand
Control for level of fluid
Grey metal
Spigot
Rubber
Flask
Center well
Manometer
White carboard
Glass tubing fine holed
Stand
Control for level of fluid

Figure 5.2

For this experiment, you will use either a Warburg apparatus or a respirometer that you will assemble. If you use Warburg flasks, follow the directions immediately after this section. If you assemble respirometers, use the instructions on page 81.

Instructions for Measuring with the Warburg Apparatus

1. Turn on the thermostat of the Warburg bath and adjust the temperature to 37°C.
2. Place some germinating wheat or barley seedlings in the bottom of two Warburg flasks, and label the flasks *A* and *B*.
3. In flask *A*, add water to the top of the center well. In flask *B*, carefully place some 10% KOH solution with a Pasteur pipette, filling no more than 1/3 of the center well. *Be sure you do not drop any KOH on the seeds.* Prepare a filter paper wick by tearing or cutting a 1 in. by 1/2 in. piece, and folding it lengthwise. Insert the wick in the center well of flask *B*. The solution KOH is used to absorb CO_2, and the wick increases the exposure of the environment within the flask to the absorbant.
4. Replace the stoppers, syringe, plunger, and open cockstop in the flasks. Place the flasks in the Warburg bath.
5. Connect the rubber tubing to the calibrated glass tube that contains the blue solution (Brodie solution).
6. After the connection has been made, examine the calibrated tube for air bubbles. Report the presence of air bubbles to your instructor.
7. Let the system equilibrate for 5 min. before closing the stop cock. Keep the syringe in place.
8. After closing the stop cock, raise and lower the syringe plunger to make sure the system is closed. If the liquid in the tube mirrors the action of the syringe plunger, you are ready to adjust the level of the solution to zero. If the system is not closed, check for loose connections, open stopcocks, and so forth.
9. Wait 10 min. and record the level of Brodie solution on Chart 5.1, indicating that the value is (+) or (–).

CHART 5.1

		Calibrated Tube Reading	Volume Measurement
–KOH	0 time		
$O_2 + CO_2$	10 min		
+KOH	0 time		
O_2	10 min		

The difference between your 10 min. reading and zero time reading represents the volume of the gas exchange taking place as the seedlings respire. Oxygen, which would tend to lower the level of the Brodie solution in the left arm of the calibrated tube, has been taken out of the closed system. On the other hand, CO_2 has been given off as the result of respiration. In the flask with the KOH solution the CO_2 has been absorbed in the center well. The reading indicates only the oxygen taken from the closed system. However, in the other flask, the emission of CO_2 would tend to raise the level of the solution.

10. Calculate the RQ in the following way:

$CO_2 + O_2$ = total gas exchange
CO_2 = total gas exchange − O_2
total gas exchange = $O_2 + CO_2$ (10-min reading − 0-time reading)
$O_2 = O_2$ (10-min reading − 0-time reading)
CO_2/O_2 = RQ

For example

$O_2 + CO_2$	0 time 10 min	2.0 3.2	−1.2
O_2	0 time 10 min	2.0 5.6	−3.6

$O_2 = 3.6$
$CO_2 = (-)\ 1.2 - (-3.6)$
$CO_2 = -1.2 + 3.6$
$CO_2 = 2.4$
$RQ = CO_2/O_2$
$= 2.4/3.6$
$= 0.66$

Since an RQ of approximately 0.7 is characteristic of lipids, the substance metabolizing in the example flask is fat.

11. From the RQ of your data, what is the major substance undergoing breakdown?

12. Check your conclusion by performing the appropriate chemical determinations on the seedlings.

Crush several seedlings in your test tubes. Add 1 cm of water to each test tube. Run the following tests to determine the major constituent of the seedlings (see procedures in Chapter 3) – Biuret (protein), Sudan IV (fats), Benedict (reducing sugar), and IKI (starch).

Instructions for Measuring with a Respirometer

To assemble the respirometer see Fig. 5.3.

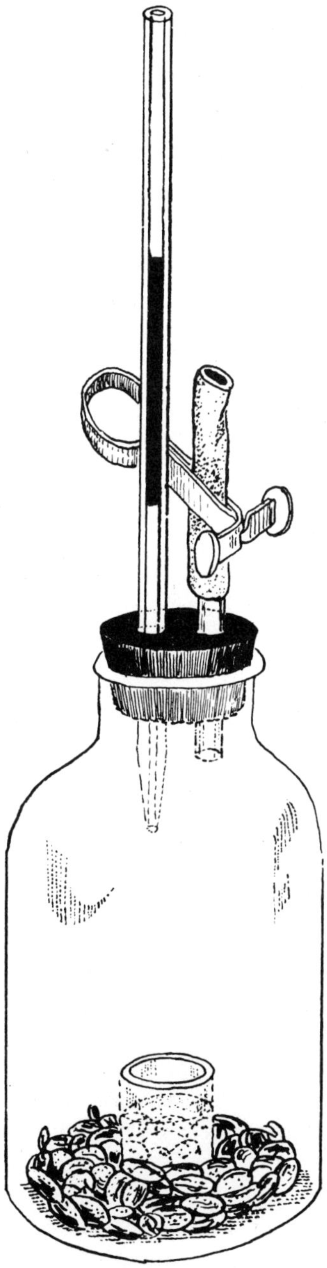

Figure 5.3

1. Place the germinating wheat or barley seedlings in the bottle ¼ in. deep. One group in the class will be delegated to run this experiment without seedlings to serve as a control.
2. Remove the pinch clamp from the auxiliary tube.
3. Fill the calibrated tube (pipette) with Brodie solution by sucking on the end. Be careful. Remember, this is not a straw. If rubber pipette bulbs are available, place on pipette and draw up the solution. If you do get any liquid in your mouth, rinse several times immediately. Fill the calibrated tube only to the 0.2-cc mark, and cover the top of the tube tightly with your finger. This may take a little practice.
4. Place the rubber stopper into the bottle, keeping your finger over the top of the calibrated tube.
5. Replace the pinch clamp, and remove your finger from the top of the tube. Why doesn't the Brodie solution flow out of the pipette?
6. Note the time when the setup is completed, and the reading of the level of the Brodie solution in the calibrated tube. Record your information on Chart 5.1.
7. Wait 10 min and record the level of the Brodie solution in the calibrated tube on Chart 5.1.

The difference between this figure and the one taken at zero time represents the volume of total gas exchange that has taken place as the seedlings are respiring. Oxygen, which would tend to lower the level of the Brodie solution in the calibrated tube, has been taken out of the closed system. On the other hand, CO_2 has been given off as the result of respiration, and this would tend to raise the level of the solution. The result of these two opposing forces is your second reading.

Since KOH (potassium hydroxide) absorbs atmospheric CO_2 and converts it into a carbonate precipitate (solid form), repeating this procedure in an atmosphere of KOH would provide a reading only of the amount of oxygen used. By appropriate calculation, the RQ and, hence, the type of fuel used by the seedlings can be ascertained.

8. Carefully disassemble the respirometer, covering the top of the calibrated tube with your finger, before removing the pinch clamp or the cork from the bottle.
9. Shake seeds to one side. Insert a vial containing a 10% KOH solution, using forceps to lower the beaker or vial into place. Be careful not to have your skin in contact with the KOH, because it may burn severely. Do not place vial on top of seeds.
10. Reassemble the apparatus as described, noting the level of solution at zero time and at 10 min. Record your information on Chart 5.1. Calculate the RQ in the following way:

$CO_2 + O_2$ = total gas exchange.
CO_2 = total gas exchange $- O_2$.
Total gas exchange = $O_2 + CO_2$ 10-min reading $-$ 0 time.
$O_2 = O_2$ 10-min reading $-$ 0-time reading.
CO_2/O_2 = RQ

For example:

$O_2 + CO_2$	0 time 10 min	2.0 3.2	−1.2
O_2	0 time 10 min	2.0 5.6	−3.6

$O_2 = 3.6$
$CO_2 = (-)1.2 - (-3.6)$
$CO_2 = -1.2 + 3.6$
$CO_2 = 2.4$
$RQ = CO_2/O_2$
$= 2.4/3.6$
$= 0.66$

Since an RQ of approximately 0.7 is characteristic of lipids, the substance metabolizing in the example flask is fat.

11. From the RQ of your data what is the major substance undergoing breakdown?
12. Check your conclusion by performing the appropriate chemical determination on the seedlings. Crush several seedlings in a test tube. Add 1 cm of water to each test tube. Run the following tests to determine the major constituent of the seedlings (see procedures in Chapter 3) – Biuret (protein), Sudan IV (fats), Benedict (reducing sugar), and IKI (starch).

Biosynthetic Reactions

Organismal growth and maintenance depends on the synthesis of structural and functional molecules, the vast majority of which are proteins, fats, and carbohydrates. Most of the energy derived from respiration is used in the synthesis of these polymers.

Biosynthetic reactions are similar to most degradative reactions in that (1) individual reactions involve small changes, (2) enzymes are required, (3) a carrier molecule is often a necessary part of the reaction, and (4) an energy source is required. The carrier molecule brings a small unit to a mother molecule; the unit attaches itself to the larger molecule through the work of an enzyme associated with high-energy molecules.

small molecule + carrier → small molecule-carrier complex
small molecule-carrier complex + ATP → activated small molecule-carrier complex
Activated small molecule-carrier complex + substrate $\xrightarrow{\text{enzyme}}$ Product + enzyme + ADP

In this exercise you will synthesize starch by mixing glucose-1-phosphate and the enzyme starch synthetase. The first step in the procedure is to isolate the enzyme.

1. Cut one quarter of a peeled potato in half.
2. Place the pieces of potato in a blender and add 65 ml of water.

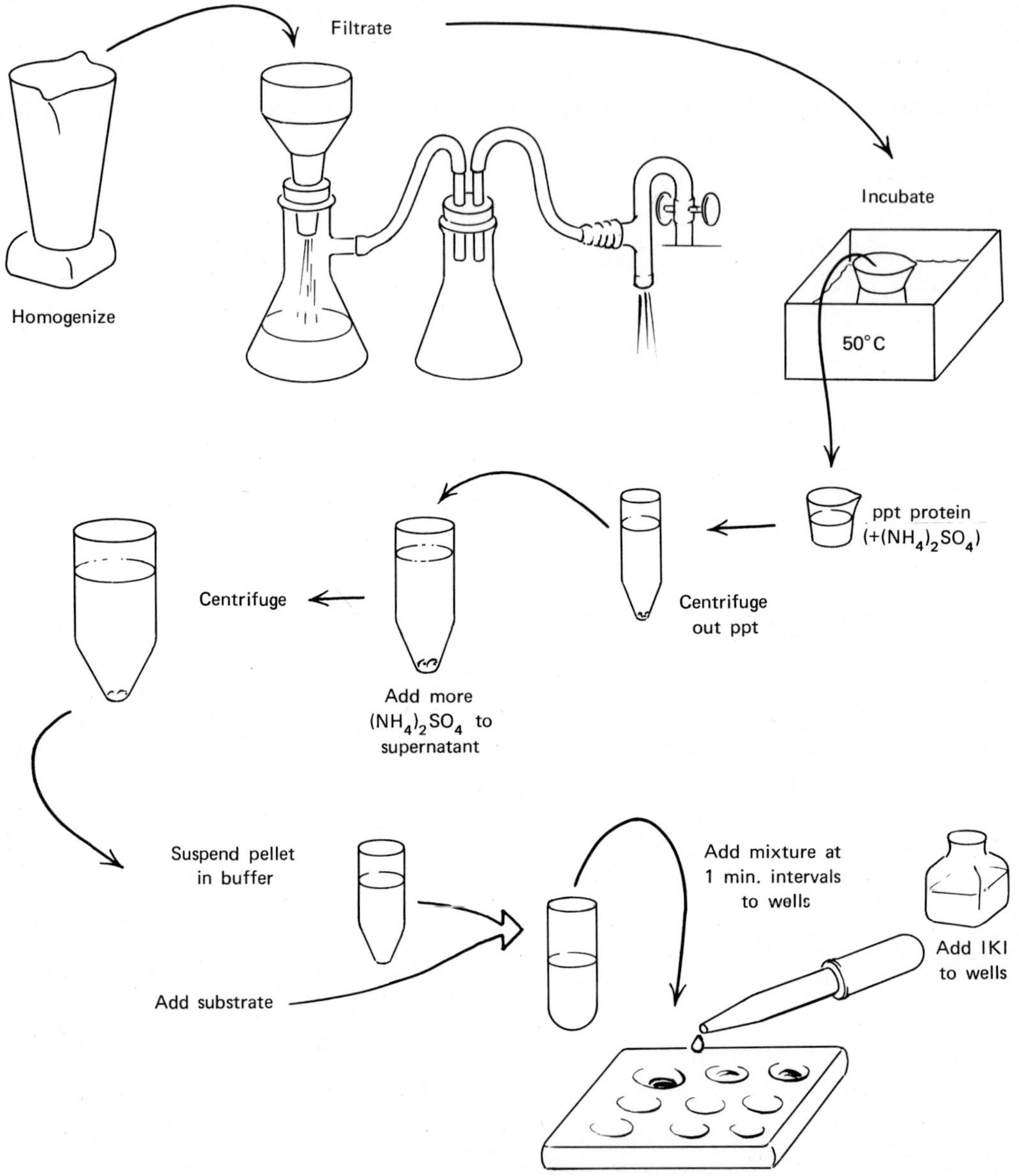

Figure 5.4

3. Pour the mixture through four layers of cheesecloth into a 125 ml beaker. The large pieces of potato have now been removed.
4. Place a piece of filter paper (Whatman #1) on a Buchner funnel and wet the paper.

5. Attach the rubber tubing from the sidearm of the flask to the sidearm of the faucet attachment, and turn on the water faucet. The rush of water past the sidearm will create a negative pressure in the sidearm and in the flask, sucking down the material in the funnel.
6. Pour the potato filtrate into the funnel.
7. After filtration is complete, remove the tubing from the faucet attachment sidearm and turn the faucet off quickly.
8. Pour the filtrate into a beaker, and place the beaker in a 50° water bath. The enzyme *amylase,* which breaks down starch to two glucose units (maltose) is destroyed at 50°C. If you want to synthesize starch, the material must be void of any factors that might break down starch.
9. After five min, measure 25 ml of the heated filtrate into a beaker, and slowly add 5 g of ammonium sulfate $(NH_4)_2SO_4$. Stir until the crystals are in solution and a brown precipitate forms. This precipitate is the unwanted protein (amylase).
10. Mark a conical centrifuge tube at 10 cm from the bottom with a wax pencil, and pour the filtrate + ammonium sulfate into the tube up to the mark.
11. Place two large beakers on the plates of a beam balance, and add toothpicks, paper clips, and so on, until the two beakers are balanced.
12. Place your filled conical centrifuge tube in one of the beakers, and have another student place his in the other beaker.
13. Add water to the centrifuge tubes as necessary to balance them.
14. Place the balanced centrifuge tubes directly opposite one another in the head of a clinical centrifuge. Wait until the head of the centrifuge is filled with balanced tubes before it turns on.
15. Turn on the centrifuge, gradually increasing the setting until it is at full speed. Continue spinning at full speed for 5 min.
16. Turn off centrifuge and let the head slow and stop by itself.
17. Remove the tube, and pour the supernatant liquid into a graduated cylinder. The pellet may not be packed too tightly so it is best to pour the supernatant with a quick sudden motion. Discard the pellet by washing the centrifuge tube.
18. Add water to the supernatant to the 25-ml level of the graduated cylinder.
19. Add 4 g of ammonium sulfate to the supernatant, stirring until it reaches solution and until another brown precipitate forms. <u>This is the precipitate you want to keep because it is mostly the enzyme starch synthetase.</u>
20. Pour the mixture into a conical centrifuge tube and repeat the balancing procedures outlined in steps #12 to 14.
21. Centrifuge at full speed for five min.
22. Remove the centrifuge tube from the head after the centrifuge has stopped, and *discard the supernatant.*
23. Add 1 ml of the buffer solution to the pellet and stir.

24. Place the tube in a rack on your desk.
25. Obtain and prepare a spot plate by placing one drop of IKI solution in one of the small wells.
26. Add 0.5 ml of glucose-1-phosphate solution to the resuspended pellet in the tube and swirl to mix.
27. Immediately after swirling, remove some of the mixture with a Pasteur pipette, and place three drops over the drop of IKI in the well of the spot plate. Record the intensity of color on Chart 5.2.
28. Place another drop of IKI in the next well on the spot plate.
29. One minute after the addition of glucose-1-phosphate, remove some of the mixture with a pipette and place three drops over the IKI drop in the well. Record intensity on Chart 5.2.

CHART 5.2

Time (min)	Intensity of Color[a]
0	
1	
2	
3	
4	
5	
6	

[a]Use +, +++, +++++, and so on for indication of color intensity.

30. Continue to take on minute readings for a total of 6 min.
31. When did starch begin to be synthesized in measureable quantities?

32. Give three ways in which the reaction could be speeded up.

33. Why would not the extract from the Buchner funnel have been as effective a source of enzyme as your final preparation?

34. Where does the energy that drives this reaction come from?

6

Photosynthesis

MATERIALS

CLASS STOCK

cultures
- 20 ml *Chlorella* grown in light
- 20 ml *Chlorella* grown in dark
- 20 ml *Schizosaccharomyces* grown in light
- 20 ml *Schizosaccharomyces* grown in dark
- *Chlorella* grown in cellaphane covered tubes (red, orange, yellow, green, and blue)

chemicals
- phenol red solution
- 0.1N HCl
- 0.1N KOH
- 95% ethyl alcohol
- IKI solution
- petroleum ether-chloroform solution

sprigs of *Anacharis* plant
leaves of *Coleus* plant
filter paper (12 in. by 1 in. strips)

STUDENT STATION

cover slips
slides
single-edged razor blade
millimeter ruler
Pasteur pipette
microscope
microscope lamp
test tubes (4)
wax pencil
aluminum foil
beaker
bunsen burner
petri plate
capillary tubes
graduated cylinder (100-ml)
stopper for graduated cylinder

INTRODUCTION

The law of conservation of matter states that matter can neither be created nor destroyed. It implies that no new carbon atoms can be created to form organized organic structures in existing organisms or in the formation of new individuals. However, organized carbon structures must be synthesized. If they were not all the carbon present in the universe would eventually accumulate in the form of CO_2 from combustion and metabolic processes. Life has continued to thrive on this planet because of the evolution of enzyme systems capable of converting carbon dioxide to organic form.

Organic synthesis through carbon dioxide fixation can occur through several processes. Photosynthesis, the synthesis of organic molecules with the use of light energy, appears to be the most efficient means of CO_2 fixation simply because it is the process used by an overwhelming proportion of organic synthetic organisms – plants.

The process of photosynthesis may be regarded as two coupled series of reactions. The first pathway requires light and involves the generation of energy (as ATP) and NADPH from this exposure to light; it is referred to as the light reaction. The other series is dependent on the energy and NADPH generated by the light reaction, but does not require light in order to proceed. This is called the dark reaction. The dark reaction involves CO_2 fixation reactions and the subsequent formation of glucose by means of reactions closely analogous to a reversal of part of the anaerobic glycolysis scheme.

The light reaction involves the excitation of chlorophyll *a* and chlorophyll *b* by light. This excitation of chlorophyll *a* produces a high-energy electron, which, while passing through a system analogous to the electron transport system, generates ATP. Excitation of chlorophyll *b* also produces a high-energy electron, but this electron plays a role in the splitting of water.

$$2\ H_2O \xrightarrow[\text{light}]{\text{electron chlorophyll } b} 2\ H^+ + 2\ \text{electron} + H_2O$$

(NADPH)

The electron from the splitting of water then proceeds through a transport system with a concomitant generation of energy. The generation of energy through the transport system associated with photosynthesis is called photosynthetic oxidation.

Chlorophyll *a* and *b* are the only photosynthetic pigments capable of undergoing this energy-yielding reaction. Other photosynthetic pigments present in the chloroplast, such as xanthophylls and carotenes, are excited by light, and may transmit this energy to chlorophyll *a* or *b*.

In the dark phase, energy and the NADPH derived from the chlorophyll *a* and *b* reactions are used to form a more complex organic product. Carbon dioxide is fixed to a five-carbon ribulose di-phosphate molecule to form a branched six-carbon molecule, which, after several intermediate steps, splits into two three-carbon molecules.

$$C_5H_{10}O_5 + CO_2 \longrightarrow\longrightarrow 2\ C_3H_6O_3$$

The following series of experiments will demonstrate the requirements for light, chlorophyll, CO_2, and the products of photosynthesis, starch, and oxygen.

THE LIGHT REQUIREMENT

Your instructor has prepared several cultures of microorganisms for your examination. The tubes contain either the green alga *Chlorella* or the yeast *Schizosaccharomyces.* Both are in liquid media containing *no* organic molecules except a nonmetabolized one necessary to stabilize the medium. The cultures were grown under light and dark conditions

Preparation of the Counting Chamber

To determine the relative number of organisms that exist under the prescribed conditions, a method of accurate counting must be devised. The standard method is to estimate the number of organisms per milliliter volume of the culture flask. Since the number of cells per milliliter often reaches astronomical figures, it is impossible to count directly. Also, 1 ml of liquid will flood a slide and will not fit beneath a cover slip. One solution to this problem is to take a known sample of the solution, and count the cells in this calibrated amount. Multiplication by the appropriate factors will yield the number of cells per milliliter of culture.

1. Obtain a plastic cover slip, a single-edged razor and a millimeter ruler.
2. Holding the cover slip over Fig. 6.1, etch the lines on the cover slip with the razor.

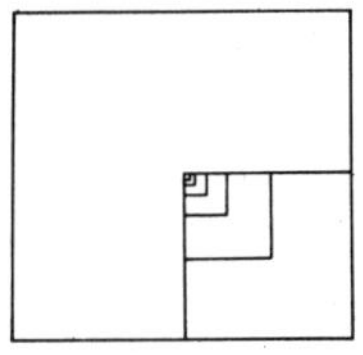

Figure 6.1

3. Each of the letters *(a, b, c, d, e, f,* and *g)* delineates a square.
   a = the entire cover slip,
   b = ¼ the area of the cover slip,
   c = 1/16 the area of the cover slip, including areas represented by *d, e, f,* and *g,*
   d = areas represented by *d, e, f,* and *g,*
   e = areas represented by *e, f,* and *g,*
   f = areas represented by *f* and *g,*
   g = smallest square portion of the cover slip.
4. When the culture fluid is transferred from a Pasteur pipette, a single drop is equal to 1/24 ml. By counting the number of cells in area *a* and multiplying by 28, the number of cells per milliliter can be calculated. (The factor for multiplication is greater because the liquid from one drop of a Pasteur pipette does not fit under the cover slip and there is a calculated seepage factor taken into account.)

Square	Factor
a	2.8×10^1
b	1.1×10^2
c	4.3×10^2
d	1.7×10^3
e	6.8×10^3
f	2.7×10^4
g	1.1×10^5

Using the Counting Chamber

1. After shaking the culture, obtain a drop of *Chlorella* suspension that was grown in the light and make a wet mount using a drop from a Pasteur pipette (held at a 45° angle to the slide).
2. Observe the slide under low power and determine which of the delineated areas contains between 10 and 30 cells.
3. Count the number in that area and multiplying by the appropriate factor to obtain the number of cells per milliliter of the culture.
4. Enter the data on Chart 6.1.
5. Repeat the procedure with the other culture of *Chlorella* and yeast, cleaning the slide and cover slip thoroughly after each count has been made.

CHART 6.1

Tube	Condition	Count	Area	Factor	Cells/ml
1	Chlorella (light)				
2	Chlorella (dark)				
3	*Schizosaccharomyces* (light)				
4	*Schizosaccharomyces* (dark)				

6. Draw a *Schizosaccharomyces* cell and a *Chlorella* cell below, indicating distinguishing features of each.

7. What is the most obvious difference between the yeast and the green alga?

8. Which organism requires light for growth?

THE QUALITY OF LIGHT

Visible light is actually a mixture of many different colors or light waves, which may be considered to be different qualities of light. In order to ascertain whether different kinds of light are equally efficient for photosynthesis, cultures of *Chlorella* have been wrapped in colored cellophane so that the light entering the tube is that of the color of the cellophane. Using the counting chamber prepared in the previous exercise, you will determine which lights best support growth.

Notice that each of the tubes is surrounded by a different number of layers of cellophane. This is done to ensure that each of the tubes is exposed to the same amount of radiant energy.

1. Obtain a drop of one of the cultures in a Pasteur pipette and prepare a wet mount using the counting chamber as your cover slip.
2. Determine which area has between 10 and 30 cells and count, multiplying by the appropriate factor to arrive at the number of cells per milliliter.
3. Enter your data on Chart 6.2.
4. Repeat for all the cultures, washing the cover slip and slide between each of the counts.
5. Which lights are the most efficient for photosynthesis?
6. The tubes were wrapped in different layers of cellophane to permit the same intensity of light throughout. How did this eliminate an important variable?

CHART 6.2

Tube Color	Cells/ml
Red	
Orange	
Yellow	
Green	
Blue	

THE CARBON DIOXIDE REQUIREMENT

1. Fill two test tubes about 2/3 full with water.
2. Add 3 drops of phenol red solution. Phenol red is an indicator dye that turns pink under alkaline conditions, and yellow with acid.
3. Tap water may be slightly alkaline and the tubes may be pink. Add 0.1N HCl drop by drop until the tubes turn slightly yellow. Both tubes are slightly acidic in character.
4. Label the test tubes *A* and *B* with a wax pencil. To tube *A* add one drop of 0.1N KOH. The liquid should turn slightly pink.
5. Repeat the procedure with tube *B*. Both tubes are now slightly alkaline in character.
6. Bubble the last part of several breaths into test tube *A* with a Pasteur pipette until the liquid turns yellow. When CO_2 is added to a solution the following reaction takes place

   $$CO_2 + H_2O \longrightarrow \underset{\text{carbonic acid}}{H_2CO_3}$$

   Now each of the test tubes is balanced on either side of neutrality.
7. Obtain two sprigs of the aquatic floating plant *Anacharis* and cut to 1½ in lengths with a razor blade.
8. Insert the sprigs upside down into test tubes *A* and *B*.
9. Cover both test tubes with aluminum foil, twisting it around the neck of the tubes. Place the test tubes near a light source (a 250-Watt bulb or a brightly lit window).
10. Check every few minutes for a color change.
11. Note the tube and the time of the color change on Chart 6.3.
12. Perform the same experiment, but keep both tubes in the dark for twice the length of time, using fresh tubes but the same sprigs of *Anacharis.*
13. Enter your data on Chart 6.3.
14. Carbon dioxide in solution yields carbonic acid. How did the results of the test tubes in the dark show that respiration was occurring?

CHART 6.3

Tube	Initial Condition	Minutes for Change in Light	Minutes for Change in Dark
A	slightly acidic		
B	slightly basic		

15. How did the reaction in the light show that CO_2 is necessary for photosynthesis?

16. Why was KOH added to both tubes?

17. Was any phenomenon other than the color change noticed in the tubes when they were in the light?

THE PRODUCT OF PHOTOSYNTHESIS

You have seen that *Chlorella* can perform photosynthesis. It would seem that chlorophyll was required, since the yeast, which did not contain any such pigment, did not photosynthesize. There is no need to demonstrate more clearly that chlorophyll is indeed required for photosynthesis, and this fact will be assumed proven for the following experiment.

The usual product of photosynthesis is starch. In order to test for the presence of starch in the plant it is necessary to remove the pigments that would mask the results of the chemical test.

1. Obtain a leaf of the variegated *Coleus* plant.
2. On the next page draw the leaf, indicating the position of colors.
3. Boil the leaf in a beaker containing 95% alcohol (½ in. depth) over a low Bunsen burner flame or a hot plate, having a wet towel handy to throw over the beaker in case the alcohol catches fire.
4. After a considerable amount of green chlorophyll is extracted and the leaf appears white, place the leaf in a petri plate and add IKI solution.
5. Draw the leaf again, indicating the areas that contain starch.
6. Boil the alcohol solution down to a few milliliters and put aside.
7. Which of the areas of the leaf seem to have the greatest concentration of starch?

(1) coleus leaf (2)

8. Does the slight accumulation of starch in the other areas indicate that a small amount of photosynthesis might have taken place, or is there some other way in which this starch accumulation might have occurred?

ANALYSIS OF THE PIGMENT-PAPER CHROMATOGRAPHY

One of the major means of analyzing complex combinations of molecules has been paper chromatography. Paper chromatography utilizes the differing migration properties of molecules to separate and distinguish between them. You have undoubtedly observed what happens when the end of a sugar cube is placed in a cup of coffee. The coffee migrates slowly up the sugar cube until the entire cube is wet. The same happens when one end of a piece of absorbent paper is placed in a liquid. The liquid will be absorbed by the paper and will migrate upward until the paper is saturated. If material that is soluble in the liquid is placed on the paper, this material will dissolve as the liquid moves upward. If this material is a combination of substances, some will be more soluble in the liquid than others and will therefore tend to move more quickly. Also, since substances vary in their attraction to the paper they will migrate at different rates. The result is that as the liquid moves up the paper, a separation of the substances of the original material will occur. The longer the migration is permitted to occur, the greater the separation of the substances will be. This explanation is an oversimplification of a very complex physical-chemical phenomenon.

1. Obtain a finely drawn out glass tube and place it in the chlorophyll-alcohol solution saved from the previous experiment. The solution will be drawn up the tube by capillary action.
2. Take a strip of filter paper 12 in. long and 1 in. wide and taper one end to a point.
3. About ½ in. above the apex of the point, touch the paper with the glass tube to release a drop of chlorophyll solution. Try to keep the drop very small.
4. Wait until this dries, and release another small drop in the same place.
5. Repeat until all the solution in the tube is used up (about 6 drops).
6. To a 100-ml graduate cylinder add 3 ml of a petroleum ether-chloroform solution. (Be careful that all Bunsen burners in the laboratory are off before working with this highly inflammable solution.)

7. Carefully place the filter paper, point downward, into the petroleum ether-chloroform solution, being careful not to touch the sides of the cylinder with the filter paper or to put the part of the paper with the spot of chlorophyll directly in the solution.
8. Plug the top of the cylinder with a rubber stopper or plastic food wrap to make the container airtight.
9. After 10 min observe the filter paper and notice the colored bands.
10. Draw the filter paper indicating the position of the chlorophyll *a* (blue green), chlorophyll *b* (yellow green), xanthophyll (yellow), and carotene (orange or red or deep yellow). (Figure 6.2 illustrates the apparatus.)
11. Which pigment is the most soluble? The least soluble? Why did you come to these conclusions?

12. Why was petroleum ether-chloroform used instead of water?

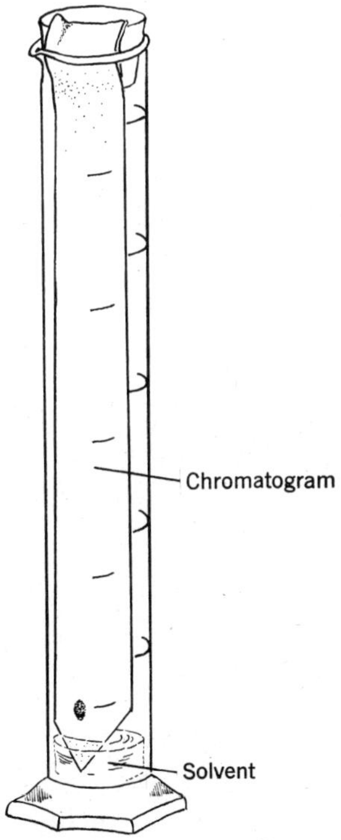

Figure 6.2

7

Nutrition

MATERIALS

CLASS STOCK

starch agar innoculated with *Escherichia coli*
starch agar innoculated with *Bacillus subtilis*
EMB agar innoculated with *Escherichia coli*
EMB agar innoculated with *Bacillus subtilis*
chemicals
  IKI solution
  Nessler's reagent
  Trommsdorf solution
  25% sulfuric acid
  $CaCl_2$
  $CaNO_3 \cdot 4H_2O$
  $MgCl_2 \cdot 6H_2O$
  $MgNO_3 \cdot 6H_2O$
  $MgSO_4 \cdot 7H_2O$
  KCl
  KNO_3
  KH_2PO_4
  $NaNO_3$
  $NaH_2PO_4 \cdot H_2O$
  Na_2SO_4

STUDENT STATION

slides (3)
Pasteur pipette
flasks (8 2000-ml)
glass rods (8)
bottles (8)
aluminum foil
culture test tubes (9)
wax pencil
counting chamber cover slip
corks for test tubes (9)
test tube rack

CLASS STOCK (cont'd)

- $FeCl_3$
- EDTA
- H_3BO_3
- $MnCl_2 \cdot 4H_2O$
- $ZnCl_2$
- $CuCl_2 \cdot 2H_2O$
- $NaMO_4 \cdot 2H_2O$
- Solution I:
  - $NaCl_3$ 3.0 g
  - $MgSO_4$ 1.0 g
  - $MgCl_2 \cdot 4H_2O$ 0.4 g
  - KCl 0.08 g
  - NH_4NO_3 0.4 g
  - $CaCl_2$ 0.15 g
  - H_2O 100 ml
- Solution II:
  - K_2HPO_4 15 g
  - glycine 2.5 g
  - glycyl-glycine 5.0 g
  - citric acid 0.3 g
  - H_2O 100 ml
- Solution III:
  - $FeCl_3 \cdot 6H_2O$ 2.5 g
  - $MnCl_2 \cdot 4H_2O$ 0.2 g
  - $ZnSO_4 \cdot 7H_2O$ 0.02 g
  - $CuSO_4$ 0.12 g
  - $CaCl_2 \cdot 6H_2O$ 0.002 g
  - $NaMO_4 \cdot 2H_2O$ 0.0001 g
  - citric acid 10.0 g
  - H_2O 1000 ml
- Solution IV:
  - vitamin B_1 10.0 mg
  - H_2O 1000 ml
- vitamin B_{12}

4% peptone tubes; half innoculated with soil; the other half sterile
pH paper
nitrite formation tubes innoculated with soil
distilled water (preferably glass distilled)
nonabsorbent cotton
eight- to ten-day-old tomato seedlings

INTRODUCTION

Enzyme systems, and hence metabolic reactions present in an organism, directly dictate the nutrition of that organism. Several proposals have been made to classify organisms according to their mode of nutrition, and most have been used at some time for nutritional descriptions of organisms. Many nutritionists presently prefer a classification system based on the type of energy source and source of electrons that generate ATP when passing through a transport system. These processes are preferred because they reflect the most fundamental aspect of the nutritional picture – energy requirement for continued existence.

The source of energy may be either light (photo) or chemical (chemo). The source of electrons may be either from inorganic (litho) or organic (organo) molecules. There are thus four basic classifications in the system, all of which end in the latin word for eat, troph.

1. Energy source–light; electron source–inorganic molecules: Photolithotroph
2. Energy source–light; electron source–organic molecules: Photoorganotroph
3. Energy source–chemical; electron source–inorganic molecules: Chemolithotroph
4. Energy source–chemical; electron source–organic molecules: Chemoorganotroph

The most common are the photolithotrophs, which include some bacteria, many protists, and all plants, and the chemoorganotrophs, which include many bacteria, protists, and all animals. Chemolithotrophic and photoorganotrophic nutrition is found only in some bacterial groups.

Although this classification is desirable because it focuses on a fundamental aspect of organisms, it is not suitable for all types of nutritional discussions. Sometimes it is necessary to use terms of other nutritional classifications. For example, one classification system employed for many years was constructed with reference to carbon sources. When discussing carbon requirements, it is necessary to use the terms autotrophy for organisms capable of producing organic molecules from carbon dioxide, and heterotrophy, for organisms that must take in organic molecules. Hence, a plant is classified as a photolithotroph according to the energy criteria, and an autotroph according to the source of carbon.

Furthermore, using either energy or carbon source criteria does not indicate the requirements for a variety of other chemicals. Venus flytrap, for instance, is a photolithotroph and an autotroph according to the energy and carbon source, respectively, but requires an organic form of nitrogen. Some algae are also photolithotrophs and autotrophs, but require certain vitamins, not as carbon sources, but as organic molecules for which they do not have the enzyme systems to produce.

Although nutritional classifications serve their purpose in providing well-defined terms, they do not present a total picture of an organism. Organisms must have sources of nitrogen, sulfur, oxygen, and phosphorus, as well as many other elements. Also organisms need organic molecules necessary for metabolism that it cannot produce itself and energy and carbon sources. This exercise will provide the study of several nutritional requirements of organisms. Since energy requirements are difficult to illustrate, the study will be limited to elemental and vitamin requirements.

THE CARBON REQUIREMENT

Regardless of the way in which carbon atoms are positioned in molecules, there is some organisms with the metabolic machinery capable of utilizing these carbons as a food source. Because of the great variety of carbon compounds utilized by organisms, the total amount of carbon in the universe can be viewed as existing in a steady state condition although being shared by all forms of life. Thus the total carbon in the universe passes from organic to inorganic states through respiration and decay, and back to organic states through photosyntheses – the carbon cycle.

The aspect of the carbon cycle involving autotrophic (photosynthetic) organisms was covered in the previous chapter. This section, therefore, is devoted to elucidating some features of heterotrophic (nonphotosynthetic) organisms, by examining different organisms and their capacity to metabolize different foodstuffs in partial fulfillment of their carbon requirement.

Your instructor has prepared four petri plates for your examination. Each contains a gelled substance, agar, which provides a semisolid surface on which microorganisms can grow. Since agar alone does not provide foodstuffs for the growing organisms, nutrients must be added. Generally, a nutrient solution containing carbohydrates, peptides, and vitamins is added to the agar to sustain the microorganisms. However, for nutritional studies special food material or vital stains may be added to test the ability of an organism to utilize a particular food.

Starch agar is a nutrient medium plus starch. Almost all aerobic microorganisms will grow on this medium because of the abundance of other carbon sources, organic nitrogen sources, and vitamins. However, only those organisms that contain the enzymes to break polysaccharide bonds are able to utilize starch as a nutrient. This may be indicated by flooding the plate with IKI soution. The absence of a black IKI reaction around the colonies indicates that starch has been broken down.

Eosin-methylene blue (EMB) agar is specialized for the detection of bacteria capable of lactose utilization of fermentation. Lactose is a disaccharide formed by the union of a glucose and a galactose molecule. Bacteria capable of lactose fermentation take on a red dye, and appear as a red brassy color on the red-green medium.

1. Obtain your four plates from the instructor.
2. Flood the two starch plates with IKI solution.
3. Examine the two EMB plates.
4. Record your observations on Chart 7.1

CHART 7.1

Agar	Bacteria	Observations
Starch	*Escherichia coli*	
	Bacillus subtilis	
EMB	*Escherichia coli*	
	Bacillus subtilis	

5. Which of the bacteria is capable of utilizing starch as food?

6. The utilization of starch as food indicates the presence of an enzyme. What is the action of this enzyme?

7. Which of the bacteria is capable of utilizing lactose as a food?

8. What is the action of the enzyme whose presence is implied by the utilization of lactose?

9. Do the results imply that a bacterium not capable of utilizing lactose is incapable of using glucose or galactose as food material? Explain.

10. Why would both of these bacteria be considered heterotrophs?

11. Where on the following cycle would you place each of the bacteria you have studied?

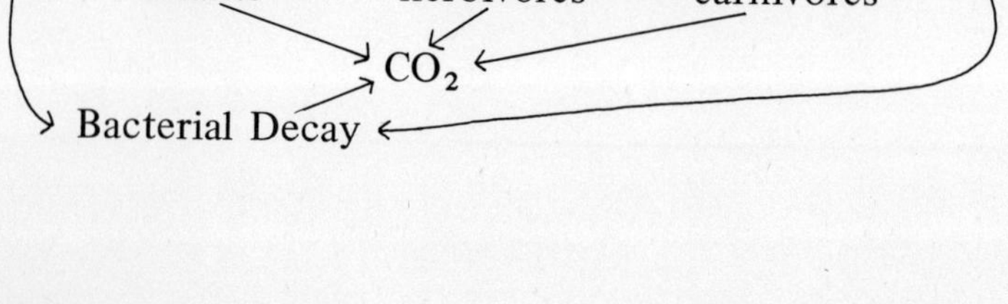

THE NITROGEN REQUIREMENT

Just as different types of organisms are capable of utilizing different carbon sources, different types of organisms are capable of utilizing different nitrogen sources. Most plants can metabolize nitrogen to aminate organic acids only when the nitrogen enters the plant in nitrate (NO_3^+) form. Animals, on the other hand, can use nitrogen only when it is already in organic form, such as in a plant or animal protein. Other types of organisms can only utilize nitrogen in the form of ammonia (NH_3), or nitrite (NO_2^+), or atmospheric nitrogen (N_2). The <u>nitrogen cycle</u> generates all of the varied nitrogen requirements. It permits wastes and elements that are bound to organic molecules in dead organisms to be returned to a pool available to living forms.

This exercise will examine two of the nitrogen conversions that take place during the cycle.

Ammonification

Two test tubes have been filled with 4% peptone solution, a medium that is rich in polypeptides (and therefore, nitrogen). One was innoculated with some soil, the other was left sterile. Both tubes were incubated for 48 hr.

1. Place one drop of Nessler's reagent at each end of a slide.
2. Add a drop of the liquid from one peptone tube to one of the drops of Nessler's reagent, and a drop from the other peptone tube to the other drop of Nessler's reagent.
3. A yellow color indicates the presence of ammonia. From which specific part or parts of the polypeptide molecules does the NH_3 group originate?

4. What is in the soil that metabolizes polypeptides?

5. Are all the bacteria in the soil taking part in metabolizing polypeptide nitrogen to ammonia? Explain.

6. Not all organisms give off nitrogen waste in the form of ammonia. Why? (Clue: Test the pH of the tube in which you found ammonia.)

7. In this test, the bacteria were provided with a protein or polypeptide source. What is the source of nitrogen in their natural habitat?

8. What part do some soil bacteria play in the nitrogen cycle shown in Fig. 7.1?

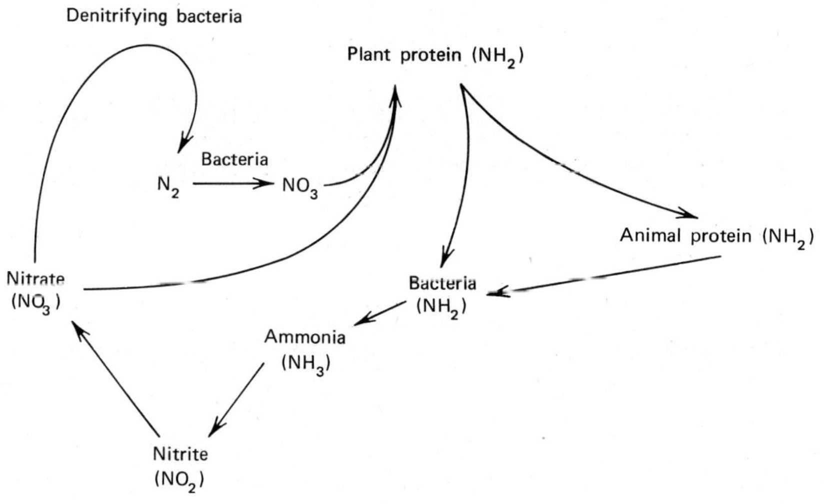

Figure 7.1

Nitrification

For your examination, the instructor has prepared five tubes containing nitrite formation medium. Each medium was innoculated with 0.1 g of soil (neutral or slightly alkaline pH) a week apart from the rest and was incubated at room temperature. The nitrite formation medium is inorganic, having no direct protein source. The only nitrogen in the medium is in the form of ammonia $(NH_4)_2SO_4$.

1. On a clean slide, mix three drops of Trommsdorf solution with one drop of sulfuric acid.
2. To the mixture on the slide add a drop of the culture that has been incubated for the *shortest* duration, using a pipette or glass rod since metal may give a false reaction.
3. A blue-black color indicates the presence of nitrites.
4. Record your results on Chart 7.2.

CHART 7.2

Weeks of Incubation	Presence of Ammonia	Presence of Nitrite
1		
2		
3		
4		
5		

5. Repeat the test for ammonia, using Nessler's reagent as described in the previous exercise.
6. Clean the slide, and repeat tests for nitrite and ammonia on the remaining tubes.
7. Enter your data on Chart 7.2 and then convert it to graph form on Fig. 7.2.
8. How long does it take before an appreciable amount of NH_3 is converted to NO_2?

9. How does this compare with the length of time necessary to convert organic nitrogen to ammonia in this experiment?

10. What does this time difference reflect regarding the relative numbers of organisms present in soil that accomplish ammonification and those that accomplish nitrification?

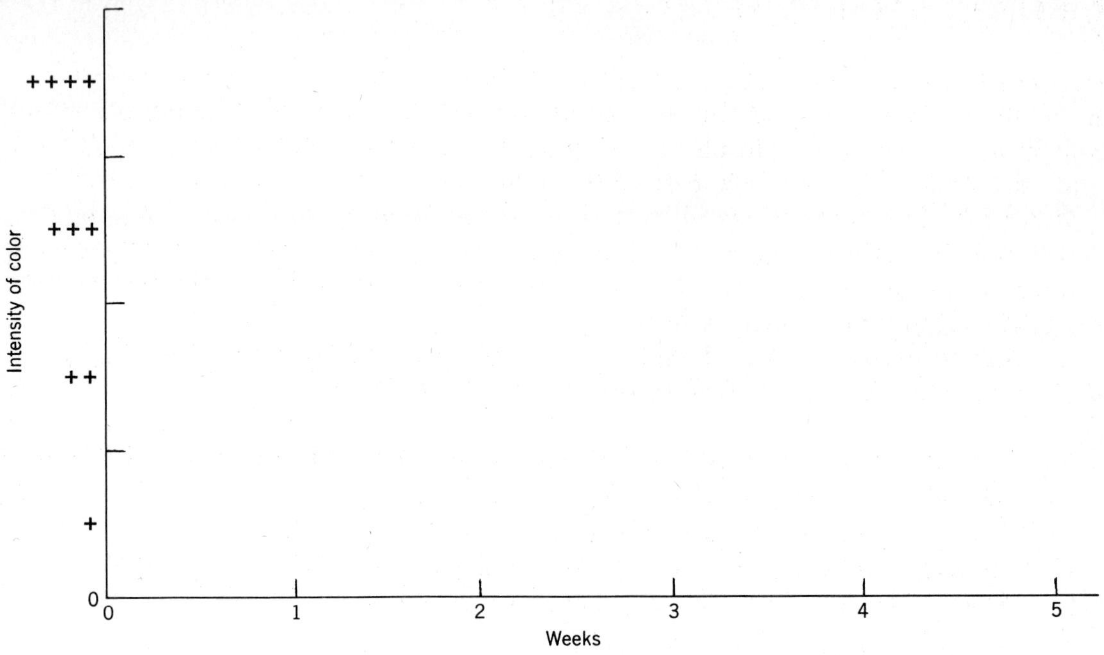

Figure 7.2

11. What is the organic nitrogen source for the ammonifying bacteria?

12. What is the significance of the relative number of organisms and their respective conversions to the nitrogen cycle?

13. What are the two possible directions that an NO_2 molecule can take after being formed by nitrification?

14. Under natural conditions, what is the source of ammonia on which nitrifying bacteria will grow?

INORGANIC REQUIREMENTS

In addition to carbon and nitrogen, organisms require a variety of inorganic elements. Some of these, such as phosphorus and sulfur, are necessary for the structure of certain organic molecular species. Phosphorus (P) in a phosphate form (PO_4) is part of the nucleotide molecule. Sulfur (S) is part of the amino acid cysteine.

Other elements are required in their ionic forms. Iron in the ferric form (Fe^{++}) is responsible for cytochrome activity. Magnesium (Mg^{++}) is incorporated into chlorophyll structure. Still other ionic forms of elements are necessary for maintaining the balance of materials and the pH in cellular systems.

In order to determine what happens to an organism when different inorganic requirements are not met, you will perform the following experiment.

1. Make up the following separate solutions. (These may be made up in groups, or divided among different students in the class.
   (a) $CaCl_2$ 11.1 g in 100 ml distilled water
   (b) $Ca(NO_3)\cdot 4H_2O$ 23.6 g in 100 ml distilled water
   (c) $MgCl_2\cdot 6H_2O$ 20.3 g in 100 ml distilled water
   (d) $Mg(NO_3)\cdot 6H_2O$ 25.6 g in 100 ml distilled water
   (e) $Mg(SO_4)\cdot 7H_2O$ 24.6 g in 100 ml distilled water
   (f) KCl 7.4 g in 100 ml distilled water
   (g) KNO_3 8.5 g in 100 ml distilled water
   (h) KH_2PO_4 13.6 g in 100 ml distilled water
   (i) $NaNO_3$ 8.5 g in 100 ml distilled water
   (j) $NaH_2PO_4\cdot H_2O$ 13.8 g in 100 ml distilled water
   (k) Na_2SO_4 14.2 g in 100 ml distilled water
   (l) Iron solution[1]:
   $FeCl_3$ 5.0 g in 1 l distilled water
   EDTA 5.0 g in 1 l distilled water
   (m) Microelement solution[1]:
   H_3BO_3 2.5 g in 1 l distilled water
   $MnCl_2\cdot 4H_2O$ 1.5 g in 1 l distilled water
   $ZnCl_2$ 0.10 g in 1 l distilled water
   $CuCl_2\cdot 2H_2O$ 0.05 g in 1 l distilled water
   $NaMo_4\cdot 2H_2O$ 0.05 g in 1 l distilled water
2. Place 1 l of distilled water into each of eight 2000-ml flasks.
3. Label the flasks as follows:
   (a) complete
   (b) without potassium (–K)
   (c) without calcium (–Ca)

[1] The iron and microelement solutions may be made by the instructor before the laboratory exercise and used for all laboratories that week.

(d) without nitrogen (–N)
(e) without magnesium (–Mg)
(f) without phosphorus (–P)
(g) without sulfur (–S)
(h) without iron (–Fe)

4. Following Chart 7.3, place the appropriate number of milliliters of the solutions you have prepared into each flask.
5. What are the sources of the following inorganic elements and ions in the complete medium?

Ca^{++}	sulfur
nitrogen	phosphate ion
oxygen	iron
nitrate ion NO_3^-	microelements
magnesium	

6. In preparation of a medium without potassium why do you eliminate KNO_3 and KH_2PO_4 but add $NaNO_3$ and NaH_2PO_4?

7. Mix each of the flasks with glass rods (use a different glass rod for each).
8. Transfer each medium from the flask to appropriately labeled bottles. Cover the bottles with each aluminum foil. The labels should include your name, the medium within the bottle, and the date.
9. Obtain four eight- to ten-day-old tomato seedlings for each bottle. You will need a total of 32 seedlings.
10. Wrap four seedlings in a strip of nonabsorbent cotton so that the stems are on one side of the wrapping and the roots on the other (see Fig. 7-3).
11. Place the roots in the medium in one of your bottles, and twist the cotton plug into place.
12. Repeat for all bottles of medium.
13. Place the medium bottles with the seedlings near a window or on lighted shelves.
14. After one week, check each group of seedlings for deficiencies. Enter your observations on Chart 7.4 according to the symptoms you observe (leaf curling, leaf coloration, stunted growth, pitted or mottled leaves, and so forth.
15. After observing and recording your data, return the bottles to the lighted shelves or window for another week.
16. Repeat your observations and recordings on Chart 7.4 the following week.
17. Are all deficiencies manifested in the same way?

CHART 7.3

Flask	$CaCl_2$	$Ca(NO_3)_2 \cdot 4H_2O$	$MgCl_2 \cdot 6H_2O$	$MgNO_3 \cdot 6H_2O$	$MgSO_4 \cdot 7H_2O$	KCl	KNO_3	KH_2PO_4	$NaNO_3$	$NaH_2PO_4 \cdot H_2O$	Na_2SO_4	Iron Sol.	Micro. Sol.
Complete	—	3 ml	—	—	2 ml	—	2 ml	2 ml	—	—	—	1 ml	1 ml
Without potassium	—	3 ml	—	—	2 ml	—	—	—	2 ml	2 ml	—	1 ml	1 ml
Without calcium	—	—	—	3 ml	2 ml	—	2 ml	2 ml	—	—	—	1 ml	1 ml
Without nitrogen	3 ml	—	—	—	2 ml	2 ml	—	2 ml	—	—	—	1 ml	1 ml
Without magnesium	—	3 ml	—	—	—	—	2 ml	2 ml	—	—	2 ml	1 ml	1 ml
Without phosphorus	—	3 ml	—	—	2 ml	2 ml	2 ml	—	—	—	—	1 ml	1 ml
Without sulfur	—	3 ml	2 ml	—	—	—	2 ml	2 ml	—	—	—	1 ml	1 ml
Without iron	—	3 ml	—	—	2 ml	—	2 ml	2 ml	—	—	—	—	1 ml

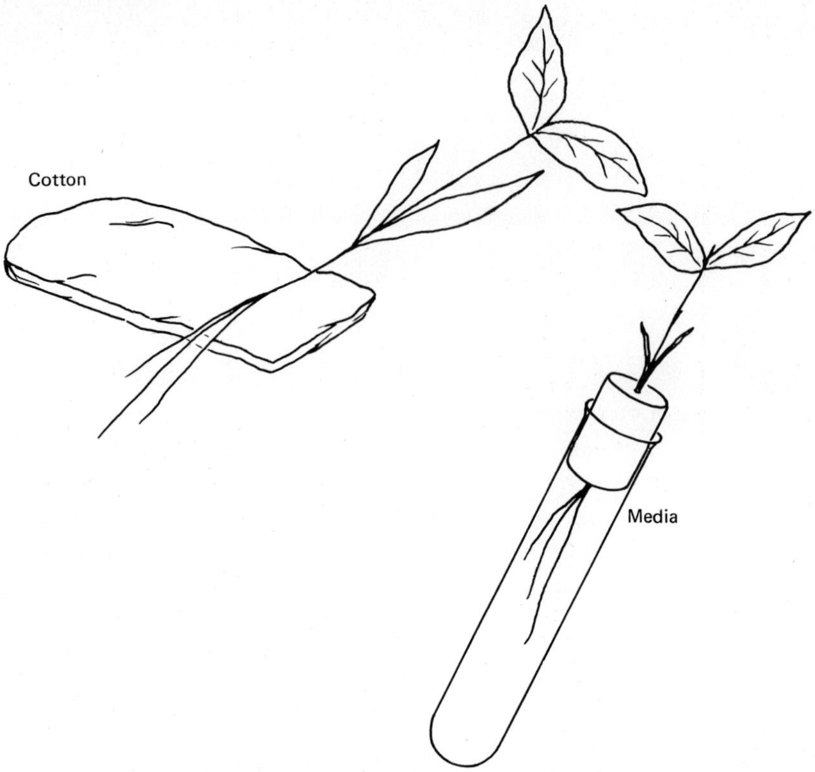

Figure 7.3 Preparation of seedlings.

CHART 7.4

Medium	Deficiency Symptoms	
	One Week	Two Weeks
Complete		
Without potassium		
Without calcium		
Without nitrogen		
Without magnesium		
Without phosphorus		
Without sulfur		
Without iron		

18. What compounds that are necessary for growth and maintenance have phosphate groups? What would happen if phosphate was limiting? How is a phosphate deficiency expressed in the growth and form of the plant?

19. Both nitrogen and sulfur are necessary for protein structure. Sulfur is also important as part of certain enzymes (for example, coenzyme *A*). If sulfur deficiency affects proteins, the symptoms should be similar to those obtained in the nitrogen deficiency series. If sulfur deficiency primarily affects metabolic reactions, the symptoms would be quite different. What is affected by sulfur deficiency?

20. What molecules in the plant are dependent on the presence of iron for their activity? Do the results of iron deficiency coincide with what you would expect if these molecules were not active?

21. Potassium, calcium, and magnesium ions (K^+, Ca^{++}, and Mg^{++}) are important for the maintenance of the ionic balance in cells. What happens when the ionic balance is upset? Are the symptoms of these three deficiencies similar?

22. Some ions are mobile, that is, they can move from old to new regions. Which ions are not mobile (deficiency expressed in growing tip and young leaves)?

THE VITAMIN REQUIREMENT – AUTOTROPHIC NUTRITION

As previously mentioned, the nutrition of an organism depends on the presence of enzyme systems, which in turn depends on the capacity of the organism to synthesize the protein portion of the enzyme and either to synthesize or to obtain the coenzyme portion of the enzyme. Although all organisms can synthesize the apoenzyme, many do not have the enzyme systems necessary for synthesis of the coenzyme. Hence, the coenzyme must be taken in and becomes added to the nutritional requirements of the organism.

Most of the water-soluble vitamins are coenzymes. Vitamin B_2, riboflavin, is easily converted to FAD (flavin adenine dinucleotide), the coenzyme active both as a deyhydrogenase in metabolism and as a part of the electron transport system. Since neither metabolism nor the electron transport system functions properly without FAD, vitamin B_2 also becomes a nutritional requirement for organisms incapable of synthesizing FAD.

Vitamin B_{12}, cobalamine, is a coenzyme active in the transfer of carbon units in the synthesis of purines, pyrimidines, and amino acids. Strain 753 of the Algal Culture Collection at Indiana University is a *Euglena gracilis* which requires vitamin B_{12}. It should be noted that both cobalamine and chlorophyll molecules are porphyrin rings. The *Euglena* species synthesizes the porphyrin ring of chlorophyll, but does not synthesize the porphyrin ring for cobalamine.

In the preparation of a defined medium for an organism, each element must be present in a concentration that will produce optimal growth. In this experiment you will determine the optimal concentration of vitamin B_{12} for the growth of *Euglena gracilis*.

1. Your instructor has prepared four solutions for you. The contents of these solutions are as follows:

   Solution I: NaCl, $MgSO_4$, $MgCl_2$, KCl, NH_4NO_3, $CaCl_2$, and H_2O.
   Solution II: K_2HPO_4, glycine, glycyl-glycine, citric acid, and H_2O.
   Solution III: $FeCl_3 \cdot 6H_2O$, $MnCl_2 \cdot 4H_2O$, $ZnSO_4 \cdot 7H_2O$, $CuSO_4$, $CaCl_2 \cdot 6H_2O$, $NaMoO_4 \cdot 2H_2O$, citric acid, and H_2O.
   Solution IV: vitamin B_1, H_2O.

2. Is *Euglena gracilis* a complete autotroph? Explain.

3. Solutions I, II, and III were combined, and KOH was added to adjust the pH to 8.0. The combined solutions were autoclaved and after cooling, Solution IV was added.

4. Why were the solutions combined prior to rather than after autoclaving?

5. With a wax pencil mark each of 9 culture test tubes 9 cm from the bottom.
6. Fill to the mark with the medium.
7. Label the test tubes as follows with paper labels. You will be examining these next week and wax pencil may rub off.
   (a) –2 (f) –7
   (b) –3 (g) –8
   (c) –4 (h) –9
   (d) –5 (i) –10
   (e) –6
8. Draw 1.0 ml of a sterile vitamin B_{12} and water 0.1 mg/100 ml solution into a pipette. Cover the top of the pipette quickly with your index finger, and release the pressure slightly to bring the liquid to the 1-ml mark. This may take some practice.
9. Tap the pipette lightly on the edge of the flask from which you have taken it to remove excess solution, and dispense the contents of the pipette into the test tube labeled –2. Do not dip the pipette into the test tube, but let the vitamin B_{12} solution flow from near the mouth of the test tube. Gently shake the tube. You have now made a 1/10 dilution. The concentration of vitamin B_{12} in the tube is 0.01 mg or 10^{-2} gm.
10. Using a fresh pipette, draw 1.0 ml of the medium in the –2 tube, tap lightly at the mouth of the –2 tube, and dispense the contents into the –3 tube. Gently swirl the –3 tube. The concentration of vitamin B_{12} in this tube is 0.001 mg or 10^{-3} mg (or 1 μ g/ml).
11. Repeat the procedure, placing 1 ml of tube –3 into tube –4, and 1 ml of tube –4 into tube –5, and so forth. Use a clean pipette for each transfer, and swirl each tube after dispensing the vitamin B_{12} solution into it.
12. You now have a series of B_{12} concentrations ranging from 10^{-2} to 10^{-10} or 0.0000000001 mg/100 ml of water. Autoclave tubes for 15 minutes.
13. Take up some of the culture of *Euglena gracillis* with a sterile Pasteur pipette. Hold the pipette in an almost horizontal position as you return to your desk so that the organisms do not collect at the bottom.
14. Holding the Pasteur pipette almost horizontal, dispense one drop of culture suspension into each of the tubes.
15. Swirl the tubes gently.
16. In order to determine the amount of growth that has taken place after a period of time, you must have an idea of how many organisms there were to begin with. Prepare a wet mount using a drop of solution from tube –10 first, and your counting chamber. Count the number of organisms, calculate the cells per milliliter, and enter your information on Chart 7.5.

CHART 7.5

Vitamin B_{12} Concentration	Initial No. of Organisms	Final No. of Organisms	x-fold Increase
10^{-2} mg			
10^{-3} mg			
10^{-4} mg			
10^{-5} mg			
10^{-6} mg			
10^{-7} mg			
10^{-8} mg			
10^{-9} mg			
10^{-10} mg			

17. You may use the same Pasteur pipette for taking samples from all cultures if you work from −10 to −9, and so on. The amount of vitamin B_{12} on the outside of the pipette from a ten-fold more dilute solution is negligible.
18. Washing and drying the counting chamber and the slide each time, make a count of each solution and enter accordingly on Chart 7.5.
19. Plug the tubes with cotton or top with cork after the counts of each tube have been made.
20. Place in a rack near a light source for one week.
21. After one week, make counts on all tubes and record the number of organisms per milliliter. Do not discard your tubes after the counts have been made.
22. Calculate the increase in cell number for each tube.
23. Plot the increase against the vitamin B_{12} concentration on the graph of Fig. 7.4.
24. According to your data, what concentration of vitamin B_{12} should be used in a defined medium for *Euglena gracilis*?

25. How does the optimal concentration of vitamin B_{12} in comparison to the concentrations of ions in the plant medium of the previous exercise indicate it is not a carbon source, although it participates in enzyme action?

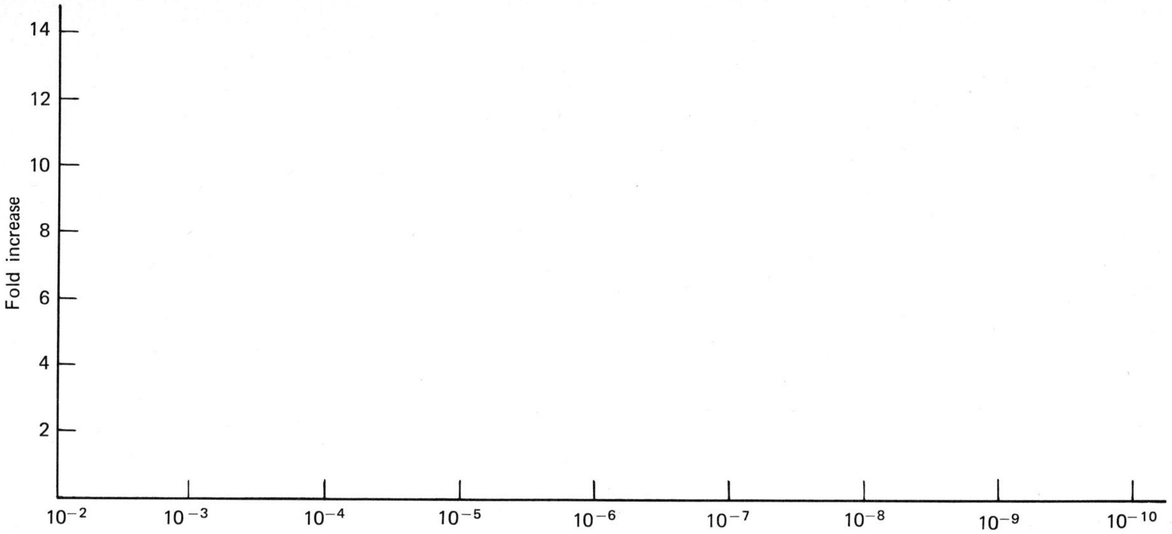

Figure 7.4

26. Pernicious anemia in man is caused by a vitamin B_{12} deficiency. How can Euglena be used to test the vitamin B_{12} level in the blood?

Part II

THE CELL

8

Cellular Organization

MATERIALS

CLASS STOCK

- chemicals
  - 0.85% NaCl solution (0.5%, 1.0%, 3.0%, 5.0%)
  - 0.85% saline solution
  - 1.0% trypsin
  - 1.0% steapsin
  - 1.0% amylase
  - 10% urea (0.5%, 1.0%, 3.0%, 5.0%)
  - 1.0% oxalic acid
  - 0.25% sucrose solution (2.0%)
  - Janus green dye solution
  - 0.25 M sucrose + 0.0018 M $CaCl_2$
  - neutral red solution
- *Anacharis* or moss leaf
- red blood cells
- cored carrot with glass tube
- blender
- cheesecloth
- balances
- clinical centrifuge
- centrifuge tube
- 40°C water bath

STUDENT STATION

- test tubes
- wax pencil
- millimeter ruler
- slides
- cover slips
- microscope
- microscope lamp
- lens paper
- scissors
- beakers
- Pasteur pipettes
- test tube rack

CLASS STOCK (cont'd)

Euglena medium with 0.1% glucose
liver or cauliflower
Serval centrifuge

INTRODUCTION

A cell is a highly integrated ensemble of functions. Each component of the cell represents a specific area of metabolic activity, and all of these areas together result in a single system capable of carrying out biological processes such as growth, reproduction, and secretion. In eucaryotic cells, each area of activity takes a characteristic form within the cell. These intracellular regions, or organelles, and macromolecular complexes, are highly specific in shape, size, and function.

The electron microscope has revealed the intricate structures and efficient activities of organelles and macromolecular complexes. For example, a cellular component that primarily stores cell nutrients or wastes usually has a large volume to surface area ratio, whereas organelles involved in chemical activity on their membrane surfaces usually have an extremely low volume to surface area relationship. The form of the latter often includes involutions or laminations.

Cytology, the study of cells, has progressed from an observational and descriptive aspect of biology to an experimental one, using the microscope as a research tool in conjunction with physical and chemical data. Coordinating observational and physio-chemical data has resulted in the elucidation of the form and function of cellular components.

A MACROMOLECULAR COMPLEX—THE CELL MEMBRANE

Macromolecular complexes are groups of large molecules resulting in a structural unit. Organelles, on the other hand, consist of two or more macromolecular complexes and are usually capable of reproduction.

Undoubtedly one of the most important macromolecular complexes in the cell is the membrane system, which includes the cell membrane, the endoplasmic reticulum (a three-dimensional layering of membranous shelves throughout the cell), the nuclear envelope (which encloses the nucleus), and the membrane components of organelles such as chloroplasts and mitochondrion. Cell membranes are the easiest to study, and the properties of the cell membrane appear to be the same as other components of the membrane system.

Chemical Analysis of the Cell Membrane

Plant cells are surrounded by a relatively thick cell wall; it is difficult to observe the cell membrane. Therefore, animal cells should be used for a study of the chemical properties of the cell membrane. A plant cell retains its shape regardless of the state of the protoplast because of the structural character of the cell wall. However, in animal cells the shape of the cell is variable because of the elasticity of the cell membrane.

The red blood cell is usually used for an analysis of the cell membrane. When the intact red blood cell is placed in an isotonic solution—a solution balanced with the internal environment of the cell—the cell retains its disc-like shape. Isotonic cell suspensions are turbid. When the red blood cell is lysed, by digestion of the cell membrane, and the cell contents pass into the isotonic medium, the solution has a clear appearance.

1. Prepare four clean test tubes by marking each 5 cm and 6 cm from the bottom with a wax pencil.
2. Fill each to the first mark with 0.85% NaCl solution, and mark the tubes A, B, C, and D.
3. Add two drops of red blood cells to each tube.
4. Add to the second mark
   (a) 0.85% saline
   (b) 1.0% trypsin (protein-digesting enzyme)
   (c) 1.0% steapsin (fat-digesting enzyme)
   (d) 1.0% amylase (starch-digesting enzyme)
   Note: Each of the enzyme solutions must be made up in 0.85% saline.
5. Mix each tube gently.
6. Illuminate the tubes from the front, either with a flashlight or by carrying the tubes to a window.
7. Place a piece of newsprint behind the set of tubes and attempt to read through the tubes. This will help you decide which tubes are turbid and have intact red blood cells, and which tubes are clear and have had the membranes of the red blood cells affected by the treatment.
8. Enter your data on Chart 8.1.

CHART 8.1

	A	B	C	D
Turbid				
Clear				

9. Which enzyme(s) caused dissolution of the cell membrane?

10. From the results, what can you conclude about the chemical nature of the cell membrane?

Physical Properties of the Cell Membrane

The cell membrane is regarded as a semipermeable structure, having the property of selectively permitting the passage of various materials. This selectivity does not seem to depend on the size of a molecule, because some large molecules easily enter the cell although others are excluded.

The Passage of Water Across a Semipermeable Membrane

A water molecule, by virtue of its properties, passes readily through the cell membrane. This passage of water through a semipermeable membrane is termed osmosis. Unfortunately, osmosis is often defined as an unequal movement of water either in or out of the cell—a definition based on the techniques used to study this phenomenon. The direction in which most of the water molecules flow is determined by the concentration of dissolved molecular material on either side of the membrane.

In a situation where the concentration of material and the subsequent pressure caused by that material is equal both inside and outside the cell, that is, where the internal and external environment of the cell have equal concentrations of water, the water will pass readily through the semipermeable membrane in both directions at the same rate. This condition of the environment is considered to be isotonic to that of the cell. Vacuolar membranes are also semipermeable, and the concentration of material within the vacuole is always isotonic to the internal cellular environment.

Under conditions where there is a higher concentration of materials outside the cell, water will move more rapidly from the inside of the cell, causing a decrease in cell volume. This transfer of water molecules will continue until a point is reached where the concentrations within the cell and within the environment are equal or until the cell shrinks too severely to function with any degree of efficiency. Under these conditions the environment is termed hypertonic to the cell.

The reverse situation may also exist. A solution that contains a concentration of dissolved molecules less than that of the cell will cause water molecules to pass across the cell membrane into the cell. As a result, the cell will swell and become turgid. Once an equilibrium is reached, the isotonic condition is restored and the passage of water molecules to and from the cell will again have equal rates. If, however, the difference in concentration is too great, the membrane may reach the limit of its elasticity before the equilibrium is reached and the cell will burst. Conditions that cause either a swelling or a bursting of cells are termed hypotonic to the cell (see Fig. 8.1).

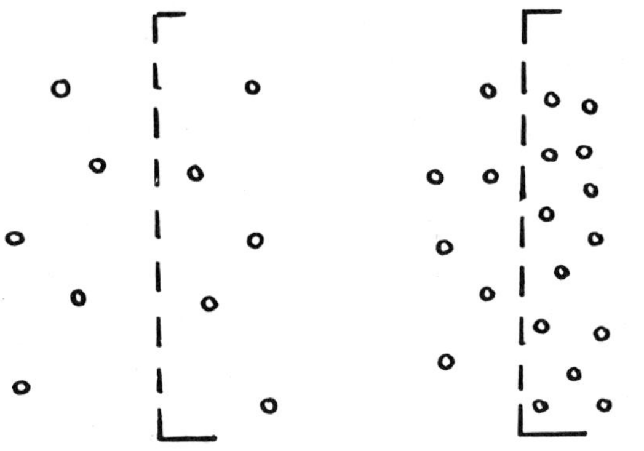

Figure 8.1

The Isotonic Condition

The shrinkage of cells as a result of exposure to hypertonic conditions is a property of both plant and animal cells. Terminology for the same phenomenon differs in reference to the two types of cells. Shrinking of plant cells is termed plasmolysis, whereas shrinking of animal cells is called crenation. Plant cells are surrounded by a cell wall, which remains unaltered on the shrinking or plasmolyzing of the cell within. The presence of a cell wall provides an indication of the original size of the cell.

When a plant cell is exposed to an environment hypotonic to the internal concentrations of the cell, the cell takes in water and begins to swell. The plant cell, however, is limited in its ability to increase in size because of the cell wall. When the plant cell membrane becomes pressed against the cell wall, the water uptake will cease, regardless of whether the cell has reached the isotonic condition. Bursting is a property only of animal cells. (Because red blood cells are standardly used for this type of experiment, the term hemolysis is generally accepted for reference to bursting of animal cells.)

EXAMINATION OF THE PLASMOLYZED CELL

1. Prepare a wet mount of *Anacharis* or moss leaf.
2. Examine under low and high power.
3. Draw a few representative cells under high power in the space provided.

4. You have used either distilled water or tap water for preparing your slide; what is the condition of the environment compared with that of the cell?

5. As the result of this environment, in which direction is there a higher rate of passage of water molecules?

6. Is this the normal cell you have drawn? Label accordingly.

7. Clean the slide thoroughly.
8. Using the same *Anacharis* or moss leaf, use 10% urea to prepare a wet mount.
9. Wait 5 min for the exchange of water to occur.
10. Examine the slide under low and high power.
11. Draw a few representative cells under high power in the space provided.
12. What state is the environment of a 10% urea solution compared to that of the cell?

13. Examine the slide carefully. What percentage of the cells have undergone physical change?

14. Clean the slide thoroughly.
15. Using the same *Anacharis* or moss leaf, prepare a wet mount with distilled or tap water.
16. Wait 5 min for exchange of water to take place.

17. Examine the slide under low and high power.
18. Is the plasmolyzed condition reversible?

Once you are familiar with the appearance of a plasmolyzed cell, proceed to the next part of the exercise.

DETERMINATION OF THE ISOTONIC POINT

You will be given ten solutions for testing; five of urea and five of NaCl.

Urea	NaCl
0.0%	0.0%
0.5%	0.5%
1.0%	1.0%
3.0%	3.0%
5.0%	5.0%

Your instructor may choose to label these solutions with letters instead of with the actual percentages so that you must rely solely on your results instead of on any preconceived notions. This work may be done individually, or it may be divided among groups in the class. A chart will be provided on the blackboard for class data to be tabulated. Each solution should be done by the class at least in triplicate.

1. Make a wet mount of an *Anacharis (Elodea)* or moss leaf, using one of the solutions of either urea or NaCl.
2. Let the slide stand for 2 min to permit water exchange.
3. Observe the slide under low and high power to determine the approximate percentage of plasmolyzed cells.

$$\frac{\text{plasmolyzed cells}}{\text{total cells in field}} \times 100 = \text{percent plasmolysis}$$

4. Enter your data on Chart 8.2.
5. Wash the slide thoroughly and with a fresh leaf of *Anacharis* or moss repeat the procedure using a different solution of either urea or NaCl.
6. Repeat the procedure until you have determined the percentage of plasmolysis for all the solutions that have been assigned.
7. Convert this data to graph form, using the graph of Fig. 8.2.

Zero percent plasmolysis occurs in those solutions that are hypotonic to the concentration of materials within the *Anacharis* cell and therefore cause swelling,

CHART 8.2

	Urea (% Plasmolysis)				
	0	0.5	1.0	3.0	5.0
Individual data					
Averaged class data					

	NaCl (% Plasmolysis)				
	0	0.5	1.0	3.0	5.0
Individual data					
Averaged class data					

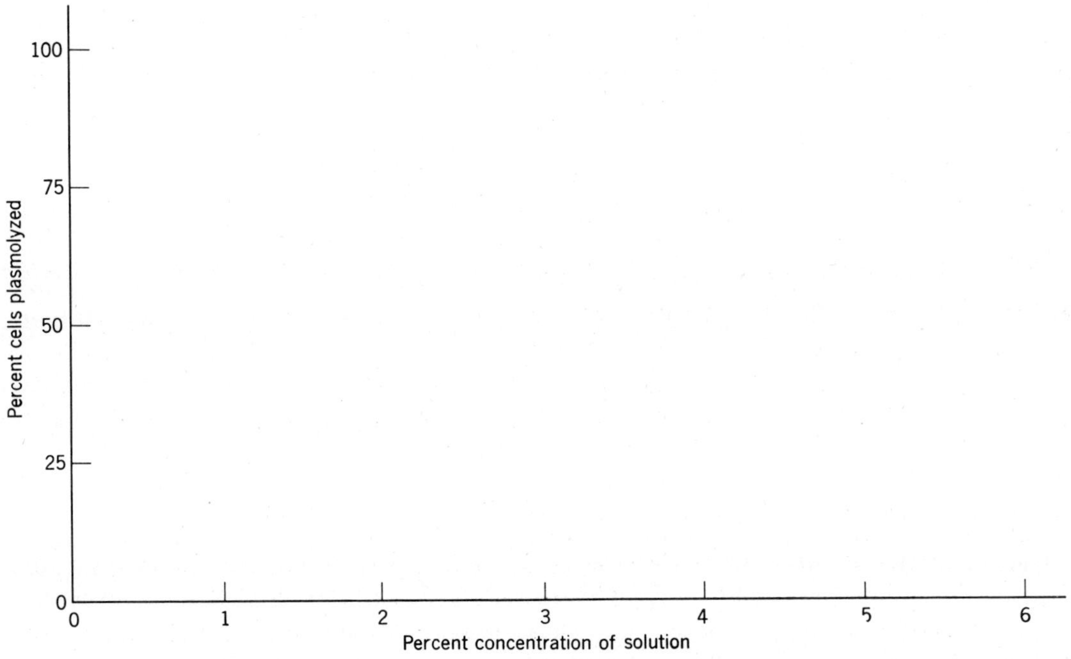

Figure 8.2

whereas 90 to 100 percent plasmolysis occurs when solutions are hypertonic to concentrations within the *Anacharis* cell. In the range between those solutions that cause total plasmolysis and no plasmolysis is the solution that is isotonic to the cell contents. Another way to regard this phenomenon is to recall the microscopic view of a plant cell. The vacuoles within the cell are necessarily in harmony with the concentration of cellular material and are isotonic to the cell. Determining the solution that is in harmony with the cell is, actually, determining the concentration of materials inside the vacuoles of the cell.

The isotonic point is an exact concentration instead of a range of concentrations. A slight change in concentration to either side of the isotonic point will cause either a shrinking or a swelling of the cell. This isotonic point, then, is the point where 50 percent plasmolysis occurs. Concentrations varying from this point will cause accompanying changes in the cell. You will note from your graph that the range between 0 and 100 is very narrow.

8. Determine the isotonic point of urea from the graph.
9. Determine the isotonic point of NaCl from the graph.
10. Urea remains in molecular form in solution, whereas NaCl dissociates into Na^+ and Cl^- ions. How does this dissociation affect the number of particles in solution of NaCl compared with urea?

11. How is the difference between the number of particles in solution in the NaCl and urea reflected in the isotonic points of each?

12. Which seems to be more important in determining the passage of water though a membrane, the type or the number of particles?

The Passage of Solute Through a Cell Membrane

The selectivity of material which passes through the cell membrane appears to be somewhat independent of the size and shape of the molecular structure. It has been hypothesized that fat solubility may be an important factor involved in the transfer of material across the membrane. It has also been hypothesized that the ability of the molecule to dissociate is inversely proportional to its ability to cross the membrane. Undoubtedly all these factors are involved. However, the selection of materials that pass into the cell is independent of their value to the cell. Many harmful compounds are admitted, and some nontoxic materials are rejected.

THE PASSAGE OF OXALIC ACID INTO THE CELL

Oxalic acid is a fat-soluble organic acid that is of little nutritive value to the cell and may indeed be harmful in large quantities. Oxalic acid, in a supersaturated condition, will form oxalate crystals because of insufficient water for solubility.

1. Prepare a wet mount of *Anacharis* with a 1.0% solution of oxalic acid.
2. Examine the slide immediately.
3. What visible sign is there that oxalic acid has entered the cell?

PASSAGE OF DYE THROUGH A CELL MEMBRANE

The instructor has prepared a demonstration to illustrate the passage of dyes through a cell membrane or a system of membranes (Fig. 8.3). A carrot has been cored and the

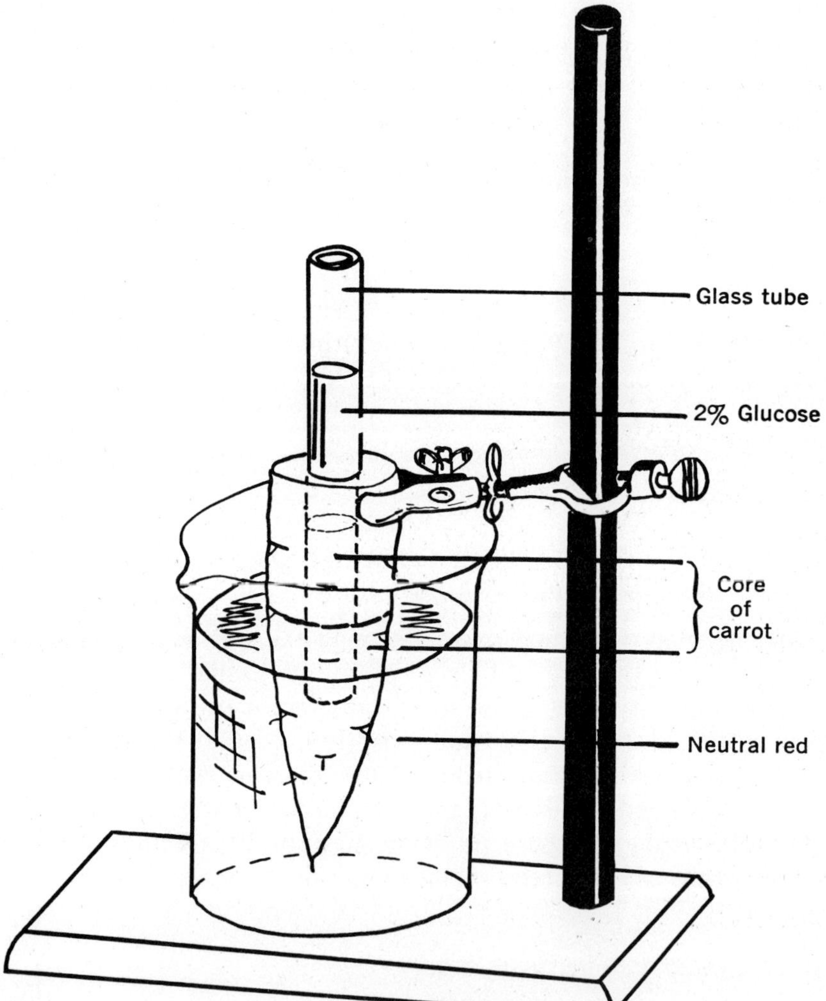

Figure 8.3

core has been partially replaced and extended by a glass tube after being filled with a 2% glucose solution. The carrot has been suspended in a solution of neutral red. The cell membranes of the carrot cells collectively act as a single membrane.

1. Test the solution surrounding the carrot for the presence of glucose (Benedict's Test).
2. What color is the solution moving up the tube? Why?

3. Indicate on Fig. 8.3 the relative concentrations of glucose and neutral red in the cells and in the core of the carrot, and, with arrows, indicate the direction of the movement of the two molecular types.
4. The cells of the carrot are obviously undisturbed by the passage of dye across them. What type of dye is this?

The Structure of the Cell Membrane

1. Examine the electron micrographs of Fig. 8.4 and 8.5.
2. From your knowledge of the position of the cell membrane in relation to other cellular components, and from what you would imagine it looks like from your studies, identify the cell membrane by drawing a circle around a portion of it in Fig. 8.4.
3. How many layers does the cell membrane have? What type of molecules comprise these layers?

4. Once you are familiar with the structure of the cell membrane, study Fig. 8.5 and identify the cell membrane by circling a portion of it.
5. Why is the surface area of the cell membrane in Fig. 8.5 so much greater than in Fig. 8.4?

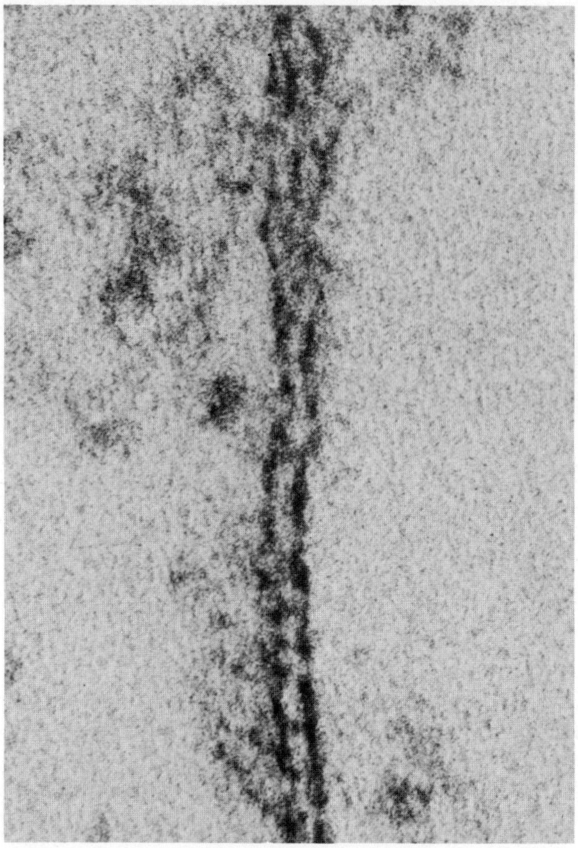

Figure **8.4** **The cell or unit membrane: rat-tongue blood capillary (760,000X).** Electron microscopy revealed the cell membrane to be a double-layered structure. (Courtesy G. E. Palade.)

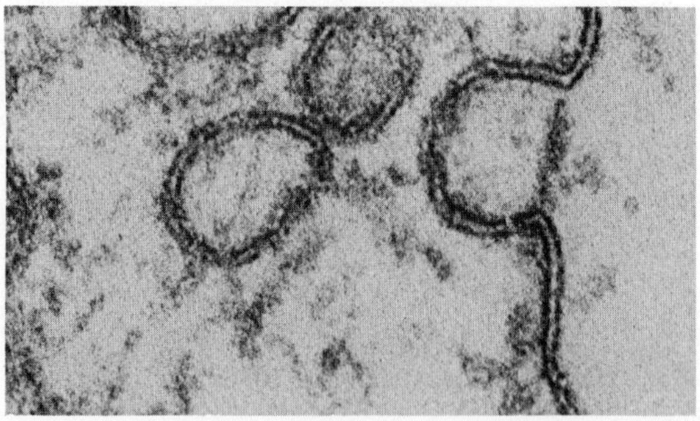

Figure **8.5** **Pinocytic vesicles of the cell: rat-tongue blood capillary (380,000X).** Cells active in the transport of material have invaginations that either are continuous with the interior of the cell or that appear to break off from the membrane and form vesicles that enter the interior of the cell. (Courtesy G. E. Palade.)

6. Knowing that metabolic reactions take place on membrane surfaces, how would this form increase the efficiency of cell function?

ORGANELLES – CHLOROPLASTS AND MITOCHONDRIA

Chloroplasts and mitochondria are the easiest organelles in the cell to study. Both of these organelles are part of the membrane system in the cell and the macromolecular complex of membranes is a major part of their structure. These membranes have the same properties as those found for cell membranes. Hence, chloroplasts and mitochondria have an internal environment quite distinct from the cytoplasm in which they exist.

Most of the recent studies on cellular components have been carried out on cell fractions. The role of the chloroplast in photosynthesis and the separation of the light and dark reactions of photosynthesis were studied on suspensions of the chloroplast fraction, that is, the chloroplasts from many cells in suspension. More accurate observations can also be made of organelles when they are outside the cellular environment.

1. Obtain several fresh sprigs of *Anacharis.*
2. Cut one into pieces. Remove the leaves and place them with 50 ml of 0.25% sucrose solution into a blender.
3. Cover the blender and run it on maximum speed for one min.
4. Pour the resulting green solution through four layers of cheesecloth into a beaker. The blender has homogenized the material, breaking cell walls and membranes, and forming a general suspension of cell membrane fragments, organelles, macromolecular complexes, and so forth.
5. Fill a small centrifuge tube about ½ full. Mark you initials on the tube with a wax pencil.
6. Place two small beakers on the pans of a balance, one on each pan, and add scraps of paper, nails, or paper clips until they balance.
7. Place your centrifuge tube in one of the beakers, and have another student place his centrifuge tube in the other.
8. Add the green suspension with a Pasteur pipette to the lighter tube until they balance.
9. Place the tubes *directly opposite* one another in a clinical centrifuge.
10. When the centrifuge head is full of tubes, turn on the centrifuge to full speed.
11. Allow time for the centrifuge to reach full speed and run it for 3 min more.

12. Slowly turn the centrifuge dial to the off position and let the head stop spinning by itself.
13. Gently remove the centrifuge tube from the head, and discard the supernatant.
14. Add 0.25% sucrose solution to the pellet, and draw liquid up and down with a Pasteur pipette until the pellet material is in suspension.
15. Take a drop of the suspension and make a water mount.
16. Examine the contents of the pellet under the microscope. What organelles and fragments do you observe? Draw.

17. Some of the cytoplasmic material may have clung to the organelles when they were brought down in centrifugation. Washing and recentrifuging usually frees them, and gives a cleaner fraction. Usually washing is done three times, but for our purposes, the single centrifugation will be sufficient.
18. Wash the slide and cover slip.
19. Make another wet mount, this time placing a drop of 2% sucrose over the specimen before placing the cover slip.
20. What macromolecular complex encloses the chloroplast contents? Give experimental evidence that confirms your answer.

21. Transfer some of the resuspended pellet material into two test tubes. Fill each about 1/3 full, and add resuspending solution until they are about 2/3 full.
22. Follow the procedure outlined in Chapter 6, under the **Carbon Dioxide Requirement** for balancing two tubes on either side of neutrality with phenol red indicator.
23. Do you observe a color change? In which direction is the change?

24. What process was instrumental in producing the change?

25. Why is research on a chloroplast fraction a valid means of studying photosynthesis?

Ultrastructure of a Chloroplast

Electron micrograph studies of chloroplasts are done on whole plant cells instead of chloroplast fractions so that the fine structure can be observed in natural conditions, undisturbed by physical treatments that might alter the true state.

1. Examine the electron micrograph of a portion of a plant cell shown in Fig. 8.6.
2. Examine the labeled chloroplast.
3. Describe the different macromolecular complex structures that make up the organelle.
4. Is the arrangement of the membrane system in the chloroplast random?

5. Packets of stacked membranes are called grana. The reactions for the dark reaction, and some of those for the light reaction take place on the membrane surfaces. Why is the structure of the granum consistent with a chemical function?

Figure 8.6 Chloroplast. An electron micrograph of a chloroplast of *Zea mays* (21,000X) reveals layered packets. (Courtesy Dr. Elliot Weier.)

Bleaching of Chloroplasts

Because of their chemical nature, chloroplasts can be altered by the same factors that affect chemical structure and reactions. Inactivation of chlorophyll causes a cessation of photosynthesis and death to most photosynthetic organisms.

1. After you have made your counts for the growth of *Euglena gracilis* in different concentrations of vitamin B_{12}, take the tube in which the most growth occurred and place it in a 40°C water bath for 30 min.
2. After removing the tube from the water bath, remove some of the cells with a Pasteur pipette and add several drops to a fresh sterile tube of medium supplemented with 0.1% glucose. Place the tube in a rack near a light source.
3. Leave the tube for one week.
4. Heat shock will bleach the chloroplasts and the *Euglena* will no longer be able to undergo photosynthesis. Will the *Euglena* die? (Hint: Is *Euglena* an autotroph?)

5. What compounds in the medium will foster growth?

6. If the chloroplasts are bleached, and photosynthesis has ceased, what use is there in placing the *Euglena* near a light source?

Isolation of Mitochondria

1. Obtain a piece of liver or several pieces of cauliflower and cut them into small pieces.
2. Place in the blender with 50 ml of 0.25% sucrose and homogenize for 2 min.
3. Pour the homogenate through four layers of cheesecloth into a beaker.
4. Fill a centrifuge tube half full with the homogenate.
5. Balance the test tube against one belonging to another student adding resuspending solution (0.25M sucrose + 0.00018M $CaCl_2$) to obtain equal weights.
6. Place the balanced tubes opposite one another in the head of the clinical centrifuge, and spin at maximum speed to bring down cells and cell fragments.
7. After running the centrifugation for 10 min, pour the supernatant into another centrifuge tube. Repeat the balancing procedures and centrifuge again in the clinical centrifuge.
8. After centrifugation, pour the supernatant into a Servall centrifuge tube. Repeat the balancing procedures and place the tubes opposite one another in a Servall centrifuge.
9. Turn the Servall to maximum speed, (10,000 rpm for a refrigerated centrifuge). Centrifuge for 10 min.
10. When the centrifuge has stopped by itself, remove the tube, pour off the supernatant, and discard.
11. Is there a visible pellet?

12. Add about 2 ml of the resuspending solution and resuspend the pellet by drawing the liquid in and out of a Pasteur pipette.
13. Place a drop of resuspended pellet material on a slide, with a drop of Janus green dye solution.
14. Prepare a wet mount and observe under the microscope at high power.

15. Janus green is specific for mitochondria. Is there any cellular material on the slide besides the mitochondria? Why? How would you eliminate other cell debris?

16. Draw a mitochondrion in the space provided.

17. On the next page is a diagram of the centrifugation series you have performed (see Fig. 8.7). Fill in the cell fractions in the spaces provided.

Ultrastructure of a Mitochondrion

1. Examine the electron micrograph of a mitochondrion in Fig. 8.8.
2. Describe the macromolecular complex structures that comprise the mitochondrion.

3. The reactions of the Kreb's cycle take place on the membrane surfaces. How is the membrane system of the mitochondrion adapted for a chemical function?

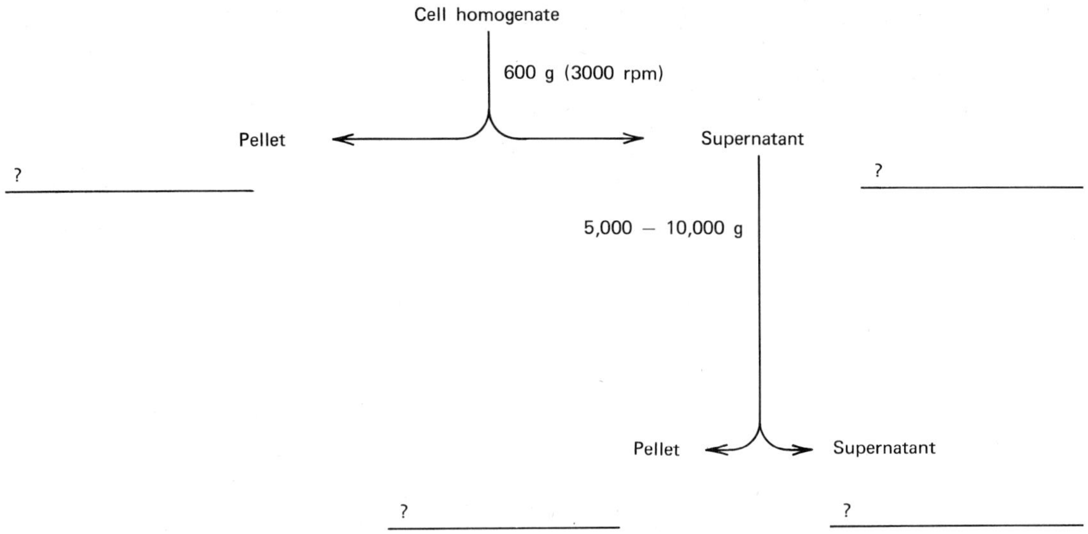

Figure 8.7

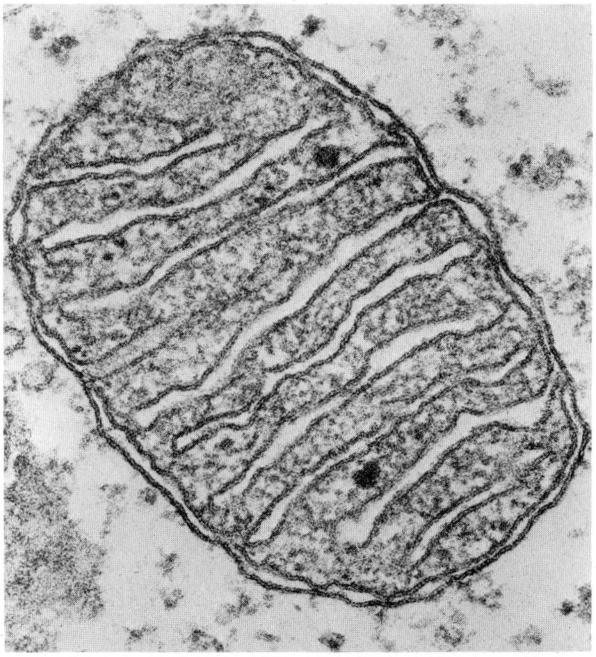

Figure 8.8 Mitochondria. The double-layered structures visible in an electron micrograph of mitochondria of a guinea pig pancreas (230,000X). (Courtesy G. E. Palade.)

4. The transfer of electrons and oxidative phosphorylation may take place in the *electron transport (ETS) particles.* What is the spacial relationship of the ETS particles to the membrane system?

5. Based on your examination of electron micrographs of a chloroplast and mitochondrion, describe two ways in which surface area to volume ratios are increased by spacial arrangement of membranes.

A CELL

A "typical" cell is an oversimplification because of the diversity of cell types in the plant kingdom. This term "typical" may be partially justified in reference to *Anacharis* cells, since they contain the elements usually associated with plant life. The chloroplasts—the site of photosynthesis—are present, and the amount of space taken up by vacuoles is typical of plant cells.

Anacharis is a flowering plant found in fresh-water lakes and ponds. The leaves, unlike those of many terrestrial plants, are only one or two cell layers thick; therefore light can penetrate for direct microscopic examination.

You have seen the cells of *Anacharis* leaf twice before in your laboratory experience, and have drawn it once before. Now, after a study of the cell organelles, you will draw it again, reflecting your comprehension of form and function in your drawing.

1. Prepare a wet mount of an *Anacharis* leaf.
2. Examine the cell under low and high power.
3. Draw a cell of *Anacharis* in the space provided, labelling the following parts:

   Cell wall—a rigid structure surrounding the cell. This structure mostly consists of cellulose, which is secreted by the cell itself. It serves for protection and support.

   Cell membrane: the same in form, function, and origin (homologous) as the cell membrane of an animal cell. *In Anacharis* it is close to the cell wall and is too thin to be identified.

   Chloroplasts: small green moving bodies in the cytoplasm which are the site of photosynthesis. The chloroplast contains the green pigment *chlorophyll.*

Cytoplasm: the streaming viscous material in which the chloroplasts flow. In *Anarchis* the cytoplasm is very close to the cell wall because of the extremely large size of the vacuole.

Vacuole: fills most of the space within the cell in an *Anacharis* cell. This large area contains salt solutions which keep the vacuole contents in equilibrium with the internal environment of the cell.

Nucleus: bubble-like structure. Close to the cell wall you may be able to see a bulge in the movement of the cytoplasm as it passes the nucleus. Look for a nucleus in one of the marginal barb cells. The nucleus has a thin covering, the nuclear envelope, and one or two small bodies within, the nucleolus (pl. nucleoli).

4. Notice the passive movement of the chloroplasts. The flowing of the cytoplasm, which causes the movement of the chloroplasts, is termed cyclosis. How does cyclosis serve to increase the efficiency of the chemical reactions on internal surfaces?

5. Notice the geometry of the cell. How is the structure of the cell adapted to obtain maximum light?

9

Reproduction of Cellular Components

MATERIALS

CLASS STOCK

Allium root, long section slide
petri plates with radish seedlings
Polytrichum with capsule

STUDENT STATION

Microscope
microscope lamp
lens paper
millimeter ruler
slides
cover slips
counting chamber
Euglena gracilis with bleached chloroplasts (from previous week)

INTRODUCTION

Internal cell control is achieved through the control of enzyme formation; the enzyme formation depends on a series a sequential systems ultimately traceable to DNA. In order to ensure that each cell contains the DNA responsible for specific enzymes, a regulated means of DNA replication and equal separation of new DNA material must occur when new cells are formed.

Previously it was thought that all the DNA of the cell was located in the nucleus, and that as long as the nucleus persisted, any part of the cell, if lost, could be replaced. Recent evidence indicates that although the great majority of DNA is indeed in the nucleus, the other organelles contain DNA that is not repeated in the nucleus. Thus, to some extent, organelle reproduction and function lies within the organelle itself. This laboratory exercise is a study of the reproduction of cell organelles. Since the nucleus contains the major portion of DNA and therefore controls the major portion of cell activity, the emphasis will be on nuclear division.

THE MITOTIC CYCLE

The mitotic cycle or mitosis refers to the chemical and cytological process by which nuclei reproduce. Since the major chemical constituent of the chromosomes within the nucleus is DNA, chromosomal duplication and segregation, the cytological events, are reflective of DNA replication and segregation, the chemical events.

In unicellular populations, each cell is capable of division, and is therefore able to give rise to two daughter cells. In multicellular organisms, specialized of differentiated cells usually do not divide , so that increase in cell number is dependent on the activity of special embryonic cells, which divide many times. The term mitotic cycle preferably describes the activity of dividing cells, because mitosis conveys the impression of a single and noncontinuing event.

The mitotic cycle consists of a series of stages that are arbitrarily established for convenience of observation. Since the genetic factors are contained on the chromosomes, the stages are based on chromosomal behavior (see Fig. 9.1).

Interphase

During interphase, the chromosomes prepare for division. Although changes cannot be detected microscopically, the cell undergoes metabolic processes and the chromosomal or genetic material duplicates. Interphase may be divided further into three stages: G_1 (prior to DNA synthesis), S (during DNA synthesis), and G_2 (after DNA synthesis). For your examination, which will be limited to microscopic observation consider interphase as a single stage, during which DNA synthesis or chromosomal duplication occurs.

Prophase

During prophase, the chromosomes condense and become individually visible. The chromosomes seem to act independently of one another. This shortening by means of condensation and coiling continues as the nuclear membrane disintegrates. Toward the end of prophase, the duplicated chromosome (or two sister chromatids) is held together only by the centromere. The disappearance of the nucleolus and the disintegration of the nuclear membrane mark the end of prophase.

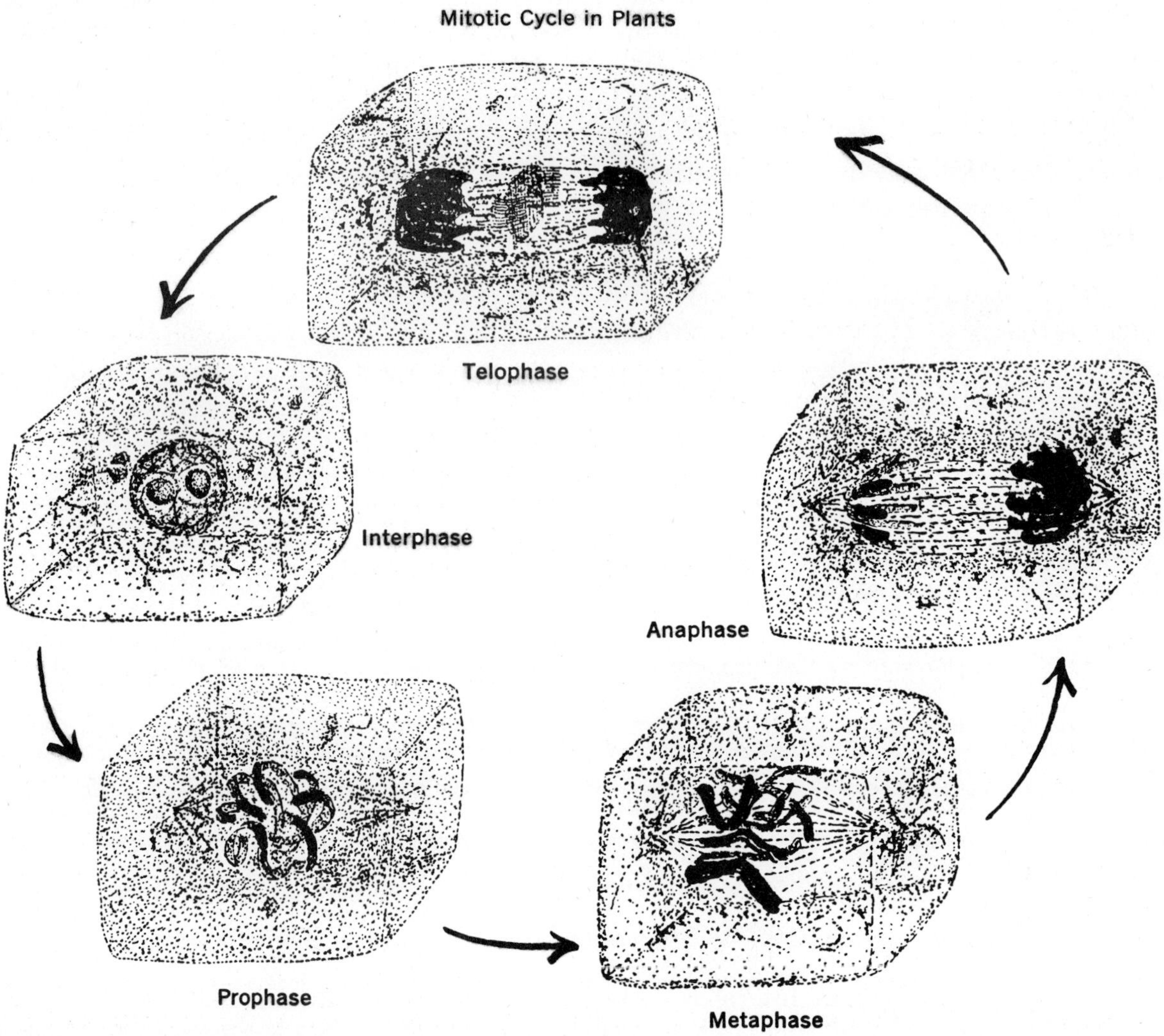

Figure 9.1

Metaphase

As the chromosomes condense, a spindle of fibers is set up across the nuclear area. Some spindle fibers stretch across the cell between two poles, while others extend from a pole to a centromere, The chromosomes line up along the central portion of the spindle, their centromeres oriented on one plane. This plane is termed the equatorial plane.

Anaphase

The chromatids become noticeably doubled at the centromere, and separate, indicating the completion of the formation of two separate and functional daughter chromosomes. The centromeres of each of the chromatids (or newly formed daughter chromosomes) migrate toward opposite poles of the spindle, pulling the rest of the chromosomes with them, giving a V-shaped appearance.

Telophase

Each of the chromatids reaches its respective pole of the spindle and the new chromosomal complement becomes tightly packed together. The new nuclear membranes are assembled around each group of chromosomes and the nucleoli are formed. In most cases, cell division or cytokinesis accompanies the nuclear division. In plants, a cell plate is formed where the equatorial plane was, and growth occurs from the center outward. In animals, the cell is pinched between two newly formed nuclei and two cells are formed.

Essentially, during the mitotic process, there has been a maintenance of the entire chromosome complement in each of the resulting cells. The duplication and subsequent divisions of each chromosome ensures the correct number of chromosomes in each of the resulting cells. The manner of separation of the chromosomes ensures the correct types in each cell. Since they contain the same chromosome complement, and consequently the same genetic complement, the cells are capable of synthesizing the same structural and functional protein and each, therefore, has the same potentials for enzyme synthesis. Given identical environments (cytoplasmic and external), the cells will be identical in their ability.

THE MITOTIC CYCLE IN PLANT TISSUE

In multicellular systems specialized cells or regions that are directly responsible for growth by means of division (increase in cell number) exist. In plants, one such region is found at the tip of the root, just above the root cap region, called the meristem or meristematic zone. The majority of cell divisions that give rise to the specialized cells of the root take place in this region (see Fig. 9.2).

1. Examine a slide of *Allium* (onion) root tip under low power.
2. Once you have located the meristematic zone near the tip, turn to high power.
3. Examine the slide, selecting good representatives of each of the stages in the mitotic cycle. Draw these in the space provided, using the standard form for biological drawings.

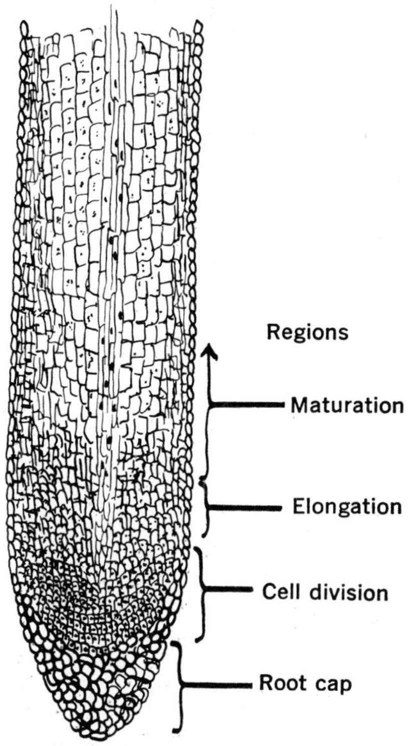

Figure 9.2

4. Each slide has three or four root sections, so you may switch from one to another in order to find one suitable for drawing.
5. Why do a number of the cells of the root appear not to have nuclei? (Hint: The method of preparing the slide involves thin slicing of the root on a longitudinal plane.)

6. In making your drawings, do *not* copy the drawings of Fig. 9.1. You will find that it will be easier to make a successful drawing if you concentrate on the chromosomal patterns of the different stages and attempt to represent these patterns instead of looking at each chromosome individually.

THE TIME SEQUENCE OF THE MITOTIC CYCLE

The relative length of time required for the completion of each stage is directly correlated with the number of cells observed in the various stages. The situation is analogous to that of a horse running. While moving, the horse repeats a series of steps. If one were to take pictures at random of a horse in full gallop one would find that some of the pictures showed the horse on one leg, some the horse on two legs, some the horse on three legs, some the horse on all four, and some the horse with all four legs off the ground. By counting the number of each type of picture, one can arrive at a relative time sequence for each stage.

For example,

1 leg	1 picture	5%
2 legs	4 pictures	20%
3 legs	4 pictures	20%
4 legs	5 pictures	25%
no legs	6 pictures	30%
	20	100%

From this information you can draw several conclusions about the various stages. The time spent on two legs is four times as long as that spent on one. The horse when galloping spends the greatest length of time with all four legs off the ground, and so forth.

Given the additional information of how long the cycle takes, the time sequence of each of these stages can be worked out. For example, if the entire running sequence takes 1 sec, then the length of time the horse is on one leg is 5 percent of 1 sec, or 0.05 sec.

This same type of reasoning may be applied to the mitotic cycle. As with the pictures in the example of the horse, the mitotic process was stopped suddenly. The number of cells that were caught in a stage requiring a greater time for completion should be correspondingly more than those cells caught in a stage that requires less time for completion.

1. Choose a single field in the meristematic zone that seems to have been actively dividing at the same time the slides were prepared.
2. Observe this field under high power.
3. Count the number of cells in each stage and enter the count in Chart 9.1.

CHART 9.1 Individual Data

	Number of Cells	Percent of Cells	Time (Hours and Minutes)
Interphase			
Prophase			
Metaphase			
Anaphase			
Telophase			
Total			

CHART 9.2 Class Data

	Number of Cells	Percent of Cells	Time (Hours and Minutes)
Interphase			
Prophase			
Metaphase			
Anaphase			
Telophase			
Total			

4. Convert each of the counts to a percentage, using the total number of cells counted as 100 percent.
5. The mitotic cycle generally takes about 16 hr. This is an approximation; the actual time may vary depending on the condition of the roots during growth. On the basis of a 16-hr cycle, work out the approximate time in hours and minutes that is spent in each stage. Enter this information on Chart 9.1.
6. The class data may be taken and put up on the blackboard for you to copy. Enter this on Chart 9.2.
7. From examination of your own data, why is it more valuable to determine the time sequence from class data than from individual data?

8. Which of the stages is the shortest? Does this comply with what you know to be the chromosomal events during the stage? Why?

9. Which of the stages seems to be the longest in duration? Does this comply with the information you have received about the events during this stage? Why?

10. State one or more events occuring at each stage. Does it seem reasonable that the time required is sufficient for the stated activities?

DETERMINATION OF THE DURATION OF THE MITOTIC CYCLE

In the experiment for the time sequence of the mitotic cycle you were told that the mitotic cycle takes about 16 hr in the onion root tip. In this experiment, you will work out the time of the mitotic cycle, and the time from metaphase through DNA replication by using appropriate inhibitors of known action, without examining the cells themselves.

1. Your instructor has prepared four petri plates of radish seedlings for your examination. These petri plates have been treated as follows:
   I. One-day-old seedlings killed and preserved with 1% formaldehyde at the beginning of the experiment. These seedlings represent the zero time of the experiment, the starting point for growth of the other seedlings in other petri plates.
   II. Five-day-old seedlings treated with 0.03% colchicine at zero time. Colchicine inhibits spindle formation, so that the mitotic cycle cannot progress beyond early metaphase.
   III. One-day-old seedlings treated with colchicine at zero time. Colchicine was washed out after 20 hours, and the seedlings were then treated with 50 mg/ml mitomycin C. Mitomycin C inhibits DNA synthesis, so that the mitotic cycle cannot progress in S stage of interphase.
   IV. One-day-old seedlings untreated, permitted to grow for 20 hours, and killed with formaldehydyde. Seedlings I and IV represent zero and 20 hours of growth respectively as references for growth of seedlings in other petri plates.
2. Place five seedlings from plate I on a piece of wet paper toweling and take it to your desk. Measure the length of the roots with a mm ruler. Enter your data on Chart 9.3.
3. Return the seedlings to plate I for other students, and take five seedlings from plate II to your desk to measure. Enter your data on Chart 9.3.
4. Do the same with five seedlings each from plates III and IV, returning the seedlings you have just measured before taking five from the next plate. Be sure to put the seedlings back in the proper plate so that students who use them after you will be measuring the proper seedlings. Enter all data on Chart 9.3 and make the appropriate calculations.

CHART 9.3

Treatment	Root Lengths	Average
0 time	1	
	2	
	3	
	4	
	5	
	Total/5 = average	
Stopped at	1	
	2	
Metaphase	3	
	4	
	5	
	Total/5 = average	
0 + 20 hr	1	
	2	
stopped at S	3	
	4	
	5	
	Total/5 = average	
20 hrs	1	
	2	
growth	3	
	4	
	5	
	Total/5 = average	

5. Calculate the rate of root growth by subtracting the averages of 0 time length from the 20 hr growth seedlings and dividing by 20.

$$\text{increase in length/hr} = \frac{\text{final length} - \text{initial length}}{\text{no. hours of growth}}$$

6. What is the rate of root growth for the radish seedlings?

Colchicine inhibits the formation of the spindle. As you have seen in the onion root tip, the cells of a meristematic region of a root are in different stages of the mitotic cycle. Some are in metaphase with the spindle already formed. These will continue through anaphase, telophase, interphase, and prophase before they are affected by the presence of colchicine. Others are in prophase and will be affected by the presence of colchicine in a short time. The amount of time it will take for a given cell to be affected depends on the stage it is in and the length of the cycle.

The situation is analogous to your going to movie houses at random times. If all the movies you went to see were two hours long, and you entered the theater at random hours, you would sometimes arrive at the beginning of a film, sometimes at the end, and sometimes in the middle. But if you averaged the number of minutes left to a film when you entered the theater, you would find that it averaged to your entering the theater when the film was half over, or one hour.

In the same way, some cells have a long time to go before being affected by colchicine, others a short time—but it would average out to half the cycle.

7. Calculate the time for the mitotic cycle to take place by subtracting the averages of the zero time from the colchicine treated seedlings, dividing by the rate of growth previously calculated, and multiplying by two.

$$\text{time for mitotic cycle} = 2 \times \frac{\text{colchicine treated time length-zero time length}}{\text{rate of growth}}$$

8. How many hours are necessary for a complete mitotic cycle in radish seedlings.

In plate III, the cells were treated with colchicine for 20 hr, so that almost all the cells were held in metaphase. After 20 hr, the roots were washed free of colchicine, and metaphase continued. Mitomycin C was added to the seedling environment to inhibit DNA synthesis at the S stage of interphase, so that the cells progressed only through metaphase, anaphase, tellophase and part of interphase.

9. Calculate the time from metaphase to the time when the cell is ready to undergo another division (as indicated by preparatory DNA synthesis) by subtracting the average lengths of 20 hr growth from the colchicine treated seedlings, from the average length of colchicine + mitomycin treated seedlings, and dividing by the rate of growth previously determined.

$$\text{time from metaphase to S} = \frac{(\text{colchicine + mitomycin C}) - \text{colchicine}}{\text{rate of growth/hr}}$$

10. How much time is necessary for a cell to go from metaphase until a daughter cell can divide?

11. How much time is necessary from the initiation of DNA synthesis to the end of late prophase?

12. Why do investigators who want to study chromosomes cytologically treat cells with colchicine?

INHERITANCE OF CHLOROPLASTS

Indirect evidence can be gathered for the contention that a chloroplast must be present for more chloroplasts to be formed. In the moss plant, *Polytrichum*, chloroplasts are inherited from one generation to the next through a single-celled reproductive body called a spore. If a chloroplast must be present for other chloroplasts to be formed, then one would expect a pattern to exist in which each spore would have at least one chloroplast. If the presence of a chloroplast is not necessary for the reproduction of chloroplasts, then the distribution of chloroplasts among spores would be random.

1. Obtain a moss plant with the enlarged tip or capsule.
2. Squeeze the capsule and notice the green spore mass emerging.
3. Make a water mount of these spores and permit the cells to absorb water for 1 min before taking a count.
4. Observe 50 spores and determine the number of chloroplasts in each.
5. Enter your data on Chart 9.4, giving the number of spores that contain for example, 0, 1, or 2 chloroplasts per spore.

   $$\frac{\text{total no. of chloroplasts}}{50} = \text{average no. of chloroplasts per spore}$$

6. Plot on the graph of Fig. 9.3 the distribution frequency of the chloroplasts per spore that you found.

CHART 9.4

No. Chloroplasts per spore	Spores	
	No.	%
1		
2		
3		
4		
5		
6		
7		
8		
Total	50	100

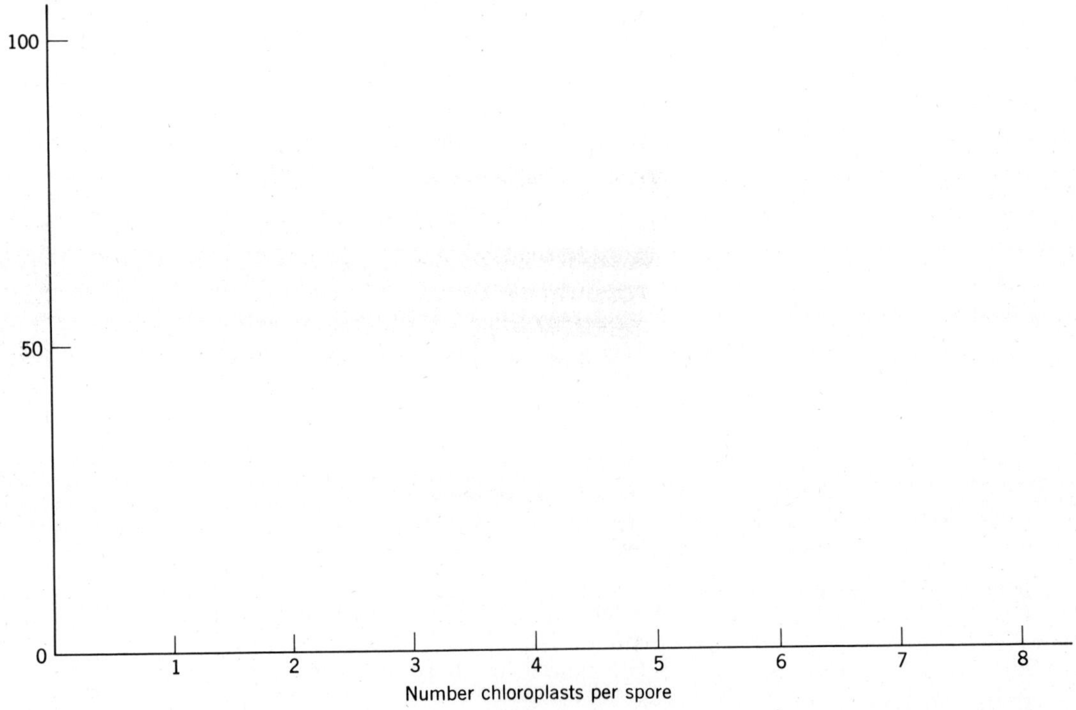

Figure 9.3

Poisson Distribution

The Poisson distribution is a mathematical computation assuming random events. If an event or series of events is wholly random, the pattern followed by these events will fit the Poisson distribution. For example, if 100 marbles are released a few feet above 100 boxes (an average of one marble per box) you would expect that some boxes would catch no marbles, others one marble each, and still others more than one marble each. According to the Poisson distribution the distribution of the marbles would be 37 boxes with no marbles, 37 boxes with one marble each, 18 boxes with 2 marbles each, 6 boxes with 3 marbles each, 1 box with 4 marbles, and no boxes with 5 marbles. If, however, instead of being dropped, these marbles were placed into boxes in some prearranged pattern, i.e., one marble per box, the pattern would immediately indicate that the events were not random. This same analytical procedure may be used in determining whether the distribution of chloroplasts in the spores is a random event or if there is some underlying mechanism that governs their distribution.

Chart 9.5 is a chart for the Poisson distribution. Given the average number of chloroplasts per spore, the percentage of spores with each number of chloroplasts may be read from the chart, assuming random distribution.

CHART 9.5

Average No. Chloro-	Percentage of Spores with Chloroplast								
plasts per spore	0	1	2	3	4	5	6	7	8
0.1	90	9	1	0	0	0	0	0	0
1	37	37	18	6	1	0	0	0	0
2	13	27	27	18	9	4	1	0	0
3	5	15	22	22	17	10	5	2	1
4	2	7	15	19	19	16	10	6	3
5	1	3	8	14	17	17	15	10	6
6	0	1	4	9	13	16	16	14	10
7	0	1	2	5	9	13	15	15	13

1. Determine the average number of chloroplasts per spore of your data.
2. Look up the average number of chloroplasts per spore that you have found on Chart 9.5. Reading across, plot the percentage of spores with chloroplasts on the graph of Fig. 9.3. For example, if you found five chloroplasts per cell as your average, you would plot the figures in the corresponding row on the graph: 1, 3, 8, 14, 17, and so on.
3. Comparing this plotted value, which corresponds to the expected value if chloroplast distribution is random, with your obtained curve will permit you to judge whether there is some mechanism underlying the distribution of chloroplasts in spores. If one curve is superimposed on the other, it indicates that the chloroplasts passing into the spores are under no special control, external or internal. If, however, the curves are

quite different, that is, if one is displaced or has a noticeably greater peak, it indicates that there is some special mechanism present for controlling the inheritance of the chloroplasts as discrete bodies.

4. The spores are contained in an opaque capsule. How does this also indicate that the chloroplasts present are primarily for genetic transmission? (Hint: could they be photosynthetic in the spore?)

5. It is generally believed that chloroplasts can be derived only from preexisting chloroplasts. What would be the consequences if a spore contained no chloroplast?

6. Recent evidence indicates that chloroplasts contain some DNA. Does this agree with your conclusions in the above experiment? Why?

Bleached Chloroplasts of Euglena

Last week, after examining your *Euglena gracilis* cultures, you submitted the tube exhibiting the highest growth to heat shock by placing it in a 40°C water bath for 30 min. As a result of this treatment, the chloroplasts gradually turned white.

1. Prepare a wet mount with one-week-old culture of bleached *Euglena*, using your counting chamber as a cover slip.
2. How many *Euglena* cells are there per milliliter?

3. Has cell division taken place in the population since the heat shock?

4. Describe the condition of the plastids in the cells you are observing (green chloroplasts, white chloroplasts, yellow chloroplasts, no chloroplasts or plastids).

5. If *nuclear* DNA alone is responsible for chloroplast reproduction you would expect to find green chloroplasts. Why?

6. If the chloroplast itself is responsible for chloroplast reproduction, and if the production mechanism was not disturbed by the heat shock, you would expect to find bleached chloroplasts in all members of the population. Why?

7. If the chloroplast is responsible for its own reproduction, and if the reproduction mechanism was disturbed by the heat shock, you would expect to find some cells with bleached chloroplasts (the original ones that underwent the heat shock) and some cells without any chloroplasts (plastids). Why?

8. Does nuclear or chloroplastic DNA seem to be responsible for chloroplast reproduction?

9. Is the mechanism for reproduction directly or indirectly affected by the heat shock?

10. Does light counteract the effect of heat shock?

10

The Theory of Particulate Heredity

MATERIALS

CLASS STOCK

Ascaris slide
ears of corn (yellow and purple, with smooth and wrinkled pericarp)
vials of *Drosophila* crosses
etherizer

STUDENT STATION

microscope
microscope lamp
lens paper
sponge
hand lens or dissecting microscope
dissecting needle

INTRODUCTION

One of the truly great achievements in biology was the discovery of several laws of inheritance by Gregor Mendel about 100 years ago. Basic to his ideas was the concept that heredity was essentially based on unit factors or genes, which behave as discrete particles in their transmission and control of cellular activities.

The Mendelian laws of heredity are based on the transmission patterns of higher plants and animals, namely diploid organisms. Each individual contains two sets of chromosomes, one from each parent, and will, in turn, pass one set on to each offspring. This one set will be a mixture of chromosomes from both parents. In analyzing hereditary patterns it must be kept in mind that two important processes are involved, (1) meiosis, or halving of the chromosome number, and subsequently (2) fertilization, or random union of eggs with sperms. Therefore, in solving genetic problems, first the types of gametes are determined and then the gametic combinations at fertilization are determined.

No treatment of deducing Mendelian laws will be attempted here, but we shall examine how it is possible to determine whether the theory is correct. The validity of Mendelian ratios are evaluated by statistical procedures to give the student experience on a more practical level. The student is referred to the textbook for a theortical consideration of Mendelian principles although a list of terms is given below to refamiliarize students with the essential considerations.

BASIC TERMINOLOGY IN GENETICS

dominant: the gene or trait expressed in the heterozygous state (A*a*); that appeared in the P_1, P_2, and F_2 generations.

recessive: that gene or trait expressed only when in the homozygous state (*aa*); not expressed in F_1 generation, but expressed in P_1 and F_2.

homozygous: both genes or alleles alike (*aa*, AA); characteristic of a pure line.

heterozygous: both genes different (A*a*); a pair of alleles.

alleles: genes contrasting or different in expression that control the same trait (blue and brown eyes); genes that always segregate at meiosis; different genes on the same site of homologous chromosomes.

homologous chromosomes, homologs: morphologically identical chromosomes, genetically similar chromosomes in a cell, one from each parent; chromosomes that pair at meiosis.

phenotype: the expression of all the genes through time (blond hair, blue eyes, female, etc.).

geneotype: all the genes or genetic elements or factors regardless of whether they contribute to the phenotype (A*a*BB*cc*D*d* etc.).

gene: the unit factor controlling a trait, a site on the chromosome, a unit of inheritance.

segregation: the separation of alleles into different gametes, the "coming-apart" of different allelic genes into different cells during meiosis (A*a* → A + *a*).

independent assortment: the way one pair of alleles segregates has no effect on how another pair of genes segregates (A*a*B*b* → AB + *ab* or A*b* + *a*B).

FRUIT FLY *(DROSOPHILA MELANOGASTER)* IDIOGRAM

Application of Terms

A = red eyes
a = brown eyes

B = long bristles
c = *cupidus interruptus* (interruption of wing veination)
y = yellow body

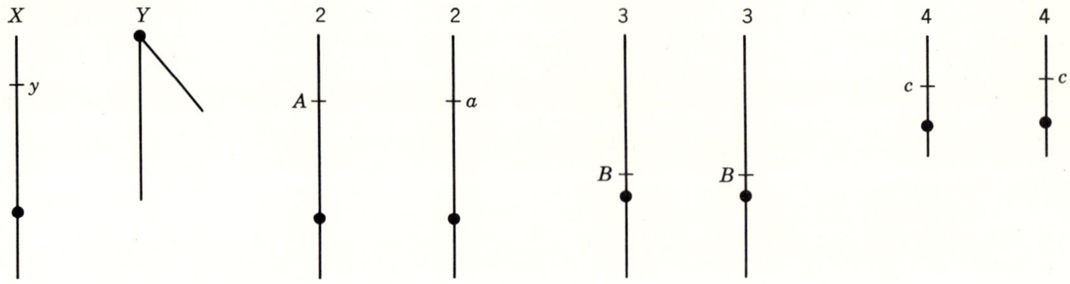

Figure 10.1

Relationships

homologous–2 and 2; 3 and 3; 4 and 4
alleles–*A* and *a*
nonalleles–*B* and *c*
heterozygous–*A* and *a*
homozygous–*B* and *B;* *c* and *c*
sex chromosome–*X*
hemizygous–*X* (only one) or *Y*
dominant–*A*,*B*
recessive–*y*, *a*, *c*
geneotype–*AaBBccy*
centromere–●
phenotype–red eyes, long bristles, *cupidus interruptus* wing, yellow body, male

MEIOSIS

The bodies of most higher plants and animals are composed of diploid cells; that is, there are two sets of chromosomes, one from each parent, in each of the nuclei. Certain specialized cells in the body undergo a process termed meiosis, which is two nuclear divisions resulting in a halving of the chromosome number to the haploid condition, ($2n \rightarrow n$) (see Fig. 10.2.)

1. Obtain a prepared slide of the parasitic worm *Ascaris* and examine under low power. The long, coiled tubules are regionally differentiated into the ovary, oviduct, and uterus. The primary oöcytes or cells that will undergo meiosis are produced in the ovary, pass into the oviduct where they are penetrated by the sperm, and enter the uterus as zygotes and young embryos.

2. Using Fig. 10.2, draw in the reproductive system, labeling the following parts.
primary oöcyte: thin-walled cells surrounded by small sperm and containing a single nucleus.
First Division: thick-walled cells containing the sperm cell at the edge of the cytoplasm. The cell contains four darkly stained pairs of chromosomes.
Secondary Oöcyte: thick-walled cells containing two nuclei at the edge of the cytoplasm, one the sperm, the other the first polar body (sloughed-off nucleus of the first division), and one functional nucleus in the center of the cytoplasm.
Second Division: same as above except cytoplasm contains chromosomes in stage of separation.
Mature Ovum: two polar bodies (first and second) are outside the cytoplasm. The cytoplasm contains two nuclei, one of the egg, and one of the sperm.
Two-Celled Embryo: thick-walled cell shows cleavage furrow dividing the zygote into two cells, each with a single nucleus.

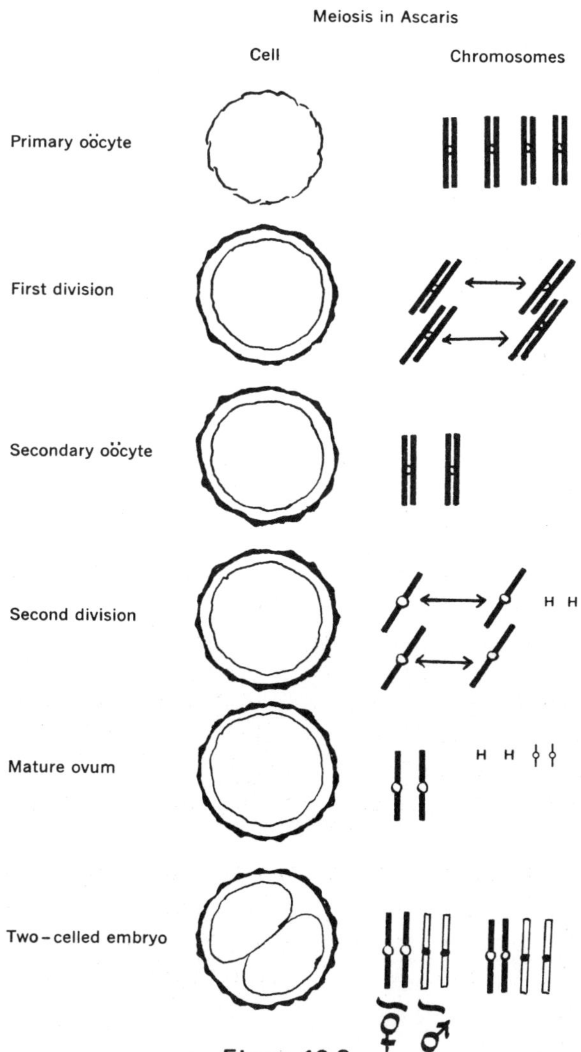

Figure 10.2

MENDELIAN INHERITANCE

You will receive an ear of corn with kernels of different colors of aleurone (outer layer of storage tissue) and different textures of pericarp (fruit wall): yellow and purple, and smooth and wrinkled, respectively. These kernels will be referred to as the F_2 generation (second filial generation) and were produced from the following crosses:

P_1	purple smooth	X	yellow wrinkled
F_1	all purple smooth		
P_2	F_1 purple smooth	X	F_1 purple smooth
F_2	to be determined		

Fill in Chart 10.1 with the phenotypes and number of each phenotypic type found on your ear of corn.

CHART 10.1

F_2 Phenotypes	Number of Each Type

No explanation of Mendelian principles of heredity will be attempted here. You will, however, attempt to determine if the counts fit the expected Mendelian ratio by employing a statistical procedure called the χ^2 (chi square) method.

Chi Square Method for Analysis of Genetic Data

Statistical procedures serve as a means of determining the probability that any given hypothesis is correct. In order to use a statistical procedure, one must (1) formulate a specific hypothesis, (2) collect sufficient data, (3) arrange the data in a manner that fits the statistical formula, and (4) make observations with great care. The Mendelian inheritance theory hypothesizes that genes assort independently of one another. According to these laws of heredity, if genes segregates independently, the patterns in which the genes will appear in the offspring should follow a predictable form. For example, if independent assortment occurs in a dihybrid cross *(AaBb X AaBb)*, a phenotypic ratio of 9 *AB*: 3 *aB*: 3 *Ab*: 1 *ab* individuals should result. See the Punnett square below.

Possible Gametes ♀

	AB	*Ab*	*aB*	*ab*
AB	*AABB*	*AABb*	*AaBB*	*AaBb*
Ab	*AABb*	*AAbb*	*AaBb*	*Aabb*
aB	*AaBB*	*AaBb*	*aaBB*	*aaBb*
ab	*AaBb*	*Aabb*	*aaBb*	*aabb*

If, in performing the actual cross, the results were as follows

Phenotype	No. of individuals
AB	900
aB	300
Ab	300
ab	100

it would be fairly certain that the hypothesis of independent assortment of these genetic factors is correct. Unfortunately, genetic ratios do not cooperate to this extent. Therefore, it is necessary to subject the data to a procedure that mathematically compares the data with the expected values based in the hypothesis. The χ^2 method is such a procedure. By using the appropriate values in the formula, a χ^2 value is obtained. By checking this χ^2 value against a probability chart for χ^2, a percentage probability value is reached. This percentage value is the probability in percent that the hypothesis is correct. A probability of greater than 10 percent is considered confirmation of the hypothesis. A probability of less that 10 percent indicates that the hypothesis must be partially or wholly rejected, in this case perhaps because some mechanism is in operation which causes assortment not to be wholly random. An alternative explanation, of course, is that the requirements for using a statistical procedure may not have been followed, that is, sufficient data may not have been collected or sufficient care taken in observations.

The χ^2 formula is expressed as follows:

$$\chi^2 = \sum \frac{(C_o - C_e)^2}{C_e}$$

C_o = number in class obtained
C_e = number in class expected
Σ = the total or sum

1. Using the data on the Chart 10.1 fill in Charts 10.2, 10.3, and 10.4.
2. The χ^2 value itself has little meaning and must be used to obtain a probability value *(P)*. This is found by checking against Chart 10.5.
   a. First look for the degrees of freedom (the number of classes minus one). For a 9:3:3:1 ratio, there are four classes; there are *three* degrees of freedom.
   b. Next move across until you come to a value equal to your χ^2 or the two numbers that are on either side of your χ^2.
   c. Move to the top of the chart and find the *P* value or the two *P* values on either side. If your χ^2 was 7.03, then with three degrees of freedom there is between 10 percent and 1 percent probability that your analysis of the situation is correct. If the data fit the expected ratio, the *P* value must be 10 percent or greater. If the data do not fit the theoretical expectation, the *P* value is less than 10 percent.

CHART 10.2

Phenotypes	C_o	C_e No.	C_e Ratio	$C_o - C_e$	$(C_o - C_e)^2$	$\frac{(C_o - C_e)^2}{C_e}$
Purple smooth			9			
Purple wrinkled			3			
Yellow smooth			3			
Yellow wrinkled			1			
Total			16		$\chi^2 = \Sigma =$	

$$\sum \frac{(C_o - C_e)^2}{C_e} = \chi^2 =$$
$$P =$$

CHART 10.3

Phenotypes	C_o	C_e No.	C_e Ratio	$C_o - C_e$	$(C_o - C_e)^2$	$\frac{(C_o - C_e)^2}{C_e}$
Purple			3			
Yellow			1			
Total			4		$\chi^2 = \Sigma =$	

$$\sum \frac{(C_o - C_e)}{C_e} = \chi^2 =$$
$$P =$$

CALCULATIONS

CHART 10.4

Phenotypes	C_o	C_e No.	C_e Ratio	$C_o - C_e$	$(C_o - C_e)^2$	$\frac{(C_o - C_e)^2}{C_e}$
Smooth			3			
Wrinkled			1			
Total			4		$\chi^2 = \Sigma =$	

$$\sum \frac{(C_o - C_e)^2}{C_e} = \chi^2 =$$
$$P =$$

CHART 10.5 Probabilities for the χ^2 Method

Degrees of Freedom	99%	90%	70%	50%	20%	10%	1%
1	.00	.02	.15	.45	1.63	2.71	6.35
2	.02	.21	.71	1.39	3.22	4.60	9.21
3	.11	.58	1.42	2.37	4.64	6.25	11.37
4	.29	1.06	2.19	3.35	6.00	7.78	13.28

3. Find the P values for Charts 10.2, 10.3, and 10.4.
4. Do the obtained data fit the expected data in all cases?

5. Can you find two pieces of information that suggest the idea of particulate heredity?

6. Which statement is correct: "Statistics give you confidence in your conclusion" or "Statistics tell you whether you are correct or not"? Why?

7. Why is it necessary to have pairs of alleles to carry out crosses?

SEX-LINKED INHERITANCE

Sex linkage follows a special pattern in heredity. In *Drosophila melanogaster,* the fruit fly, the male has one *X*-chromosome and a smaller nonmatching *Y*-chromosome, while the female has two *X*'s. This is the same pattern found in man. Under these conditions any gene located on this *X*-chromosome will be expressed in the male, whereas the female recessive traits will be expressed only in the homozygous state. Any gene on the *X*-chromosome can be detected by reciprocal crosses, for example

red-eyed ♀ X white-eyed ♂
white-eyed ♀ X red-eyed ♂

Handling of fruit flies

1. Obtain a vial of *Drosophila,* an etherizer, and a piece of sponge.
2. Lift the top of the etherizer and place a few drops of ether on the cotton on the bottom of the bottle. Replace the etherizer top.
3. Tap the vial on the sponge to knock the flies to the bottom and quickly remove the cotton plug as you invert the vial over the etherizer so that the flies fall to the gauze just above the ether-filled cotton.
4. Hold the two bottles together to prevent the flies from escaping and tap the etherizer on the sponge to keep the flies near the ether.
5. Keep the flies in the etherizer for about 30 sec until you can detect no gross movement at the bottom of the etherizer. Be careful not to over-etherize the flies and kill them.
6. Pour the flies out on a piece of white paper.
7. If the flies begin to move, invert your etherizer over them for 5 to 10 sec.

Identification of the flies

The first trait to recognize is the sex of the fly. This may be determined by the naked eye, but is easier if a hand lens or dissecting microscope is used. Turn the flies ventral side up gently with a dissecting needle. The male is smaller than the female and has a dark and rounded posterior of the abdomen. The female is larger with a broad abdomen of light color. The abdomen narrows to the pointed ovipositor at the posterior portion of the abdomen (see Fig. 10.3).

Determination of Phenotype

The eye color will be either red or white. This is easily determined with the naked eye.

1. Count the phenotype of the two reciprocal crosses.

2. Return the flies to the vial from which they came after counting them by tipping the vial in such a way that the flies do not fall on the sticky food. Replace the cotton plug and invert the vial, tapping lightly so that the flies will rest on the dry cotton plug.

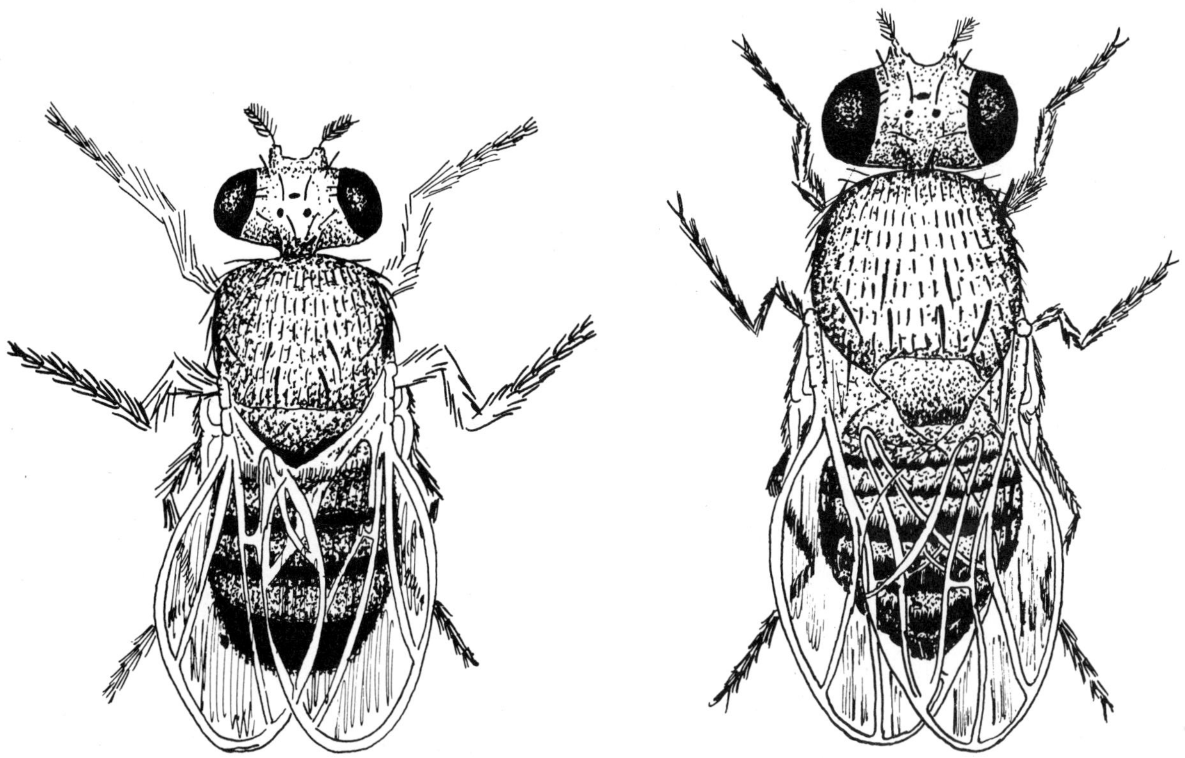

Male

Female

Figure 10.3

3. Enter the data on Chart 10.6.

CHART 10.6

Sex	Phenotype	Red ♀ X White ♂	White ♀ X Red ♂
♂	Red		
	White		
♀	Red		
	White		

4. In Fig. 10.4, draw how the chromosomes and genes were inherited, using the following symbols:

ω^+ = red allele
ω = white allele

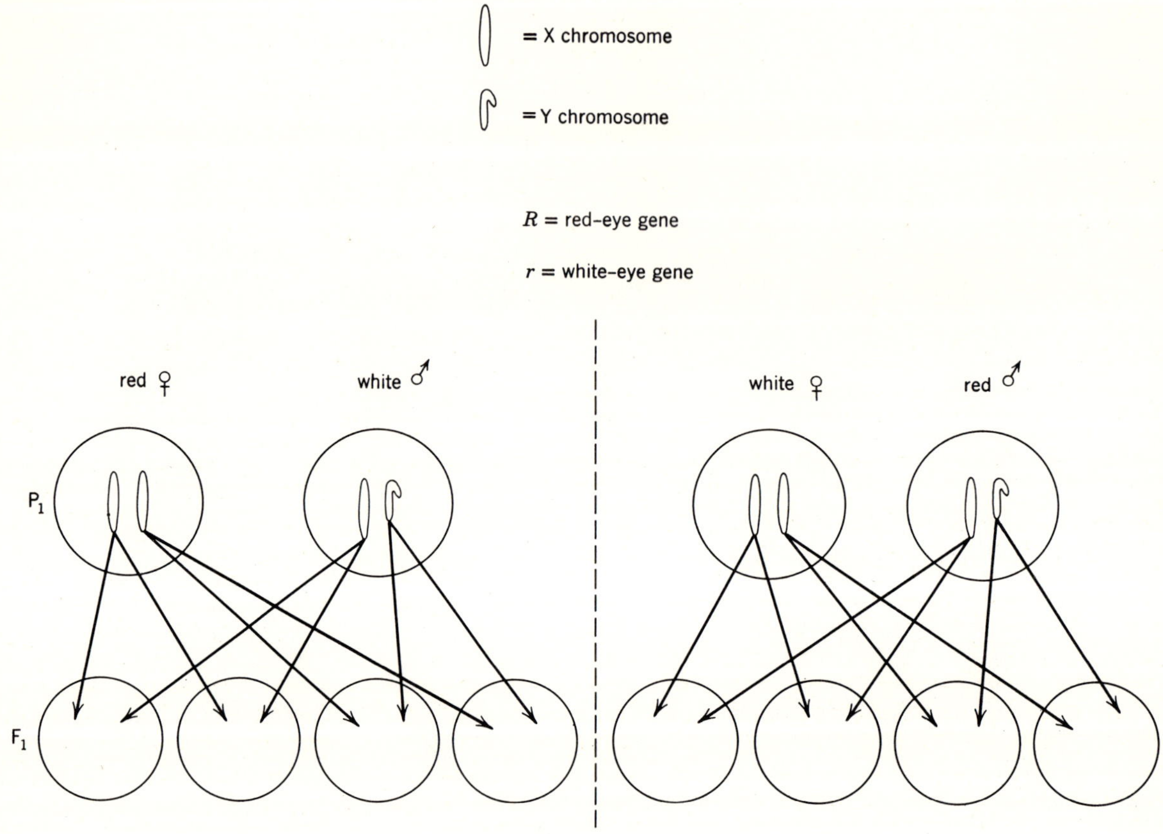

Figure 10.4

5. This pattern of inheritance is often referred to as "criss-cross" inheritance. Explain what is meant by this.

6. How does this system of inheritance ensure an approximate 50:50 ratio of males to females?

7. Color blindness in humans is a recessive sex-linked trait. Explain why more males exhibit this trait than females.

8. Give one reason why a geneticist will perform reciprocal crosses in order to determine the hereditary pattern of a new genetic trait?

11

Genetic Regulation of the Cell

MATERIALS

CLASS STOCK	STUDENT STATION
petri plates with radish seedlings	millimeter ruler
paper messages in code	pencil with eraser

INTRODUCTION

Each protein molecule, with its set sequence of amino acids and specific three-dimensional shape, lends specificity to the enzyme reaction, macromolecular complex, organelle, cell, and ultimately the organism in which it resides. Hence, those factors that control protein synthesis control the form and function of every level of biological organization.

The sequence of events leading from DNA in chromosomes to protein synthesis was finally elucidated in the early 1960s; it has become known as the Central Dogma. In its most simple form, the Central Dogma is DNA ⟶ RNA ⟶ protein, that is, DNA sequences of nucleotides are (1) transcribed into RNA sequences of nucleotides, which leave the nucleus, proceed into the cytoplasm to ribosomes (which also consist of RNA) and are then (2) translated into protein which then folds into its three-dimensional shape. Thus the gene, or cistron, which is composed of the nucleotides for a given polypeptide, and which can be read or not read, in accordance with a host of other factors (inhibitors, activators) controls RNA synthesis, and subsequently protein synthesis.

THE CENTRAL DOGMA

As always, it is impossible to repeat the sophisticated modern experimentation in recent biological research. Hence, only some of the salient features of proteins, RNA, and DNA can be demonstrated in ways that illustrate the basic features, but not the specific steps and all types of molecules involved.

1. Your instructor has prepared five petri plates for your examination. Each contains radish seedlings which have been treated in the following ways:
   (a) Formaldehyde 1% added to one-day-old seedlings. This plate is a control and represents zero time.
   (b) Formaldehyde 1% added to one-day + 20 hr-old seedlings. This plate is the control representing 20 hr of growth after zero time.
   (c) Cyclohexamide (100 μg/ml) added to zero-time seedlings. Cyclohexamide inhibits protein synthesis but does not destroy existing protein.
   (d) Actinomycin D (20μg/ml) added to zero-time seedlings. Actinomycin D inhibits RNA synthesis (transcription from DNA to RNA sequences) but does not destroy existing RNA.
   (e) Mitomycin C added to zero-time seedlings. Mitomycin inhibits DNA synthesis.
2. Take five seedlings from plate I place them on a piece of wet paper toweling and take them back to your desk. Measure the length of the roots of the seedlings with a millimeter ruler and enter your data on Chart 11.1.
3. Return the seedlings to plate (a) and take five seedlings from plate (b) to your desk. Measure the roots and enter your data on Chart 11.1.
4. Repeat with seedlings from plates (c), (d), and (e), making sure you return the seedlings to the proper plates for other students to use. All data on Chart 11.1.
5. Take the average of each group of seedling root lengths.
6. Calculate the rate of growth by subtracting the average zero-time length from the average zero time + 20 hr-growth length and dividing by 20 hr.

$$\text{millimeter growth per hour} = \frac{0 \text{ time} + 20\ (\bar{m}) - 0 \text{ time}\ (\bar{m})}{20 \text{ hr}}$$

7. What is the rate of growth?

Individual protein molecules spontaneously deteriorate and are replaced by newly synthesized ones. Although this may at first seem wasteful, it is actually an important feature of the efficient function of cellular systems. The proteins active during cell division for dissolution of the nuclear membrane and formation of the spindle cannot be present in

CHART 11.1

Treatment	Root Lengths	Average
0 time	1	
	2	
	3	
	4	
	5	
	Total/5 = average	
0 time + 20 hrs	1	
	2	
	3	
	4	
	5	
	Total/5 = average	
Cyclohexamide	1	
	2	
	3	
	4	
	5	
	Total/5 = average	
Actinomycin D	1	
	2	
	3	
	4	
	5	
	Total/5 = average	
Mitomycin C	1	
	2	
	3	
	4	
	5	
	Total/5 = average	

interphase when the cell is actively metabolizing. By the same token, a large number of metabolic enzymes synthesized during cell division would unbalance the cellular equilibrium. As you demonstrated in the exercise in Chapter 9, little protein synthesize other than that for spindle formation is active during the division process. Therefore, deterioration of protein molecules is as integral a part of the system as their synthesis.

Cyclohexamide inhibits protein synthesis and does not allow new proteins to be formed. This means that the cells will continue to grow only as long as the previously synthesized protein remains functional. Once that protein deteriorates, growth will cease.

It is possible to calculate the half-life of the population of protein molecules by using the same type of reasoning in the experiment on radish seedlings found in Chapter 9. Since some of the protein functional at the time cyclohexamide was added was just synthesized, this protein would remain functional for the longest time. Other protein was in the last stages of functionality, and these molecules would deteriorate soon. The rest of the protein is in stages varying from almost newly synthesized to approaching the last stages of functionality. As increasing numbers of molecules deteriorate, growth will slow, and finally cease. The time between zero-time and the time growth ceased is the half-life of a protein.

8. Calculate the half-life of a protein by subtracting the average zero time from the average cyclohexamide root lengths, and divide by the rate of growth.

$$\text{half-life of a protein} = \frac{\text{cyclohexamide } (\overline{m}) - 0 \text{ time } (\overline{m})}{\text{rate of growth/hr}}$$

9. How long is the average full life of a protein? (Since there are many kinds of proteins, this represents an approximation.)

10. According to your calculations for the mitotic cycle in Chapter 9, would spindle fibers and enzymes responsible for spindle fibers deteriorate before the S stage of interphase?

11. In view of your answer for the average life of a protein, is it logical that most of the cell's energy is expended in protein synthesis? Explain.

Actinomycin D inhibits the transcription of base sequences from DNA to RNA, but does not directly affect protein synthesis. After actinomycin D has been added, no new RNA is formed, but the already existing RNA continues in its role in protein synthesis. When the existing RNA deteriorates, the cell is dependent on existing protein. And when the protein deteriorates, growth ceases. Therefore, actinomycin D indirectly affects protein synthesis by inhibiting the formation of the RNA on which it is dependent.

12. Calculate the half-life of RNA and protein by subtracting the average zero time from the average actinomyocin D root length and dividing by the growth rate.

$$\text{half-life of RNA and protein} = \frac{(\overline{m})\ \text{Actinomycin D} - (\overline{m})\ 0\ \text{time}}{\text{rate of growth}}$$

13. What is the average full life of RNA and protein?

14. Since you know the average full life of protein, what is the average full life of RNA?

15. How does this experiment with cyclohexamide and actinomycin D illustrate that the sequence of synthesis is RNA → protein, and not protein → RNA?

16. Of what advantage is a short half-life for RNA?

Mitomycin C, as you are aware from your experiment in Chapter 9, inhibits DNA synthesis. Unlike RNA and protein, DNA does not deteriorate with time. In fact, one of the major advantages to DNA as an information molecule is its great thermal stability. Therefore, it is difficult to calculate a half-life for DNA, simply because thermal denaturation at 37°C occurs over a span of decades. What would you be calculating if you followed the above procedures used for actinomycin D and cyclohexamide for mitomycin C?

When DNA is about to replicate, no RNA is formed, so that for a short period of time, there is no RNA synthesis. When mitomycin C is added, the cells are suspended in the stage where no RNA synthesis takes place, and the cell is dependent on already formed RNA and proteins for growth. Growth will stop when these molecules deteriorate.

17. Calculate half the mitotic cycle by subtracting the average zero-time from the average mitomycin C heated root lengths and dividing by the growth rate.

$$\tfrac{1}{2}\ \text{mitotic cycle} = \frac{(\overline{m})\ \text{mitomycin C} - (\overline{m})\ 0\ \text{time}}{\text{rate of growth}}$$

18. What is the average length of the whole mitotic cycle?

19. DNA is *directly* responsible for synthesizing what two types of molecules?

20. Would some protein destruction destroy the cell permanently? Explain.

21. Would some RNA destruction destroy the cell permanently? Explain.

22. Would some DNA destruction destroy the cell permanently? Explain.

23. Would you conclude from this experiment that the sequence is DNA → RNA → protein, and not RNA → DNA → protein? Why?

THE CELL AS AN AUTOMATON

The common way of studying cells is to describe their structure and function and to follow these descriptions with experimental investigations. Another and lesser used means of studying cells and organisms is to construct models that simulate the properties of living systems. By this approach the entire individual can be evaluated, and the interplay between various parts, which create a functional organization, can be inspected. This approach assumes that the cell is an automaton, that is, a machine capable of carrying out all its activities from within when supplied only with the necessary building blocks and energy.

This series of exercises is designed to illustrate how an automaton operates, especially with reference to living forms. An automaton, whether living or man-made, must be capable of (1) deciphering instructions into actions, (2) counting, (3) turning off and on various activities and thereby regulating rates, and (4) possessing some form of memory. These capacities must be achieved by designing the automaton so that no external force is necessary for control of operation, and that any control necessary is derived from qualities of the individual components, and not, as one would assume, from some central control.

The living automaton must be capable of two functions: self-regulation and self-reproduction. This series of exercises will lead up to these two criteria.

Reading Instructions and the Genetic Code

One feature necessary for an automaton is that it be capable of deciphering instructions into action. The striking similarities between the events of the Central Dogma and the translation of a code did not escape biologists. As a result, the language of cryptology and information theory is used to describe certain features of the DNA-RNA-protein sequence. Typical phrases arose, such as, the genetic "code," "transcription" of DNA to RNA.

The genetic code is based on four nucleotides. In DNA these are adenine *(A)*, thymine *(T)*, guanine *(G)*, and cytosine *(C)*; in RNA the thymine is replaced by uracil *(U)*. In the 1950s, when Watson and Crick proposed their model for DNA, they included a provision by which certain bases attract one another. In the double helical structure, an adenine on one strand was always paired with a thymine on the other strand, and a cytosine on one strand was always paired with a guanine on the other. This attraction of bases, which led to base pairs *(A-T, C-G)* provided the foundation for DNA transcription.

It is believed that one strand of DNA is "read" or transcribed at a time. When the strand is to be read, the *RNA* bases *(A,U,G,C)* are attracted to it. The adenine will pair with a uracil, the cytosine with an quanine, and so forth. The DNA sequence

AATTCGCGACGTATA

would be transcribed into the RNA sequence

UUAAGCGCUGCAUAU

The transcription of the coded message to an RNA sequence was a minor achievement compared to the translation of the code, which was achieved in 1965 by Nirenberg. Nirenberg postulated a three-base code, in which three sequential bases determined a single amino acid. Thus nine bases in sequence determined three specific amino acids in sequence. Much of the work was filled in by other workers, leading to the almost complete translation of the genetic code shown in Chart 11.2.

CHART 11.2 THE GENETIC DICTIONARY

First Base	Second Base U	C	A	G	Third Base
U	phenylalanine	serine	tyrosine	cysteine	U
	phenylalanine	serine	tyrosine	cysteine	C
	leucine	serine	terminal	tryptophane	A
	leucine	serine	terminal	tryptophane	G
C	leucine	proline	histidine	arginine	U
	leucine	proline	histidine	arginine	C
	leucine	proline	glutamine	arginine	A
	leucine	proline	glutamine	arginine	G
A	isoleucine	threonine	asparagine	cysteine	U
	isoleucine	threonine	asparagine	cysteine	C
	methionine	threonine	lysine	arginine	A
	methionine	threonine	lysine	arginine	G
G	valine	alanine	asparagine	glycine	U
	valine	alanine	asparagine	glycine	C
	valine	alanine	glutamic acid	glycine	A
	valine or met	alanine	glutamic acid	glycine	G

The sequence of cellular events leading to protein synthesis is now fairly well known. In a somewhat simplified form the sequence would be as follows: Different genes on the DNA transcribe messages for several types of RNA, among which are transfer RNA (tRNA), ribosomal RNA (rRNA), and messenger RNA (mRNA). Ribosomal RNA proceeds into the cytoplasm, and forms a "table" for the mRNA. (Messenger RNA cannot be translated unless it is associated with rRNA). There are about 60 different types of tRNA molecules, one for each amino acid. The tRNA's also proceed to the cytoplasm.

In the cytoplasm there are about 20 enzymes, each specific for a particular tRNA and for a single amino acid. The tRNA, enzyme, and amino acid combine to form a tRNA-enzyme-amino acid complex. The enzyme is apparently further capable of recognizing the appropriate RNA sequence on the messenger, and is thought to be responsible for the attraction between the complex and the messenger at the appropriate site.

After being aligned in a specific sequence of bases, the messenger proceeds into the cytoplasm and associates with the ribosomes. Translation of the messenger begins with a protein at one end of the messenger. The first three bases of the messenger associate with the appropriate three bases of a tRNA. For example:

mRNA *UUUUUAUCA*
tRNA *AAA*
|
enzyme
|
phenylalanine

So that the amino acid phenylaline is brought to the beginning of the messenger. Next the tRNA containing the anticode *AAU* would associate with the messenger. The tRNA carrying this code would be for the amino acid leucine. The phenylalanine and the leucine would then form a peptide link. The next tRNA brought to the messenger would carry the anticode *AGU,* and would be the tRNA for the amino acid serine. Serine would then form a peptide link with leucine, and the amino acid sequence would be phenylalanine-leucine-serine. The protein progresses down the length of the messenger as it is formed. The now free terminus will attract another tRNA with the anticode *AAA*, and another protein will commence forming. Several proteins are formed on the same messenger at the same time, each in a different stage of completion.

In order to gain some insight into certain features involved in translation you will play the role of an enzyme.

1. On the laboratory tables are a number of messages encoded similarly to what is found in DNA and RNA. Each student will receive a card with a simple instruction indicating the operation to be performed.
2. If the card reads "Place A below UUU" you will move about the room, from message to message, placing an *A* beneath a *UUU* triplet. But you must make sure that the *UUU* is a triplet sequence. For example, the sequence *CAUUUÇGC* does not contain a triplet *UUU;* the triplets are *CAU, UUC* and *GC-*. The only time *UUU* is clearly a triplet is when the other triplets to the left have been translated, and *UUU* is at the extreme left end of the undeciphered portion.
3. Continue your migration around the room to different messages until you see your triplet at the extreme left end of a message. Write your letter of the alphabet beneath the triplet and then cross the triplet out.
4. Notice that for the purposes of this exercise you are not a letter of the alphabet (which would be analogous to an amino acid in the cell). You are a device capable of recognizing a sequence on one hand, and having an endless supply of one particular letter on the other. To what enzyme in the cell are you analogous and why?

5. After all the messages have been deciphered, read them. Why is it necessary for a cell to make devices such as this before protein can be produced?

6. Why is it necessary that all the students take part in the deciphering?

7. Does an automaton need a chart or master plan containing the code and its translation?

8. Do the individual units "know" what the final result will be?

9. Can the new messages replace the cards? Where did the cards come from in the first place?

10. Is the repetition of the messages on the cards the only translation possible?

11. Explain how the following postulate is demonstrated by the exercise. "The complexity of a system does not reside in the ability to build, but in the length of instructions."

12. From the postulate above, explain how an egg can give rise to a complex mammalian body?

Efficiency of Translation

The ease with which translation depends on several factors, some of which relate to the code itself. The object of a code is always to transmit information. Hence, the code must be able to contain the information desired. For example, a code of one symbol, for example, *A*, could convey only one piece of information—whatever *A* stands for. A code of two symbols, *A* and *B* could contain two pieces of information if only two letter words are used, namely *AB* and *BA*. A code with three symbols using only three letter words would contain 27 pieces of information. The formula for the number of symbols in a code that contains words of the same length is

$n^{n'}$

where n = the number of symbols in a code, and n' = the number of symbols used per word.

1. How many pieces of information would be carried by a code of four symbols, using only four-letter words?

2. How many pieces of information would be carried by a code of four symbols using only two-letter words?

3. Construct a statement relating the number of symbols used in a code, the number of symbols per word, and the information carrying capacity of the code.

4. Protein is the meaningful translation of information in terms of the cell; amino acid units are the basic components of protein. The cell must carry information for all the amino acids and for a punctuation mark to terminate a reading sequence. What is the minimum number of pieces of information the DNA must code in order to carry the necessary information.

   $n^{n'} = ?$

Another feature encompassed in the code of an information system is a certain amount of redundancy. Redundancy is a necessary feature because it eliminates the possibility of a single alteration rendering the code useless. If, for example, only *UUU* translated to phenylalanine, and the only gene that produced the tRNA of that anti-code was mutated, no phenylalanine could be incorporated into any protein, and all protein synthesis would cease. On the other hand, too much redundancy is wasteful. An information system must contain enough redundancy to safeguard the system, but not too much, so as not to waste valuable time and energy.

In its evolution, DNA has incorporated a certain degree of redundancy. The system itself is redundant—there are many genes that transcribe each tRNA. The code itself has evolved to contain redundant elements in that more than one triplet codes for a single amino acid.

5. If, in order to contain sufficient redundancy, the DNA code must carry each piece of necessary information *twice,* what is the minimum number of pieces of information for which the DNA must code?

We shall return later to the ideas developed in this part of the chapter.

Another factor on which translation of the code depends is the simplicity of the code. This can be demonstrated by translating messages from three different codes to test the efficiency and degree of correctness.

1. Translate the messages in Fig. 11.1.

2. Compare your deciphered message with that of two other students in the following way: One student will read his message aloud. When there is a disagreement regarding a symbol, the two students agreeing are arbitrarily considered correct. If all three disagree on a symbol, all are considered incorrect.

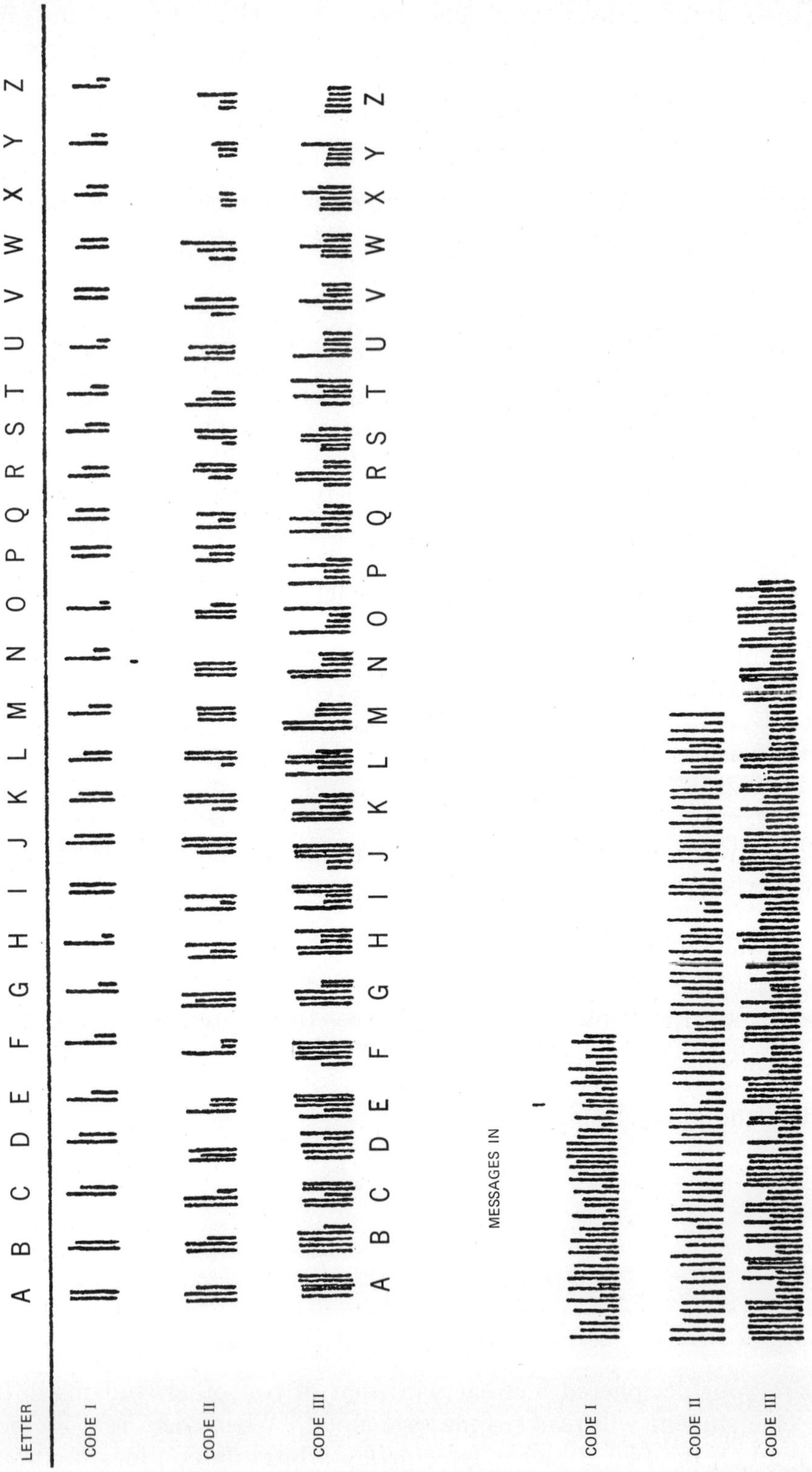
LETTER
A B C D E F G H I J K L M N O P Q R S T U V W X Y Z
CODE I
CODE II
CODE III
A B C D E F G H I J K L M N O P Q R S T U V W X Y Z
MESSAGES IN
CODE I
CODE II
CODE III

Figure 11.1

3. Enter the number of mistakes according to individual students on Chart 11.3.

CHART 11.3

	Number of Errors		
	Code I	Code II	Code III
Individual			
Group			

4. The instructor will then read off the correct translations of the messages. How many errors did your group have? Enter your information on Chart 11.3.

5. On the basis of efficiency of translation and degree of correctness, which is the best code?

6. How many different symbols were there in Code I? How many symbols were used per word?

7. How many different symbols were there in Code II? How many symbols were used per word?

8. How many different symbols were there in Code III? How many symbols were used per word?

9. Construct a statement relating the number of symbols used in a code, the number of symbols used in a word, and the efficiency of translation and degree of correctness of a code.

An information system must be a balance between a simple code and the amount of information it must contain. Examine the efficiency of the DNA code by answering the following questions:

10. The simplest code would be one of two symbols. How many symbols per word would there have to be in a DNA code to carry at least the necessary information (including redundancy) if the code were based only on adenine and thymine?

$2^{n'}$ = necessary pieces of information

11. How many symbols per word would there have to be in a DNA code to carry at least the necessary information (including redundancy) if the code were based on adenine, thymine, and guanine?

$3^{n'}$ = necessary pieces of information

12. How would the situations described in Questions 10 and 11 affect the length of the DNA molecule necessary to carry this information?

13. How many symbols per word would there have to be in a DNA code to carry at least the necessary information (including redundancy) if the code were based on five nucleotides? How would this affect the amount of redundancy?

$5^{n'}$ = necessary pieces of information

14. How is the existing DNA code, based on a four-symbol code in triplets (a 3-symbol "word") a balance between the code simplicity and information carrying capacities?

Regulation

As you have seen earlier in this exercise, the average life of a protein is relatively short. It has been stated that the proteins for spindle formation are not within the cell during interphase stages. The implication of such a statement is *that no more protein for spindle fibers is produced,* which implies that there is some mechanism in the cell for regulating protein syntheses.

The regulation of processes by turning certain ones on and off is an important feature of automatons. It would be wasteful, and in some cases detrimental, for a machine to have all its parts operating all the time. A component that can be regulated can be used to control operation and rate of operation. Such a device works according to a negative feedback system, in which the product from system A stimulates system B to act, but the product from system B can inhibit system A from operating.

Negative feedback is an important means of control in a cellular system. Activator genes (System A) produce substances that enable other genes (System B) to be transcribed. The RNA or final protein product of System B may serve to repress the activator system. The activator production ceases, and ultimately the gene analogous to system B ceases production. When the molecules produced by system B fall below a specific level or threshold, the molecular balance dictates that the repressor dissociate from the activator gene, and the process begins again.

The operation of a negative feedback system is perhaps best understood from a mechanical model, such as that illustrated in Fig. 11.2. Mexican jumping beans are crowded on a

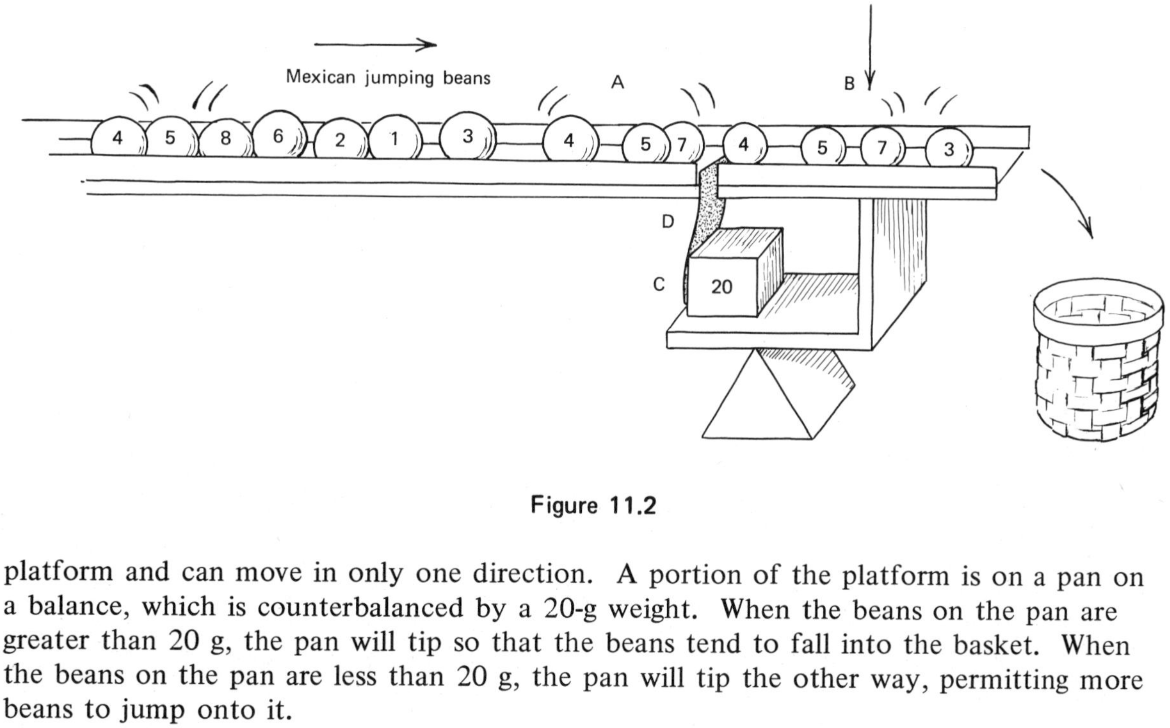

Figure 11.2

platform and can move in only one direction. A portion of the platform is on a pan on a balance, which is counterbalanced by a 20-g weight. When the beans on the pan are greater than 20 g, the pan will tip so that the beans tend to fall into the basket. When the beans on the pan are less than 20 g, the pan will tip the other way, permitting more beans to jump onto it.

1. Determine the weight on the pan each time a bean falls off or jumps on, assuming (a) that the beans jump on in the given order and (b) that the changes occur every second. For example: The weight now on the pan is 19 g; the pan will tip to per-

mit another bean to jump onto it. Since the next bean weighs 7 g, when it is on the pan, the pan will weigh 26 g, and tip in such a way that the bean closest to the basket will drop into it. This bean weighs 3 g, so the weight of the pan after 2 seconds is 23 g. Since the pan is still tipped toward the basket, another bean will drop into it—this time the 7 g bean. The pan again weighs 16 g, and will tip in a direction to permit another bean onto the pan. The plotted weights for the first three seconds would be 26 g, 23 g, 16 g.

2. Plot the values for the pan weights on the graph below (Fig. 11.3).

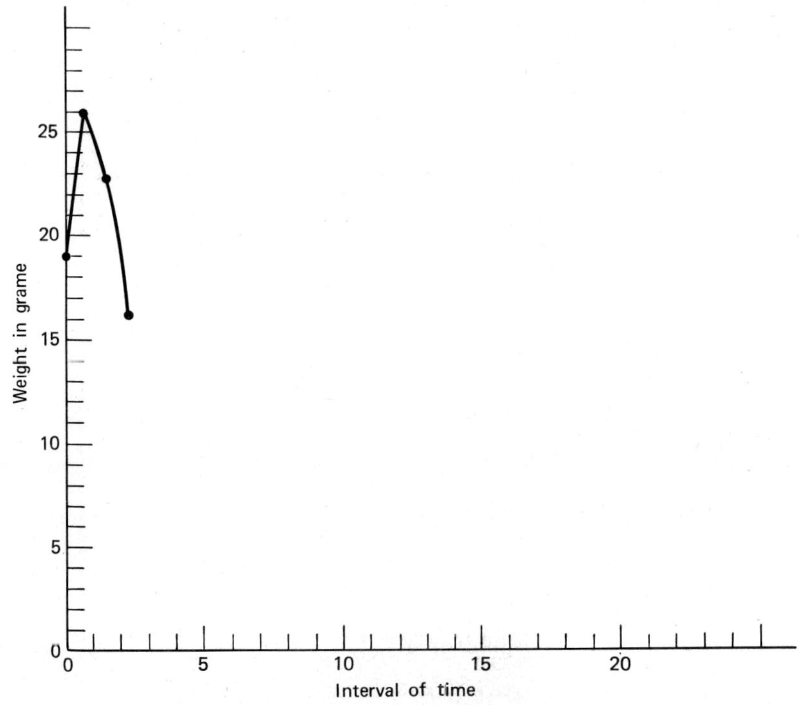

Figure 11.3

3. Explain how this device can control either off-on regulation or the rate of the process.

4. Explain how an enzyme and its product could form a feedback system?

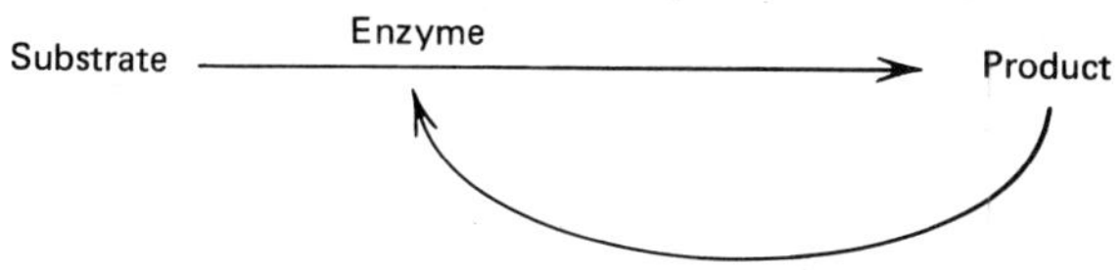

Figure 11.4

Counting

It is valuable for a system to identify the number or concentrations of some particular part or product, such as how many mitochondria or enzymes are in the cell. Below is a simple means of counting, although it is not necessarily the system found in cells. It is based on the idea that each member of a group contributes to the counting procedure, and no outside component does the counting.

1. Consider a series of circles.

   0 0 0 0 0 0 0 0 0 0 0 0 0 0 0

2. Add a +1 to each circle on the right of it, and a −1 for each circle on the left of it.
3. Add the values for each circle.
   The number of circles is equal to the largest sum plus one.
   Thus, if the largest value is +2, then +2 +1 = 3 circles.
4. According to this counting system, how many circles are there?

5. By what modification can this system count the number of circles in a 3 X 3 arrangement?

6. How is it possible to divide the circles into two equal groups?

7. What configuration of circles cannot be counted by this method?

A Recipe for a Cell

Although the devices previously illustrated can do little by themselves, with slight modification they can be integrated to produce an automaton that simulates a cell, that is, a machine that has the capacities for self-correction and self-reproduction.

1. An instruction list follows for changing a cell. Before following the instructions, draw *two* copies of Fig. 11.5. Label one *A*. This will be your reference drawing so that you may see what you have accomplished when you are finished.

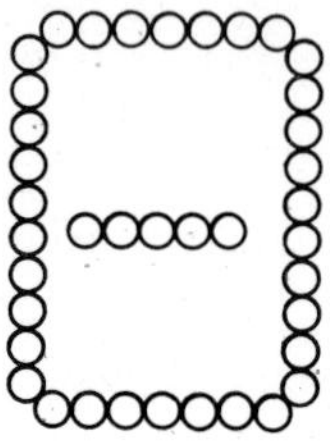

Figure 11.5

2. Begin reading the instructions, modifying drawing *B* accordingly. Notice that you do not have to know what is happening. After all, you are only an external sourcc of energy.

Instructions

(a) Measure the lengths of all four sides.
  (1) If the longest side is less than 20, continue with (b).
  (2) If the longest side is 20, continue with (c).

(b) Consideration of longest sides:
  (1) If you have an even number of circles, erase the top horizontal circles, add one circle to each vertical side, and redraw the horizontal circles adjoining the newly drawn circles. Return to (a).
  (2) If you have an odd number of circles, erase the lowest horizontal circles, and add a circle to each vertical side on the bottom. Redraw the horizontal circles adjoining the newly drawn ones. Return to (a).

(c) Consider all five rows of circles, and count them according to the method discussed earlier in this section. Mark the +1 and −1 circles on the vertical rows with Xs, and fill in the 0-valued circles on the three horizontal rows. Continue with (d).

(d) Draw a circle below any circle belonging to a row with exposed ends. Continue with (e).

(e) Draw a row of filled-in circles between the filled-in circles on the horizontal rows. Continue with (f).

(f) Count the number of closed circles between each of the horizontal rows.
 (1) If there are more than three, erase one and reconnect the central horizontal rows by moving them toward the peripheral horizontal rows. Return to (f).
 (2) If there are three, erase all closed circles and continue with (g).

(g) Draw a row of X'd circles between the X'd circles on the vertical rows.

4. Identify the steps involving counting, negative feedback, identification and correction of errors, and genetic transmission.

5. By following the instructions, what process have you simulated?

6. In what steps was the situation analogous to synthesizing cellular material?

7. In what steps was the situation analogous to DNA replication?

8. In what steps was the situation analogous to spindle formation?

9. If the nucleus of a human cell has two million genes, how long would the recipe be for simulating development and physiological adjustment of man?

Part III

GROWTH AND CELLULAR DIFFERENTIATION

12

Growth

MATERIALS

CLASS STOCK

yeast cultures grown and frozen at different times
colorimeter tubes
pea seedlings, 1-10 days old
gourds

STUDENT STATION

plastic cover slip
blank slide
razor
millimeter ruler
microscope
microscope lamp
lens paper
Pasteur pipette
string

INTRODUCTION

Growth is probably one of the most fundamental processes of life. Growth has been defined in many ways, but the most useful definition for our purposes is an increase in active protoplasmic mass. This protoplasmic mass may be represented by a new individual within a population or by the increase in size of a single organism. In a reproductive sense, growth of a population ensures the survival of the species. The maintenance of any species is based on a balance between the reproductive rate and the death rate of its members. A population is considered to grow when its rate of reproduction exceeds the death rate, resulting in an overall increase in the size of the population. Thus the success of any species is based on its reproductive rate. In more structurally complex organisms, growth is often associated with differentiation; that is, an increase in protoplasm is coupled with a change in biochemical and morphological (structural) processes.

Growth is the result of the sum of all metabolic activites of the system. Hence, growth may be studied as a biological phenomenon itself, or as a process reflective of other biological phenomena. As the latter, growth has been used several times in this laboratory. For example, in the exercise on photosynthesis the amount of growth of the *Chlorella* was an indication of the degree of operation of the photosynthetic mechanism. This experiment implied light → photosynthesis → glucose → degradation through respiratory processes → energy → biosynthetic reactions → growth.

The study of growth as a biological phenomenon has merit. In more complex organisms, it is difficult to separate growth and differentiation and to study growth as an isolated phenomenon. Accordingly, microorganisms, notably bacteria, that undergo little or no differentiation are usually used for growth studies.

There are two major means of measuring growth; the total count, and the viable count. The total count is the number of living and dead organisms. The viable count, on the other hand, is a count of only the live organisms. The definition of death for microorganisms is based on their ability to reproduce; an organism capable of dividing is considered to be alive, and one that is not, regardless of the operation of its metabolic machinery, is considered to be dead. It is impossible to distinguish between a living and dead organism by microscopic examination. The only way of obtaining a viable count then is by the plate count method. When certain bacteria are placed on the agar surface of a petri plate they remain in one place and grow. Growth of a localized bacterium gives rise to a localized population, called a colony. Each colony represents a single living bacteria originally placed on that plate. By taking samples from a growing culture, appropriately diluting, plating on an agar surface, incubating, and then counting the colonies, the number of live bacteria placed on the plate, and finally, the number taken out of the culture can be determined.

For certain types of studies, viable counts are not necessary, and the simpler methods for total counts may be employed. The counts you have made in earlier experiments with a counting chamber constitute total counts. Judging growth on the basis of the turbidity of the culture is also a total count method.

Regardless of the method used, plotting the growth of organisms on a graph generates a standard curve known as the growth curve. Initially, cells pass through a lag phase, before cell division ensues. This is possibly a period during which the cells adjust to a new environment. Gradually, the rate of cell division increases to a maximum (for the given environment), which is called the log (logarithmic) or exponential phase, during which time the organisms increase according to a pattern of 2^n, where n = the number of cell generations. At this time the metabolic machinery is fully adapted to conditions of the environment. Gradually, the rate of cell division tapers off, and the population reaches the plateau or stationary phase. Since the rate of growth equals the rate of death, no net change occurs.

The above phases can be discerned by using either total or viable counts. When viable counts are used, a fourth phase—death phase—can be determined. During this phase, the rate of death exceeds the rate of growth. This is thought to result from a limiting food supply and/or the accumulation of waste products of metabolism that act as toxic agents.

When the growth cycle is graphed, an S-shaped (sigmoid) curve is given (see Fig. 12.1).

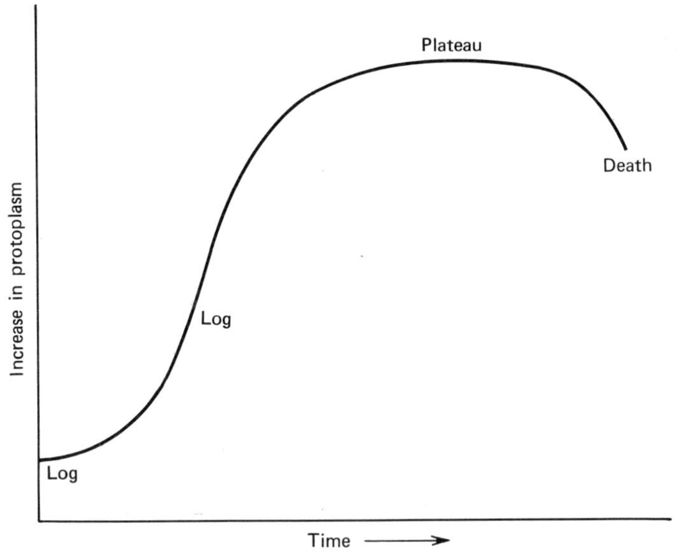

Figure 12.1

In multicellular systems, cells or organs grow at different rates. Nevertheless, the increase in total size follows the sigmoidal expression with respect to time.

GROWTH OF UNICELLULAR SYSTEMS

Your instructor has prepared cultures of yeast for your examination. Yeast is a unicellular plant that reproduces vegetatively by budding off small areas of cytoplasm while the nuclear material divides equally (mitosis). Each flask of yeast was initially innoculated with the same number of cells and grown under the same conditions. Flasks were aerated by shaking or bubbling air into them. At different time intervals (0, 4, 8, 12, 16, 20 hrs) flasks were removed from the growing chamber and frozen. In this way, you are able to observe a growth cycle that would normally take longer time than that permitted by the length of you laboratory period.

Preparation of the Counting Chamber

In order to determine increases in the size of the population, a means of accurate counting must be devised. The standard means of counting is by estimating the number of organisms (yeast cells) per milliliter volume of the culture flask. Since the number of cells per milliliter often reaches astronomical figures, it is impossible to count directly. Also 1 ml will flood a slide and will not fit beneath a cover slip. One solution to this problem is to take a known sample of the solution and count the cells in this calibrated amount. Multiplication by the factors involved will then yield the number of yeast cells per milliliter of culture.

1. Obtain a plastic cover slip, a razor, and a ruler.
2. Holding the cover slip over Fig. 12.2 etch the lines with the razor.

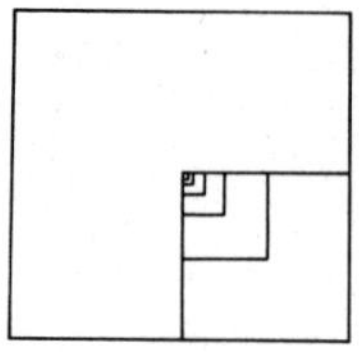

Figure 12.2

3. Each of the letters (*a, b, c, d, e, f,* and *g*) delineates a square.

   a = the entire cover slip,
   b = 1/4 the area of the cover slip,
   c = 1/16 the area of the cover slip, including areas represented by *d, e, f,* and *g,*
   d = areas represented by *d, e, f,* and *g,*
   e = areas represented by *e, f,* and *g,*
   f = areas represented by *f* and *g,*
   g = smallest square portion of the cover slip.

4. When the culture fluid is transferred from a Pasteur pipette (held at 45 deg) a single countable drop is equal to approximately 1/24 ml. By counting the number of cells in area *a* (the entire cover slip) and multiplying by 28, the number per milliliter is found. (The factor for multiplication is greater because the liquid from one drop of a Pasteur pipette does not completely fit under the cover slip and there is a calculated seepage factor taken into account.)

Square	Factor
a	2.8×10^1
b	1.1×10^2
c	4.3×10^2
d	1.7×10^3
e	6.8×10^3
f	2.7×10^4
g	1.1×10^5

Determination of the Growth Curve

1. Shake the culture flask marked "0 time" and remove a sample with the Pasteur pipette.

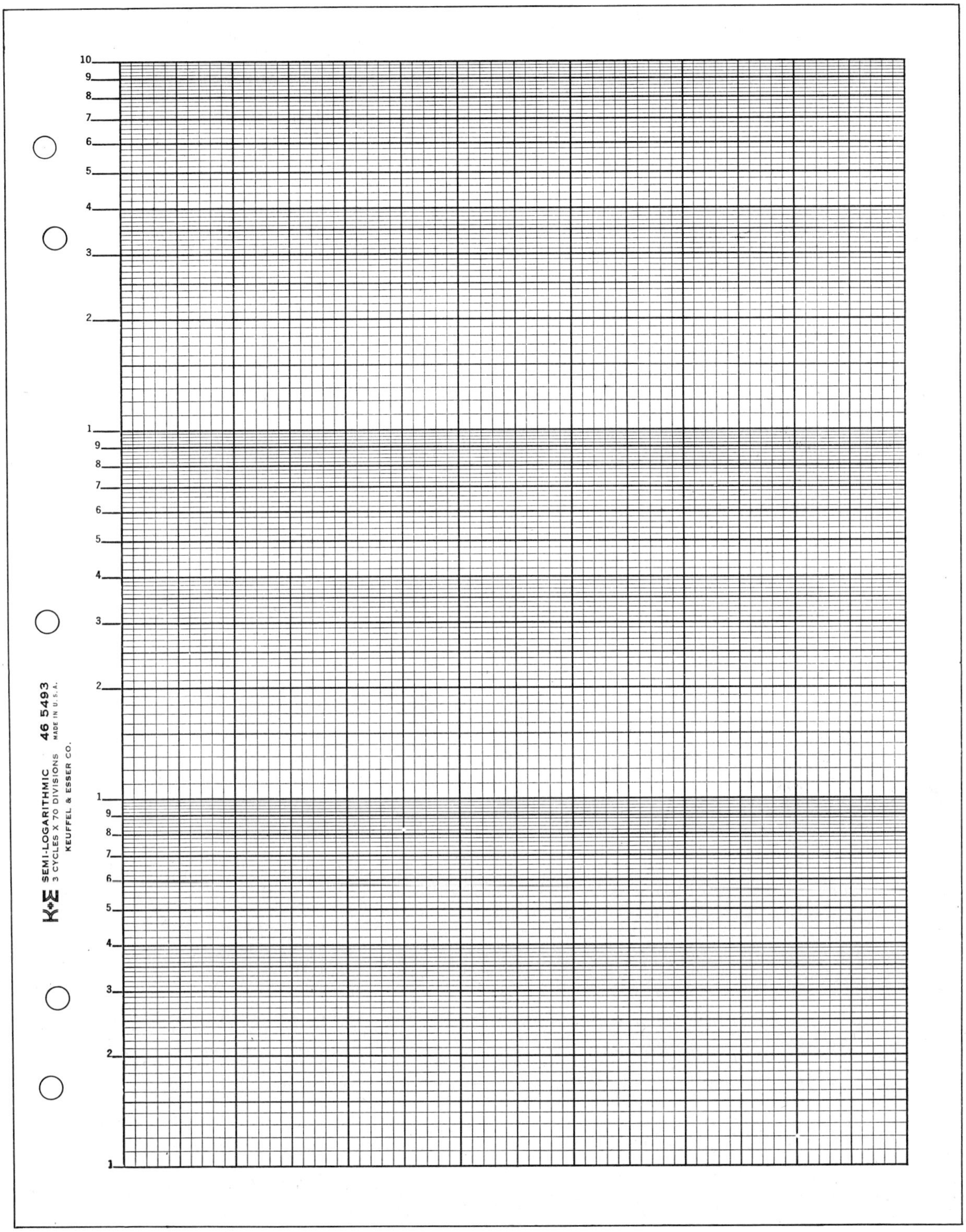
K+E SEMI-LOGARITHMIC 46 5493
3 CYCLES X 70 DIVISIONS MADE IN U.S.A.
KEUFFEL & ESSER CO.

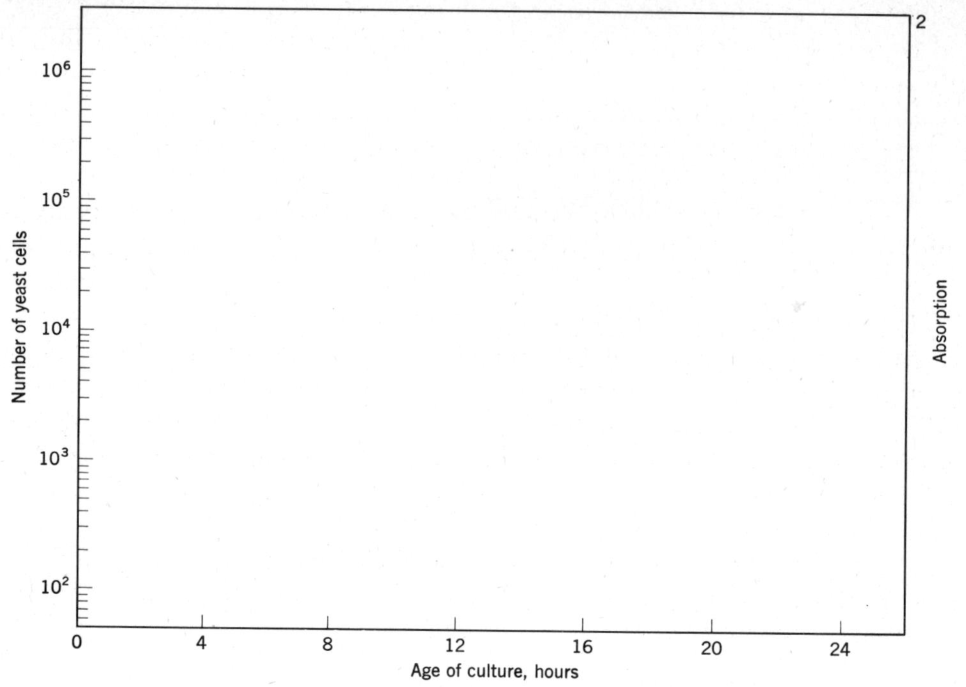

Figure 12.3

2. Add only one drop to a clean slide and prepare a wet mount with the etched cover slip. Be careful to keep the etched surface on top.
3. Place the slide on the microscope and observe the small spherical yeast cells under high power.
4. Determine which of the squares of the counting chamber contains between 10 and 30 cells.
5. Count the number of cells in one of the squares and multiply by the appropriate factor to give the number of cells per milliliter of culture.
6. Record the data on the graph of Fig. 12.3.
7. Follow the same procedure for each of the other cultures, plotting the attained values.
8. Remove a sample from the zero-time culture with a Pasteur pipette and place the sample in a clean colorimeter tube.
9. Read the sample against a tube of media with no organisms in the "spectronic 20," making sure that you wipe the bottom 2/3 of each tube to remove fingerprints.
10. Record the data on the graph of Fig. 12.3 using the right-hand ordinate.
11. Follow the same procedure for the other cultures, plotting the attained values.
12. The two curves act as a check on one another. If your curves do not agree, repeat the necessary counts.
13. Although the death phase may have been reached by 12 hr, why is it not detectable by the means of measurement?

The most important features of the growth curve are that it lends itself to data manipulation and it provides quantitative comparisons. For example, one growth study may be done with organisms in a complete medium, and another in a complete medium minus some element. Or one growth study may be done at one temperature, and another at a different temperature. From manipulation of data, the investigator can say that the growth rate under one condition is x times that of another condition.

Manipulation of growth data begins with translating the values for log phase to a log graph. When a curve can be expressed mathematically as x^y, it means that no three points on that curve lie on a straight line. When the values of such a curve are translated to a log graph, a straight line is generated. A straight line graph is much easier to read, and more importantly, the slope of the straight line is the value for the rate of the process.

Translation of the values to a log graph can be done in two ways. The existing data can be plotted on semi-log paper or the values can be converted to logarithms, and plotted on straight graph paper. For purposes of this exercise, the data will be plotted on three-cycle semi-log paper, such as that found in Fig. 12.3.

1. Replot the data of the log phase only on the graph of Fig. 12.3, being careful to assign correct values to the ordinate. Remember 2×10^4 is twice as much as 1×10^4, and 4×10^4 is twice as much as 2×10^4, and so forth.
2. Draw a straight line, averaging the points. This is done by drawing the line equidistant from all points instead of drawing a staggered line through each point.
3. Determine the generation time of the culture by selecting any point along the line, recording the corresponding time, and then finding the point that corresponds to double the original value chosen. Record the corresponding time and subtract the time values. For example:

   1×10^6 corresponding time: 2 hr
   2×10^6 corresponding time: 2.75 hr
   Generation time = 0.75 hr, or 45 min.

CHART 12.1

Value Chosen	Corresponding Time	2 × value Chosen	Corresponding Time

Generation time =

ROOT GROWTH

The instructor has prepared a series of petri plates containing pea seedlings for your examination. Seeds were placed under conditions for germination at intervals prior to the laboratory period.

Petri Plate	No. Days Germinated	Measurement
A	10	
B	9	
C	8	
D	7	
E	6	
F	5	
G	4	
H	3	
I	2	
J	1	

1. Using a millimeter ruler, measure the length of the root of five seedlings from petri plate *A*.
2. Average the five resulting measurements, and plot the average on the graph of Fig. 12.4.

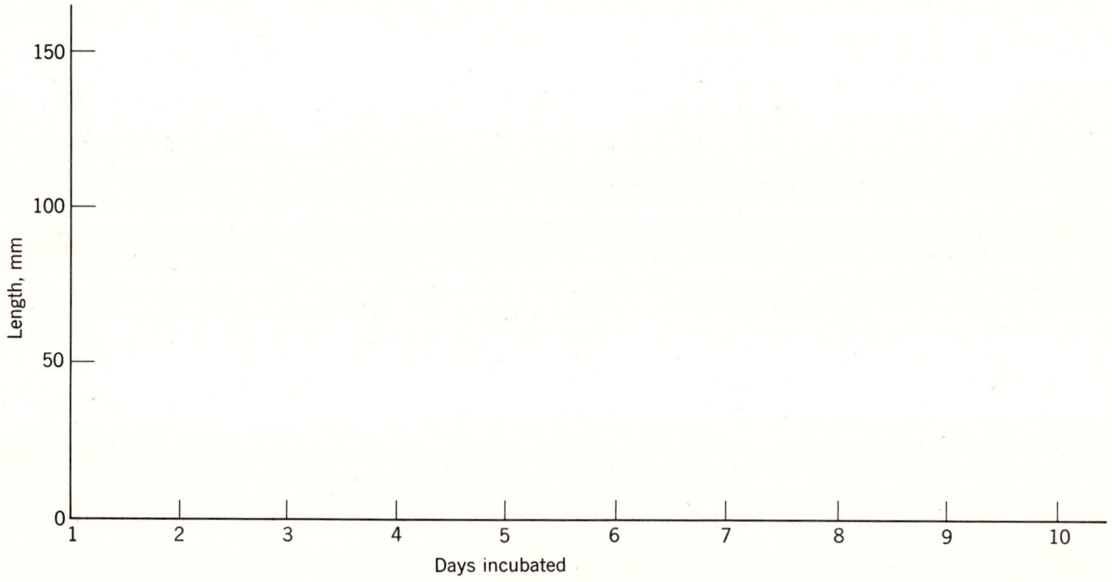

Figure 12.4

3. Repeat this procedure for all ten plates, plotting averages on the graph and drawing the curve that is formed by the plotted points.
4. Why is the time period for doubling of length of little meaning here?

5. Is cell division the only means of growth here? Why?

6. Is this the growth curve you would expect if the seedlings were planted in soil? Why?

ANIMAL GROWTH

In humans, growth from conception follows the basic sigmoidal pattern, but with an interesting deviation. Five stages may be recognized:

conception to birth	= prenatal
birth to 10 mo	= infantile
10 mo to 4 yr	= early childhood
4 yr to 14 yr	= juvenile
14 yr to 24 yr	= adolescent

1. Listed below are the percentages of weight compared to that at maturity for various stages.

−9/12 (conception) = 0%	10 yr = 39%
−4/12 (4 mo gestation) = 1%	14 yr = 60%
0 (birth) = 5%	16 yr = 79%
1 yr = 12%	18 yr = 87%
2 yr = 16%	20 yr = 92%
4 yr = 23%	24 yr = 97%

2. Plot the above data on the graph of Fig. 12.5.
3. During what stages is growth accelerated?

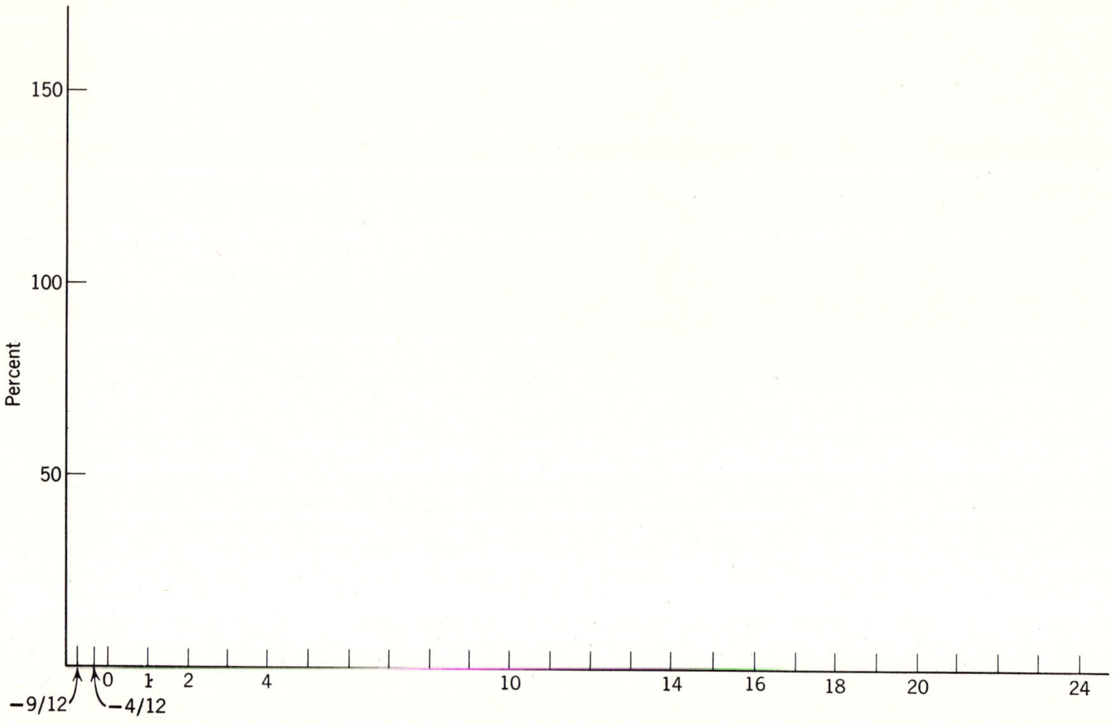

Figure 12.5

4. During what stages is growth retarded?

5. In other mammals only one accelerated period exists. Which of those in humans do you think are absent in other mammals?

ALLOMETRIC GROWTH

Structural conformation can be the result of cells growing at different rates (allometric growth) as well as differentiation.

1. Obtain a gourd and a piece of string or yarn.

Figure 12.6

2. Place the string or yarn over the line that is the circumference of the circle in Fig. 12.6.
3. Starting with the circle of string, move parts of the string appropriately to form the shape of your gourd.
4. Place arrows on the circle indicating the direction of growth.
5. As you moved part of the string outward, other parts of the string moved inward. Draw the circumference of a cross section of the sphere from which your gourd could have developed.
6. If you were to cut the gourd in half and count all the cells in the two dissimilar halves, you would see that there are approximately the same number of cells. Does allometric growth in the gourd refer to different rates of cell division?

7. If there are not more cells in the larger half, then how can you account for the difference in shape through growth processes alone?

13

Differentiation

MATERIALS

CLASS STOCK

prepared slides
- spermatogenesis in chicken
- frog neural tube
- longitudinal section of onion or bean root
- young *Tilia* stem

E. coli suspension

chemicals
- 5.0% sucrose
- 5.0% lactose
- actinomycin D (50 μg/ml) + 5.0% lactose
- 0.005% methylene blue

fern spores grown for 10 days in red light

fern spores grown for 10 days in blue light

fern spores grown for 6 days in white light, 6 or more days in red light

fern spores grown for 6 days in white light, 6 or more days in blue light.

Planaria

STUDENT STATION

microscope
microscope lamp
lens paper
test tubes
wax pencil
glass rod or Pasteur pipette
slides
cover slips
cork
razor
dishes

INTRODUCTION

During the past few years, greater emphasis has been placed on the nuclear control of various activities of the cell in response to the environmental and parental cell influences. For inheritance patterns to be transmitted, the gene must remain essentially unchanged, regardless of the changes in cell processes concerned with structure or metabolism. The nucleus may be seen as passing unchanged, through cell lineages, despite the difference in appearance of the cells in which it may be found. Compare this to a man who wears work clothes during the day, a tuxedo in the evening, and pajamas at night. In the same manner the nucleus synthesizes and attires itself in different cytoplasms or cells in order to achieve different functions.

PATHWAYS OF DIFFERENTIATION

There are two aspects to the study of different`ation—the awareness of the types of changes cells undergo to form specialized cells and the way in which genetically identical nuclei produce unlike cells. The first exercise will deal with the pathways of differentiation to study of the types of cell changes.

The Formation of Spermatozoa

The spermatozoa are one of the highly differentiated types of cells in the animal body. They function only in the reproduction of the organism, acting as gametes. As such, the most important feature of the cell is the chromosomal complement of the nucleus. For this discussion, however, the primary concern is the condition or the cellular type.

1. Obtain a slide of spermatogenesis (formation of sperm) in chicken.
2. Observe under low power.
3. In the space provided, draw the section of the testes under low power, labeling the following structures:

   seminiferous tubule—the cylindrical tubules that comprise the testes.
   lumen of tubule—the hole running through the seminiferous tubule. The sperm are produced in the tubule and released into the lumen for storage and eventual transport.
4. Draw a section of a seminiferous tubule under high power, identifying and labeling the following cell types:

   spermatogonia—found at the periphery of the tubule. As these cells undergo division, more spermatogonia are formed. At times, the spermatogonia divide and give rise to primary spermatocytes.
   primary spermatocyte—found closer to the central core, in the layer above the spermatogonia. In morphology, these do not differ very much from the spermatogonia.

Once a primary spermatocyte is formed, the fate of the cell is determined. The primary spermatocyte cannot divide to form more primary spermatocytes. Division of this type of cell yields only secondary spermatocyte.

secondary spermatocyte—found closer to the central core in the layer above the primary spermatocytes. They do not differ in gross morphology from the former types. Division of the secondary spermatocytes yields spermatids.

spermatids—found quite close to the central core of the seminiferous tubule. These show a marked morphological change, including the presence of a rudimentary flagellum and an interphase nucleus. By the process of differentiation, the spermatid is converted to a functional spermatozoan, which enters the lumen for its elimination from the body.

nurse cells—found radiating from the central core. The function of the nurse cells is undetermined, but they may provide some nutritive value to transmit to the stored spermatozoa.

Differentiation in the Embryo

The differentiation process is important in its application to embryological development. Without differentiation cell division would result in the formation of an amorphous multicellular blob of protoplasmic material. Differentiation does not occur in the embryo until after the gastrula stage. Prior to this the embryo has not undergone any increase in size or shape, as divisions segregate the already existing protoplasmic mass that originally made up the egg cell. Gastrula is marked by the delineation of the primary germ layers, endoderm, mesoderm, and ectoderm. These tissues show little morphological difference from one another and are more important in their destiny as cell types. One of the first tissues to show marked differentiation is the ectoderm, which forms the neural canal, the primary foundation of the brain and spinal cord in the mammal. This change in cell type can be readily seen in the course of the development of the neural canal.

1. Obtain a prepared slide of frog neural tube (2.60-mm embryo) and observe under low power.
2. On the outline drawing of the frog embryo in Fig. 13.1 identify, draw, and label a few examples of the cell types given below, placing them in their proper position.

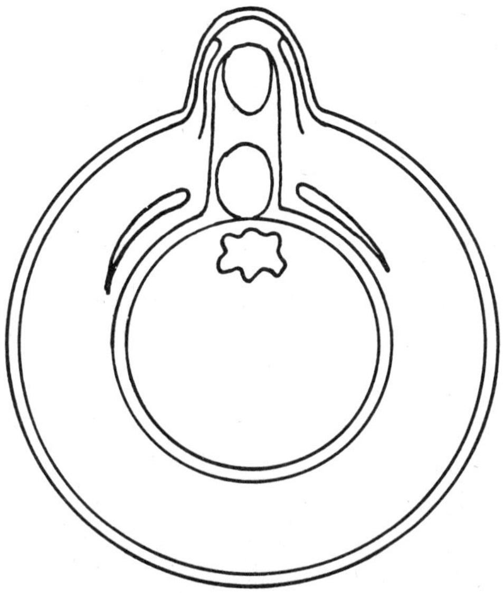

Figure 13.1

endoderm—the innermost layer of cells. Some of these cells are quite large and contain yolk material. The endoderm surrounds the cavity that eventually becomes the digestive tract of the frog.

mesoderm—the middle layer of cells. These are generally smaller than those of the endoderm and are separated from the endoderm by a small space. On the dorsal side of the embryo (where the neural tube is forming) the mesoderm may show a separation in the central part of this tissue. This is the beginning of the formation of the body cavity or coelom.

ectoderm—small cells that surround the embryo. This will eventually form the skin of the frog.

neural tube—found on the dorsal side of the embryo. It comprises cells that have differentiated from ectoderm. Notice the difference between the typical ectoderm cell, which will form the skin and the cells of the neural tube.

3. In the space provided, draw and label a typical ectoderm cell from the outer covering of the embryo and a typical neural tube cell from around the neural canal, observed under high power.

Root Formation

In plants, both the stem and the roots increase in length mainly by cell division at the apexes or tips. Cells proliferate at the apex and subsequently undergo a series of changes into mature cells. Many lines of differentiation originate from the dividing meristematic cells, each of which can be localized in a specific region of the root.

1. Obtain a slide of a longitudinal section of an onion or bean root.
2. Identify the following areas of the root and draw a cell representative of each area in the space provided.

meristematic cells—actively dividing cells that contain the mitotic figures described in Chapter 10. These cells have a somewhat square appearance with dense cytoplasm and nuclei at some stage in the mitotic cycle.
root cap cells—comprise the protective tip of the root. Notice that the cells are oval to irregular in outline, have little cytoplasm, and have nuclei that are either peripheral or absent. These cells are in various stages of degeneration.
elongation cells—adjacent to the meristematic region. The cells in this region are about twice as long as they are wide, and have small central vacuoles and a central or peripheral nucleus.
mature cells—After the cell elongates, it undergoes a final series of changes, depending on its position in the root. Move farther up the root and select one cell, indicating its characteristics, length of the cell, size of the vacuole, thickness of the cell wall, position of the nucleus, and so on.

3. Is the root elongation carried out by cell division exclusively? Why?

Vascular Bundle

As the plant increases in length by apical meristems, it increases in diameter by lateral meristems. One lateral meristem is the vascular cambium, which gives rise to water-conducting cells or xylem and food-conducting cells or phloem. The change in rate of production of xylem cells results in the well-known annual rings observed in the cross-sectional cuts of woody stems.

1. Obtain a slide of 2- to 3-year-old basswood *(Tilia)* stem and observe under low power.
2. On Fig. 13.2, draw and label the following cell types in their proper position. Only a few of each of the cell types need be drawn.

   xylem cells—located on the inside of the stem from the cambium. The vessels are large, thick-walled, and red-stained, having neither cytoplasm nor nucleus.
   xylem ray cells—short cells lined up on the radii (lines radiating from the center of the stem).
   cambium cells—flat irregularly shaped cells that lie on the periphery of the xylem. Remember that these cells are seen in cross section. In longitudinal view, they are very long and pipelike. There are several tiers of these cells.
   phloem cells—found on the outside of the cambial layer. The pholem is composed of several different kinds of cells.
   phloem ray cells—extensions of the xylem rays and found in the same position on the distal side of the cambium.
   phloem element and adjacent companion cell—found side by side. The element is a large thick-walled cell having no cytoplasmic content except a deep-staining plug of nonliving material. The companion cell is relatively small, with cytoplasm and nucleus.
   bast fibers—groups of thick-walled cells occurring in clusters of approximately 20 cells. These are found in the phloem tissue.
3. Would you expect apical and lateral elongation to be under similar controls? Why?

4. What would be the origin of the cambium cells in the root?

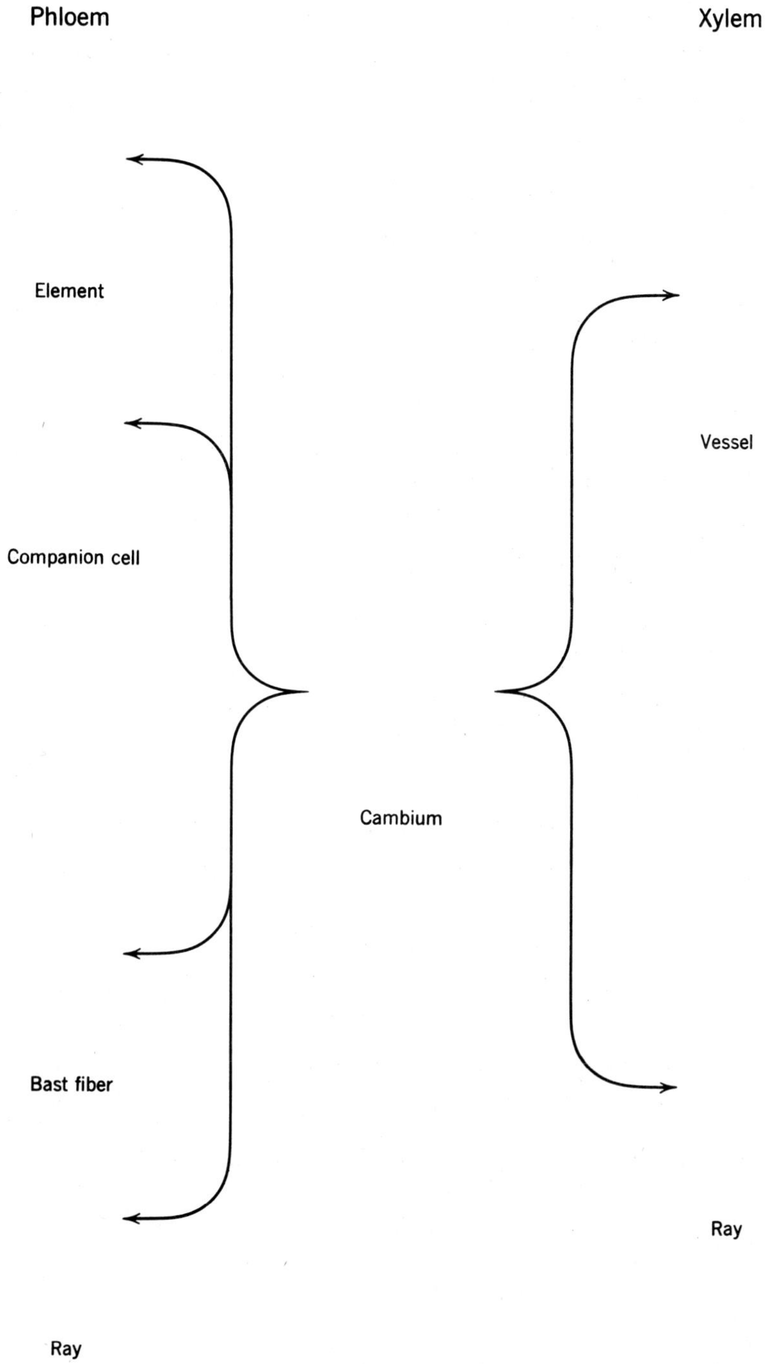
Phloem
Xylem
Element
Companion cell
Bast fiber
Ray
Cambium
Vessel
Ray

Figure 13.2

INDUCTION OF DIFFERENTIATION

Patterns of differentiation are predictable, with respect to the stage achieved and the time of occurrence. Not only will the chrysanthemum plant flower after the vegetative organs develop, but that flowering will occur when the days are short, as in autumn. When the chick has developed three primary germ layers, the heart will begin to develop and establish a beat (about 20 to 24 hours after fertilization). This predictability is based on observations of many individuals, and the stable quality of differentiation patterns suggest that some regulatory mechanism are the basis of organismal potential for differentiation.

The mechanisms that underlie organismal or cellular potentials for differentiation are active at a variety of levels within the cell, including transcription, translation, and conformation.

Induction at the Transcription Level of Cell Activity

The bacterial cell *E. coli* is able to utilize many types of molecules for carbon sources. However, the enzymes required for the breakdown of each substrate may not always be present. If the sugar lactose becomes available to the cell, the enzyme β-galactosidase will gradually increase from as few as twenty molecules to thousands per cell over a short period of time.

The production of the enzyme is induced by the entrance of lactose into the cell. Lactose combines with and thereby inactivates a repressor substance (protein) that masks the gene and prevents transcription of DNA to RNA. The combination of the repressor substance and lactose leaves the gene free to produce β-galactosidase.

Not only is β-galactosidase produced but a permease is formed that increases permeability of the cell membrane to lactose, so that more lactose molecules can enter the cell.

In this experiment you will be ascertaining the breakdown of carbon sources with methylene blue, a dye that is capable of accepting hydrogen. Methylene blue is colorless in its hydrogenated state. When oxygen is not readily available, methylene blue may act as a final hydrogen accoptor in aerobic respiration, thus substituting for oxygen:

oxygen + hydrogen (from electron transport system) $\longrightarrow$ H_2O
methylene blue + hydrogen (from electron transportation system)
$\longrightarrow$ methylene blue-hydrogen (colorless)

Thus, when breakdown of the only available carbon source is achieved, and oxygen is limiting, methylene blue will turn colorless. As long as the carbon source is not broken down, hydrogens will not be released, and methylene blue will retain its color.

1. Mark each of four clean test tubes, *A, B, C,* and *D* 5 mm, 6 mm, 7 mm, from the bottom, using a wax pencil.
2. To each of these tubes, add *E. coli* suspension (2×10^8) up to the first mark.
3. Add to the tubes up to the second mark as follows:
   A–5% sucrose
   B–5% lactose
   C–5% lactose + 10 drops actinomycin-D
   D–water

4. To each tube add 0.005% methylene blue up to the third mark and shake.
5. Record the time when the methlyene blue was added as being zero time, and record the color in each of the tubes every 15 min for 2 hr, shaking before each reading.
6. A set of standards will be available to aid in judging color change. These will correspond to full ¾, ½, ¼, and) concentrations of methylene blue. Compare your tubes with these before recording the data.
7. On the graph of Fig. 13.3 plot the color intensity of each tube for the time scored.

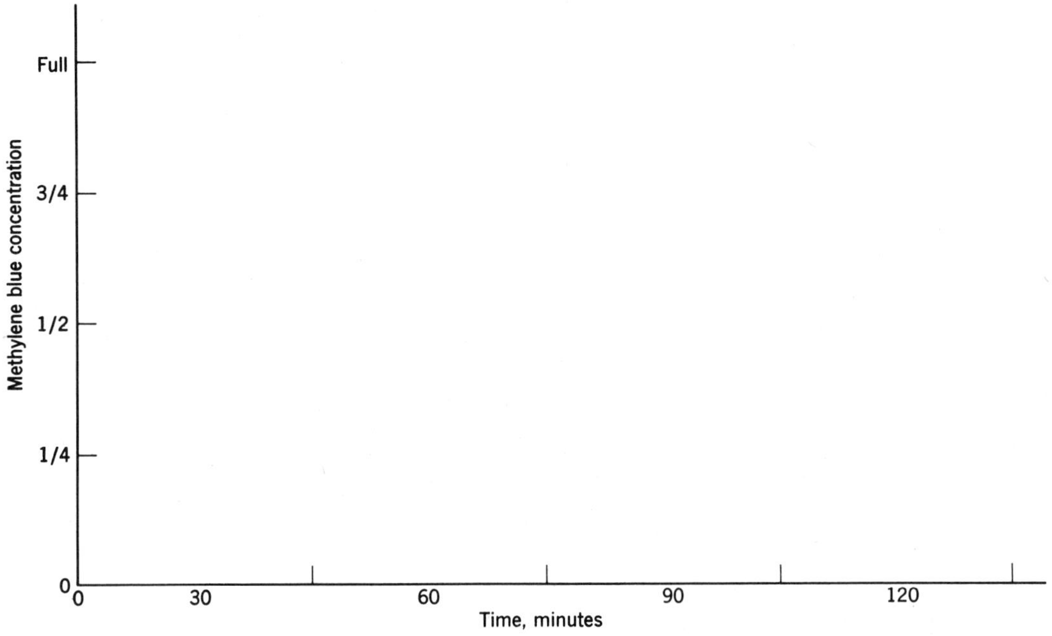

Figure 13.3

8. The enzyme that utilizes sucrose as a substrate is called sucrase, and it is a constitutive enzyme. Beta-galactosidase is an induced enzyme in some *E. coli* strains. Based on the experimental data, define constitutive and induced enzymes.

9. What would you expect to happen if cells were added to a tube of fresh medium containing lactose as the sole carbon source after being in a medium containing lactose?

10. Actinomycin-D inhibits DNA transcription to RNA. How do you know from this experiment that β-galactosidase is induced at the DNA level (transcription) of cellular organization?

11. Cyclohexamide inhibits protein synthesis. What would you conclude if the tube with only lactose as a carbon source and cyclohexamide behaved like the tube with actinomyocin-D?

Induction at the Conformation Level of Cellular Activity

The conformation of certain proteins is affected by light. Phytochrome, for example, takes on one configuration in red light, and an alternate configuration in far-red light. Each of the two three-dimensional shapes of phytochrome are necessary for certain activities prerequisite for plant growth. In many lower plants this system is coupled with a blue-light system.

1. Your instructor has prepared the following flasks for your examination:
   I. Fern spores grown for ten days in red light.
   II. Fern spores grown for ten days in blue light.
   III. Fern spores grown for six days or more in white light, six days or more in red light.
   IV. Fern spores grown for six days or more in white light, six days or more in blue light.
2. Prepare a wet mount of the material in flask I by inserting a glass rod or Pasteur pipette into the culture. Some of the material on the surface will cling to the glass surface of the rod or pipette. Hold the rod or pipette over a slide, and wash the material onto the slide with a drop or two of water dispensed from a dropper or another Pasteur pipette.
3. Examine the wet mount and draw the spore or gametophyte in the space provided.
4. Did the spores that were grown in red light for ten days germinate? Enter your answer on Chart 13.1.
5. Wash the slide and prepare a wet mount of the material in flask II.
6. Draw the spore or gametophyte from flask II in the space provided.
7. Did the spores that were grown in blue light for ten days germinate? Enter your answer on Chart 13.1.
8. Did germination of the spores result in one-dimensional (filament), two-dimensional (sheet), or three-dimensional (cubic mass) growth?
9. What type of light is required for germination?

CHART 13.1

Treatment	Development			
	Germination	Filament (1-D)	Sheet (2-D)	Cubic Mass (3-D)
Red Light 10 Days				
Blue Light 10 Days				
White Light 6 Days, Red Light 6 Days				
White Light 6 Days, Blue Light 6 Days				

10. Wash the slide and prepare a wet mount of the material in flask III.
11. Draw the gametophyte from flask III in the space provided.
12. Does the gametophyte from flask III exhibit one-dimensional, two-dimensional, or three-dimensional growth? Enter your answer on Chart 13.1.
13. Flask III was grown in white light for six days. What type of light would you deduce that white light includes?

14. Wash the slide and prepare a wet mount of the material in flask IV.
15. Draw a gametophyte from flask IV in the space provided.
16. Does the gametophyte exhibit one-dimensional, two-dimensional, or three-dimensional growth? Enter your answer on Chart 13.1.
17. What type of light is necessary for the conversion of the filament into a sheet of cells?

18. For what growth phase is the following reaction necessary?

$$\underset{\text{(far-red)}}{\text{Phytochrome}_{fr}} \longrightarrow_{l} \underset{\text{(red)}}{\text{Phytochrome}_{r}}$$

20. For what growth phase is the following reaction necessary?

$$\underset{\text{(red)}}{\text{Phytochrome}_r} \longrightarrow \underset{\text{(far-red)}}{\text{Phytochrome}_{fr}}$$

Induction at Unknown Levels of Cellular Activity

In many cases, the factors that trigger differentiation are unknown. This is particularly true for complex systems where individual events are not easily studied.

The phylum Platyhelminthes consists primarily of parasitic forms. One exception is the free-living form Planaria, which is found in running streams or rivers at 50° to 60°F, and may be collected quite easily. A small piece of liver tied to the end of a string will attract many *Planaria.* The liver may be shaken into a water-filled jar in order to release the *Planaria.* You may be asked to collect the *Planaria* for this experiment or they may be provided for you.

Planaria are rather inconvenient in size, being too large to be viewed under the microscope but too small to be seen with any accuracy by the naked eye. Ideally, a dissecting scope with a magnification of 5X should be used. If this is not available, hand lenses may be used.

The *Planaria* contains a great deal of embryonic or undifferentiated tissue. Under certain environmental conditions, this tissue may be stimulated to differentiate.

1. Gently pick up a *Planaria* and place on a cork. A medicine dropper or glass rod will serve for the transfer.
2. Using a single-edged razor, cut the *Planaria* into four parts, using the guide in Fig. 13.4.

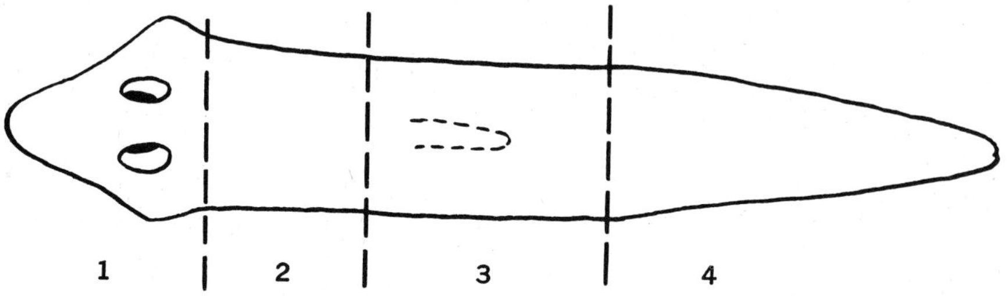

Figure 13.4

3. Place each of these parts in the specified dish. Place the most anterior part in the dish labeled 1; the most posterior in dish 4. Each dish should contain the same part of the animal from everyone in the class.

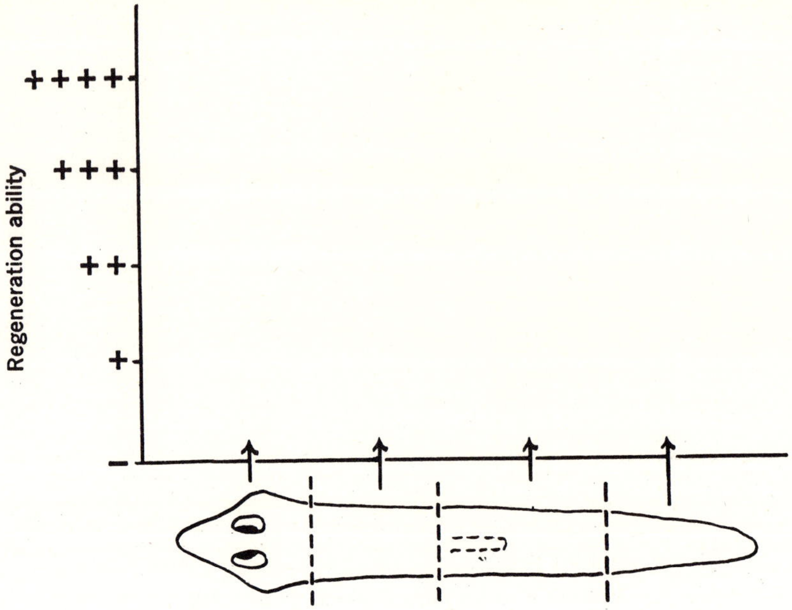

Figure 13.5

4. Examine these dishes once again in two weeks, and if necessary, in three weeks. During this time they will be kept at 10°C.
5. Record regeneration information on the graph of Fig. 13.5.
6. Are all parts of the *Planaria* equally capable of regeneration?

7. Is the regenerated *Planaria* smaller than the original animal?

8. Could you deduce whether regeneration would take place if these sections of the *Planaria* were put in a solution of colchicine? In Actinomycin-D? In mitomycin-C?

9. What level or levels do you think are involved in redifferentiation?

10. What types of factors, based on your knowledge of a cell or organism as an automaton, could induce regeneration?

11. What environmental factors might include the differentiation ability of normal embryonic tissue?

Part IV

EVOLUTION

14

Sources of Variation

MATERIALS

CLASS STOCK

EMB agar petri plates
E. coli mutabila
Drosophila crosses
- brown-eyed, yellow-bodied, curly-winged females x wild type males
- back cross of above
- yellow-bodied, white-eyed females x wild type males
- back cross of above

etherizer
Lilium anther slide

STUDENT STATION

glass spreader
dish of 95% alcohol
Bunsen burner or alcohol lamp
millimeter ruler

INTRODUCTION

Each individual contributes to the unfolding history of a species and serves as custodian of the priceless germplasm of genes that took millions of years to acquire. The passing of the hereditary material by sperm and egg is precise and controlled; but it is also tenuous—that is, the inheritance of a specific genetic factor is uncertain. To state that an organism gambles with its priceless germplasm is in a way very true. The possibility of improvement has always existed, and means have been retained to permit changes to occur. Life is controlled to a great extent, but not completely because complete control is thermodynamically impossible. The analogy to how organisms "gamble" is a game with many players betting small and constantly against the house. If an organism fails, only one of thousands dies; and if an organism wins, offspring for generations will profit.

All components of an organism, including genetic material, eventually break down and decay (entropy). At times, in their replacement, errors are made. This inexorable process of change is usually disastrous to the organism; but rare cases (one in a million) may be advantageous and may give the organism a better chance for survival. Through the unique process of reproduction life can accumulate untold numbers of beneficial changes to the extent that it may take on different appearances or functions than it originally had and give rise to many different branches of this living continuum.

The major sources of variation are:

gene mutation—a chemical alteration of a gene so as to produce a different product (such as an altered enzyme).

segregation—Mendel's first law describing the separation of alleles into different offspring. This may give rise to new combinations.

independent assortment—Mendel's second law, stating that pairs of genes are unrelated with respect to how each undergoes segregation. This may give rise to new combinations.

fertilization—combination of egg and sperm to give new associations of genes, even if the parents are somewhat related.

crossing over—exchanges of chromosomal segments by breakage and reunion between homologous chromosomes.

chromosome anomalies—parts of or whole chromosomes can be lost, doubled, or rearranged.

Other phenomena include genetic drift, heterozygosity, and certain processes in the bacteria.

The subsequent effects these changes have on the organism's survival are concerned with adaptation, which will be discussed in the next chapter. Several ideas must be introduced to bring these mechanisms into proper perspective. First, the change is unpredictable, that is, there is no direction or known pattern of these changes. Organisms, including man, do not know what challenges lie ahead, and therefore cannot foresee what genes may be needed to cope with new challenges. Second, advantages gained are usually very rare. Only a few well-documented examples are known today. Third, the functional unit of action in evolutionary prospective is the population instead of the individual organism. Fourth, populations are constantly migrating, splitting, and fusing; therefore the potential of an organism to mate with any member of its own species is greatly reduced.

GENE MUTATION

The rate of spontanous gene mutation is low, and hence it is very difficult to detect and study spontaneous mutations. A major advance in genetics was the discovery by H. J. Muller that mutations could be induced by irradiation, and that these induced mutations had effects identical to those mutations that occurred spontaneously. Thus, induced mutations could be studied to ascertain basic features of the mutational event.

Mutagenesis in Escherichia Coli

Escherichia coli is a rod-shaped bacterium found in the lower intestine of many mammals. It is capable of numerous metabolic reactions, among which is the fermentation of lactose by the enzyme β-galactosidase. In the absence of the enzyme, lactose cannot be fermented.

A strain of *E. coli*, called *E. coli mutabila,* cannot ferment lactose, but can readily mutate to a form that can ferment lactose. Eosin-methylene blue agar (EMB) contains lactose and dyes. Organisms growing on the agar and utilizing lactose as a carbon source grows as a brassy purplish colony. Bacteria, such as *E. coli mutabalia*, that cannot ferment lactose (lact$^-$) produce a white colony.

You will each receive a sterile plastic petri plate. Do not lift the cover until you are ready to plate the bacteria because the dish must remain sterile. The instructor will distribute the bacteria. Read the instructions thoroughly before you begin, and practice flaming the spreader (Fig. 14.1).

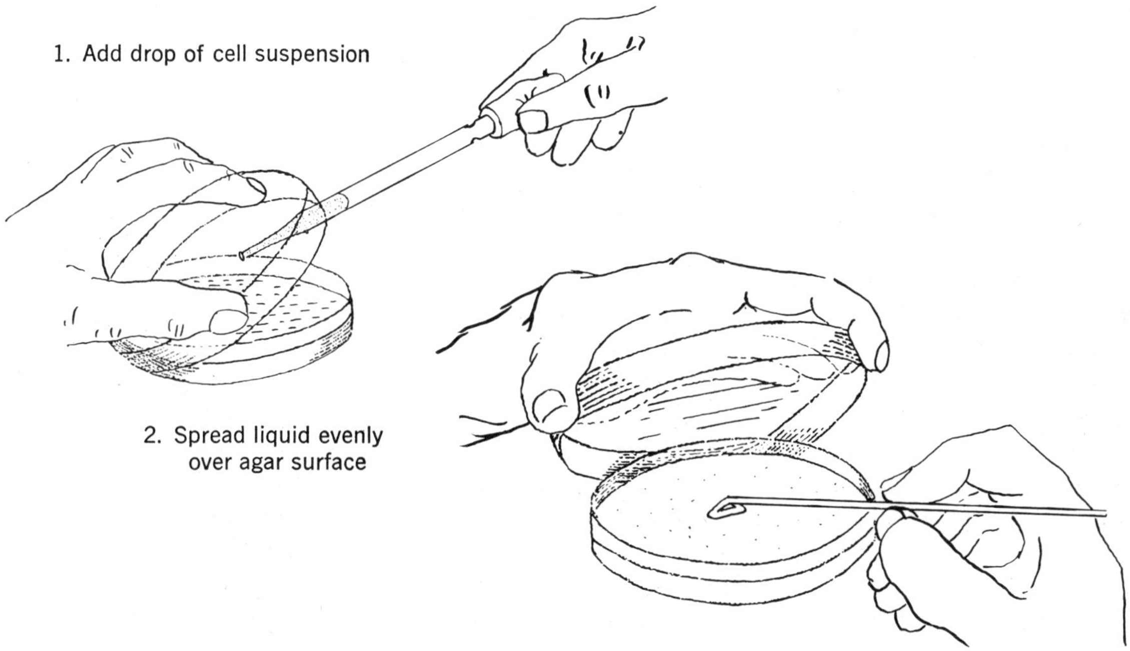

Figure 14.1

1. Obtain a glass spreader and a dish of 95% alcohol.
2. Light your Bunsen burner or alcohol lamp.

3. Obtain two drops of *E. coli mutabalia* of a 10^3 culture from the instructor. Keep the plate closed until the last moment; open it only to permit the drops to be placed on the agar. Close the plate immediately.
4. Quickly return to your desk.
5. Dip the spreader in the alcohol. Pass it through the flame of a Bunsen burner. Do not hold the spreader over the flame because it will break. The alcohol will light and will burn itself out. Do not put the spreader back into the alcohol dish, or you will ignite the alcohol there.
6. Using the flat side of your spreader, as shown in Fig. 14.1, uniformly spread the drops of bacteria to cover the entire plate, turning the plate and keeping it partially covered with the cover as you do this.
7. Recover the plate when you have finished spreading and invert it. Write your name and the dilution on the bottom of the plate. Tape the two halves together.
8. After three or four days (one week is too long) examine the plates for white colonies with purple papillae.
9. Count the number of purple papillae in 10 colonies and enter your data on Chart 14.1.
10. Measure the diameter of each colony in which you counted papillae by noting its position and measuring with a millimeter ruler against the bottom of the petri plate. Enter your data on Chart 14.1.
11. Using Fig. 14.2 and your data for the diameter of the colonies, enter the number of cells in a colony on Chart 14.1.

CHART 14.1

Colony No.	No. of Papillae	Diameter of Colony	No. of Cells in Colony
1.			
2.			
3.			
4.			
5.			
6.			
7.			
8.			
9.			
10.			
Total			
Average	T/10=		T/10=

12. Calculate the mutation frequency in the following way:

$$\frac{\text{Average number of mutational events}}{\text{Average number of cells}} \times 100 = \quad \%$$

or

$$\frac{\text{Average number of papillae}}{\text{Average number of cells in colony}} \times 100 = \qquad \%$$

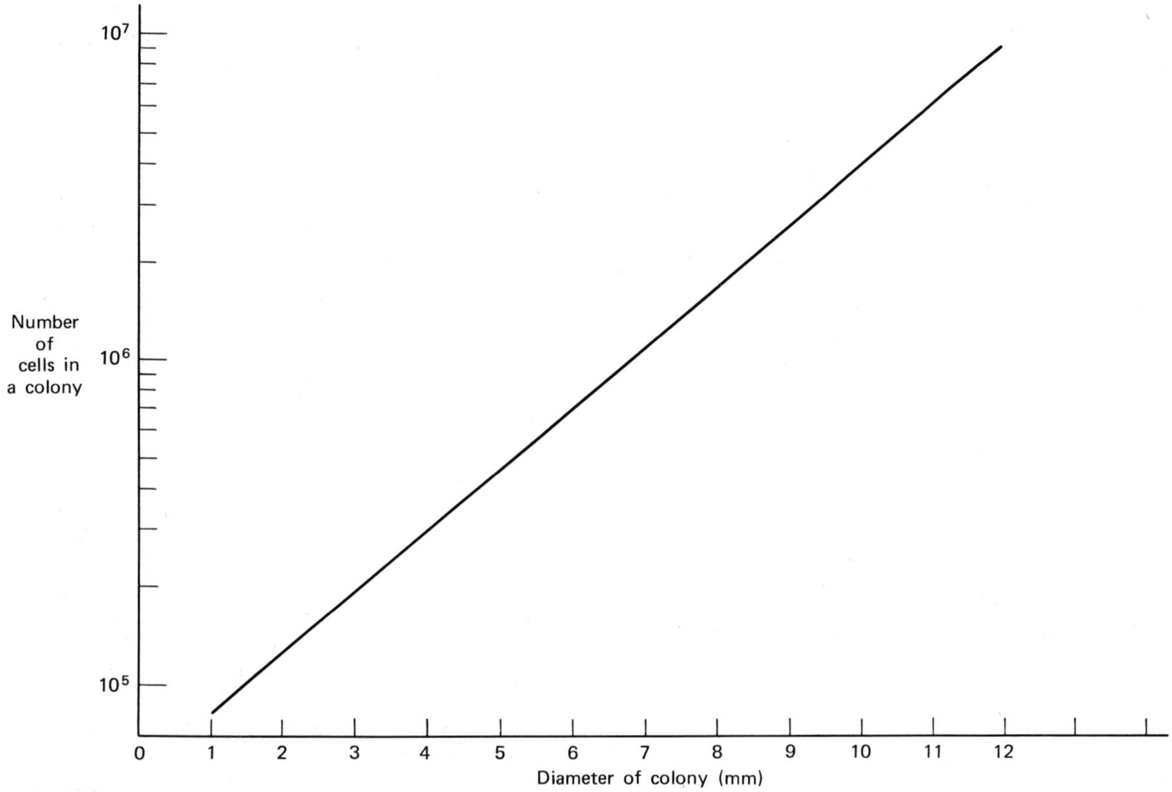

Figure 14.2

13. What is the mutation frequency for lac$^-$ to lac$^+$ in *E. coli mutabalia?*

14. What would have happened to *E. coli mutabila* if there had been no sugar other than lactose in the medium? On the basis of your answer, do you consider the mutation to lac$^+$ as detrimental or beneficial?

15. How does the presence or absence of lactose in a medium act as a selection pressure for *E. coli mutabila?*

16. Why are mutations considered to be building blocks as well as sources of variation?

17. How would you demonstrate that the change to lac^+ is a genetic change?

RECOMBINATION AS A SOURCE OF VARIATION

Independent assortment of chromosomes leads to a recombination of genetic factors, which results in a variety of gene combinations. These combinations are tested during evolution for their physiological balance within the organism, and for the advantage or disadvantage they give the organism in a particular environment.

1. Your instructor has performed the following *Drosophila* genetic crosses for you.
   (a) brown-eyed, yellow-bodied, curly-winged females X wild type males
   (b) F_1 females from cross (a) X brown-eyed, yellow-bodied, curly-winged males
2. Obtain a vial containing flies resulting from cross (b). Etherize them according to instructions given in chapter 9, and count the number of each class of flies.
3. Enter your data on Chart 14.2.

CHART 14.2

Class	Number
Wild type	
Wild type body and eyes, curly wings	
Wild type body and wings, brown eyes	
Wild type eyes and wings, yellow body	
Wild type body, curly wings, brown eyes	
Wild type eyes, yellow body, curly wings	
Wild type wings, yellow body, brown eyes	
Yellow body, brown eyes, curly wings	
Total	

4. Draw a Punnet square for the cross between F_1 female from cross (a) × brown-eyed , yellow-bodied, curly-winged male. The F_1 from cross (a) were wild type. Are the traits mentioned dominant or recessive?

5. Give the genotypes for the new gene combinations.

6. How do these new gene combinations provide a means for testing genes?

7. Calculate the ratios of each class you have counted in the following way:

$$\frac{\text{Number in each class}}{\text{Total number counted}} = \text{fraction of total}$$

8. What is the obtained ratio?

9. Determine the expected ratio from the Punnet square.

10. Of which class or classes were there more than expected, and of which were there less?

11. Which gene combinations are favored by the selection pressures available in the growing flasks?

CROSSING OVER

At the time of meiosis (see chapter 9) homologous chromosomes contributed by each parent pair lengthwise and wrap around each other like a cable. At times the tension becomes so great that the chromosomes snap and rejoin, resulting in an exchange of regions. These cytological exchanges are called chiasma and lead to a chromosomal or genetically detectably crossover. There are other equally satisfying explanations for crossing over, but the results are the same. It is by this process, the frequency of which is controlled by the cell, that parts of the maternal and paternal homologues can recombine to form new chromosomes with a new combination of genes.

1. Obtain a prepared slide of first division of meiosis in *Lilium* anther.
2. Find and draw the following in the sporogenous cells

   bivalent—pairs of chromosomes
   chiasma—the crosses or twists in each bivalent

3. What is the total number of bivalents in the *Lilium* anther nucleus?

4. What would be the 2N number of the nucleus?

5. Since each chromosome has replicated, how many strands (double helices) are there in a bivalent?

6. What is the total number of chromosomes in a cell on completion of meiosis?

7. What is the range of the number of chiasma for a bivalent?

The Genetic Effect of Crossing Over

Crossing over permits two genes on homologous chromosomes to recombine. However, the frequency of recombination is a function of the distance between alleles on the homologous chromosomes.

1. Your instructor has performed the following crosses for you.
   (a) yellow-bodied, white-eyed ♀ X wild type ♂
   (b) F_1 females from cross (a) X yellow-bodied, white-eyed ♂
   The genes for both yellow bodies and white eyes are on the *X* chromosome. There are no corresponding alleles on the *Y* chromosome.
2. Obtain a vial of F_2 flies and count each class. Enter your data on Chart 14.3.

CHART 14.3

Class	Number	
	Individual Data	Class Data
Wild type		
Yellow body, white eyes		
Wild type body, white eyes		
Yellow body, wild type eyes		
Total		

3. Draw a Punnet square representing each class based on the assumption that the genes for yellow bodies and white eyes are on different chromosomes and are recessive.

4. Draw a Punnet square assuming that the genes for yellow bodies and white eyes are on the same chromosome (the *X*) and are recessive.

5. Do your results agree with what is expected in either Punnet square?

6. How can you explain your results on the basis of crossing over?

7. The distance between two genes can be calculated in terms of crossover units, where 1 percent crossing over = one crossover unit. For this calculation, it is best to use class data.

8. Enter the data from your chart on the chart provided on the blackboard.
9. Copy all the data onto Chart 14.3.
10. Calculate the percentage of crossing over between the yellow locus and the white locus in the following way:

$$\frac{\text{No. with wild type body, white eyes + No. with yellow body, wild type eyes}}{\text{Total number flies counted}}$$

$\times$ 100 = percent crossing over

11. How many crossover units apart are the locus for yellow bodies and white eyes?

12. What happens to the meiotic products in the formation of the egg that justifies adding the number of wild type bodied, white-eyed, flies to the number of yellow-bodied, wild type eyed flies in order to obtain the number of crossovers that have taken place?

13. Draw the *X* chromosomes of the F_1 ♀ from the (a) cross and indicate why a crossover would lead to recombination.

14. Why is crossing over considered to be a source of variation?

CHROMOSOME ANOMALIES

During the process of nuclear division, some chromosomes experience accidents that lead to a loss or gain of one or more chromosomal segments. This type of event is rare, but a cell receiving an incomplete or over-complete set of chromosomes can sometimes survive and reproduce. Similar to gene mutations, the process is usually deleterious, especially if it involves the loss of a chromosome. These kinds of changes are most easily detected by a change of phenotype, which is particularly pronounced if the change involves the sex chromosomes. The same sort of accident may occur involving an entire chromosome.

In humans, a number of syndromes have recently been discovered to be the result of chromosome anomalies. Individuals with atypical traits are first chemically diagnosed. A bone marrow puncture is made to obtain marrow cells, which are then cultured. Cells are then smeared on a slide, stained, and examined. Photographs of the mitotic phases are made and the metaphase or late prophase sequence is cut up so that the chromosomes may be matched in pairs. Each chromosome pair in the human has been given a number (one to 23). Any deficiency or duplication will result in an inability of the investigator to match these chromosomes with the standard that has been set up for human chromosomes.

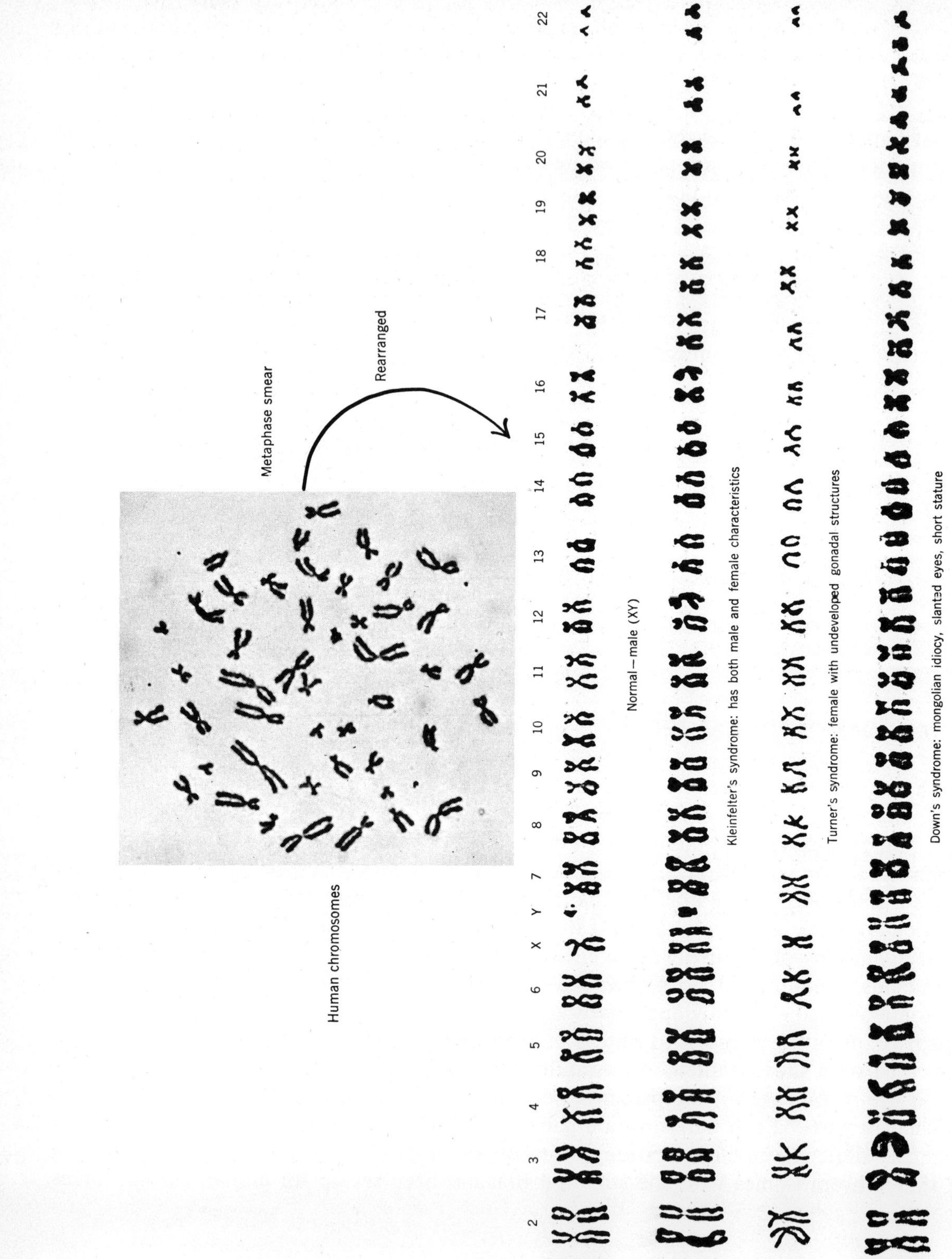
Human chromosomes
Metaphase smear
Rearranged
1 2 3 4 5 6 X Y 7 8 9 10 11 12 13 14 15 16 17 18 19 20 21 22
Normal—male (XY)
Kleinfelter's syndrome: has both male and female characteristics
Turner's syndrome: female with undeveloped gonadal structures
Down's syndrome: mongolian idiocy, slanted eyes, short stature

Figure 14.3

1. Fig. 14.3 shows several sets of chromosome pictures called idiograms. The first illustrates how they appear when photographed. The second shows how the chromosomes are matched and provides you with a standard of the normal chromosome complement in humans.
2. Study the three abnormal conditions: Turner's Syndrome, Kleinfelter's Syndrome, and Mongolian idiocy, in the chromosome idiograms below the standard.
3. Enter your observations on Chart 14.4.

CHART 14.4

Syndrome	Chromosome Number	Chromosomes that Are Added or Deleted
Turner's		
Kleinfelters		
Mongolian Idiocy		

4. What type of chromosomal anomaly might prove beneficial to an organism?

5. Why are chromosomal anomalies considered sources of variation?

15

Adaptation

MATERIALS

CLASS STOCK

culture tubes containing *E. coli*
prepared slides
- homozygous sickle-cell anemia RBC
- cross section of privet leaf
- cross section of corn leaf
- cross section of pine needle

skeletons of frog, cat, and human
hearts of dogfish shark, frog, and sheep
undissected dogfish shark
undissected bullfrog
flowers of garden pea, snapdragon, wild rose, and grass

STUDENT STATION

50 green peas
50 white peas
beaker
gram scale
hand lens or dissecting microscope
microscope
microscope lamp
lens paper

INTRODUCTION

The interaction of genetic variation with the total environment, which ranges from the physiological complementarity of genetic elements within the cells of the organism to climatic conditions of the atmosphere, brings about the emergence of new species.

A perspective of levels is very important to an accurate concept of evolution. Although variation arises on the genetic level it is expressed on the phenotypic level, so that organisms, instead of genes, are eliminated or multiplied. Speciation, on the other hand, is a population phenomenon. The genetic composition of the population (all the genes in the population that comprise the gene pool from which the next generation will derive its genetic material) is the reference point for the emergence of new species.

The Gene Pool Concept

The frequency of alleles in a gene pool can be figured out when the number of homozygous recessive members of a population are known. For example, given a population of 160 flies, of which 20 are brown eyed (bw^-bw^-) and, therefore, homozygous recessive, the number of wild type (bw^+) genes can be calculated in the following way.

all brown-eyed flies = bw^-bw^-
all wild type eyed flies = either bw^+bw^- or bw^+bw^+
Let A = frequency of bw^+ allele
Let a = frequency of bw^- allele

The population may be represented as

$$AA + 2Aa + aa$$

where aa represents the frequency of *organisms* exhibiting the brown-eyed trait (organisms having bw^-bw^-)

$$aa = \frac{20}{160} \text{ or } .166$$

$$a = \sqrt{aa} = \sqrt{.166} = .412 \text{ or } 41.2\%.$$

Thus the frequency of the bw^- gene is 41.2%. And since 1 – (the frequency of the bw^- gene) = the frequency of the bw^+ gene, the frequency of the bw^+ gene is 58.8%.

1. Calculate the frequency of the gene for curly wings in a population of 720 flies, 65 of which exhibit the curly-wing trait.

2. What is the frequency of the wild type gene (cy^+) in the gene pool of this population?

3. Analyze the following traits from members of your class:
free ear lobe (dominant), attached ear lobe (recessive)
ability to curl tongue (dominant), inability (recessive)

4. Enter the data on Chart 15.1 and determine the gene frequency for each gene in the pool comprised of the class population.

CHART 15.1

Phenotype	No. of Students	Decimal Fraction Students	Gene Frequency Dom.	Rec.
Free ear lobe				
Attached ear lobe				
Curling of tongue				
No curling of tongue				

THE HARDY WEINBERG LAW

The Hardy-Weinberg law states that a pair of alleles will not change their respective frequencies in a population as long as

1. One allele has no advantage over the other (no selection).
2. There is random mating.
3. Population size is infinitely large.
4. The mutation rates ($A \rightarrow a$; $a \rightarrow A$) are equal.
5. No migration occurs (either immigration or emigration).

To prove the law, use the gene frequency data and calculations from the problem on gene frequency of curly-winged flies. Assume that all matings can occur, cy^+cy^+ x cy^+cy^+, cy^+cy^+ x cy^+cy^-, and so forth. Obtain the frequency of each genotype by multiplying appropriate frequencies. For example, the number of heterozygous individuals would be the frequency of cy^+ (A) times the frequency of cy^- (a), or $2Aa$. Then for each cross, multiply the frequencies of the participants of the cross. Add all the results; the same gene frequency of the two alleles occurs after another generation.

Maintenance of gene frequencies would lead to stable populations, not speciation. The move toward speciation is, therefore, through violations of the assumptions of the Hardy-Weinberg law. In this laboratory, we shall discuss only a few of the ways in which violations can occur.

Selection

Two allelic genes impart slightly different phenotypes to the individuals bearing them. One of these individuals may have a greater survival advantage, although very slight, over the other, and will tend to bear more offspring. The relative fitness of each individual will then be a reflection of the ratio of the two genes. A gene frequency is meaningful because it is related to the survival advantage of its owners. One can relate this idea to the study of population shifts, antibiotic resistance, and sickle cell anemia.

Consider a population of the bacterium *E. coli* that has a gene $strep^s$ (streptomycin sensitive) which does not impart resistance to streptomycin to the cell. Suppose that a mutation of $strep^s \longrightarrow strep^r$ (streptomycin resistant) can occur. If these cells were grown in an environment that contained a high enough concentration of streptomycin (20 μg/ml) to inhibit growth, any cells that mutated to $strep^r$ would multiply and increase the frequency of the $strep^r$ gene.

1. Observe a set of culture tubes containing *E. coli.* In series P each tube contains 20 μg/ml streptomycin but a different number of cells (10^4, 10^5, 10^6, 10^7, 10^8, 10^9, 10^{10}).
2. On Fig. 15.1, shade in the density of the cells to correspond with the cell density in each tube. Bacterial growth becomes visible at a cell density of about 10^7; translucent cloudy growth is indicative of a cell density of about 10^8 cells/ml.
3. Observe and shade in the cell density of the F series on Fig. 15.1. The F series also contains 20 μg/ml streptomycin. The number on the tube will indicate the number of cells from the 10^{10} tube of the 2-day-old P series that were added. This is also indicated on the drawing.

Note: This experiment has actually been carried out in larger flasks, and each tube you have observed is a sample of the original culture. The numbers of bacteria transferred in certain cases are not possible using test tube quantities.

4. Did the mutations of $strep^s \longrightarrow strep^r$ occur before or after the cells were added to the streptomycin medium in the flask? Why?

5. Would you expect streptomycin to be present in nature? Why?

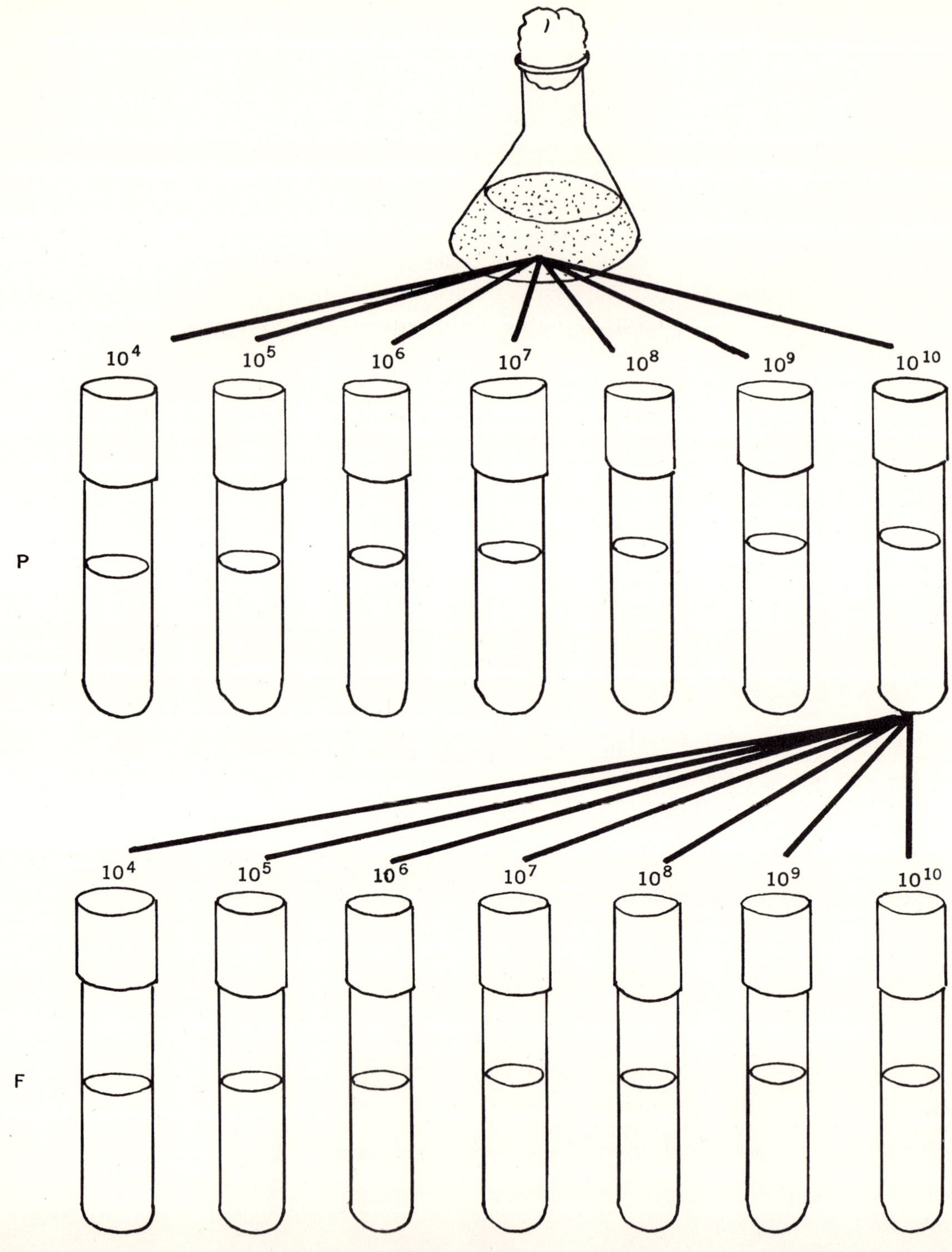

Figure 15.1

6. Why is it advantageous for the resistance to have a genetic basis?

Heterozygosity

Selection operates according to phenotype. But phenotypes do not always reflect the genetic state of an individual. For example, homozygous dominant and heterozygous individuals often have the same phenotype; selection would affect both in the same way.

There are some cases in which there are three phenotypes; one for the homozygous dominant, one for the homozygous recessive, and another for the heterozygous individual. In peas, genes for height can be dominant for tallness *(T)* and recessive for dwarfness *(t)*, but the heterozygous condition will result in taller plants than the *TT* individuals. One pair of alleles in humans that is of considerable interest has an effect on the red blood cells. Individuals with the genotype $H^A H^A$ have normal red blood cells because of the presence of hemoglobin *A*. Individuals of type $H^S H^S$ have many of their red blood cells shaped like sickles because of hemoglobin *S*, which causes an anemia commonly called sickle-cell anemia. Under this condition, the hemoglobin cannot carry as much oxygen as normal hemoglobin, and the cells tend to rupture under increased oxygen pressure. The homozygous condition of $H^S H^S$ causes death prior to maturity. Individuals with $H^S H^S$ are resistant to malarial infection, whereas $H^A H^A$ individuals are sensitive to the invasion of this parasite. Heterozygous individuals ($H^A H^S$) have normal and sickle red blood cells and also have resistance to malaria.

1. Obtain a prepared slide of a blood smear from a person homozygous for $H^S H^S$ genes.
2. In the space provided, draw and label the following:

   normal red blood cells—round, flat cells shaped somewhat like discs. These cells have no nuclei.
   sickle-celled red blood cells—sickled-shaped cells.

3. Why is the H^S gene predominantly found in West Africa and the United States?

4. Why in West Africa are 40 percent of the black people heterozygous whereas in the United States only 9 percent of the black people are heterozygous?

5. What is the selection value in West Africa compared to the United States?

Migration and Population Size

The frequency of a gene is expressed in terms of (1) a ratio involving its allele(s), and (2) many individuals of a populations. Populations frequently spread out geographically into subdivisions. This spreading is accomplished by one or more individuals migrating to establish a new population. The gene frequency of this new population will be the result of the genes in those individuals who migrated. Suppose that three individuals, all *AA*, migrated from a population of 90 percent *A*, 10 percent *a* (0.81 *AA*; 0.18*Aa*; 0.01 *aa*). The new population would be significantly deviant from the parental population by a non-representative combination of genes.

1. Obtain a beaker containing 100 peas, 50 of which are green and 50 white. This represents a parental population of genes.
2. Sample 10 peas with your eyes closed in order to make your selection random. These will represent 5 diploid individuals.
3. Calculate the gene frequency of this sampling and enter your data on the graph of Fig. 15.2.
4. Repeat the sampling 19 more times, calculating the gene frequency of each of the 20 subpopulations and tabulating on the graph. Do not connect the graph points.
5. How can chance play a significant role in determining gene frequency?

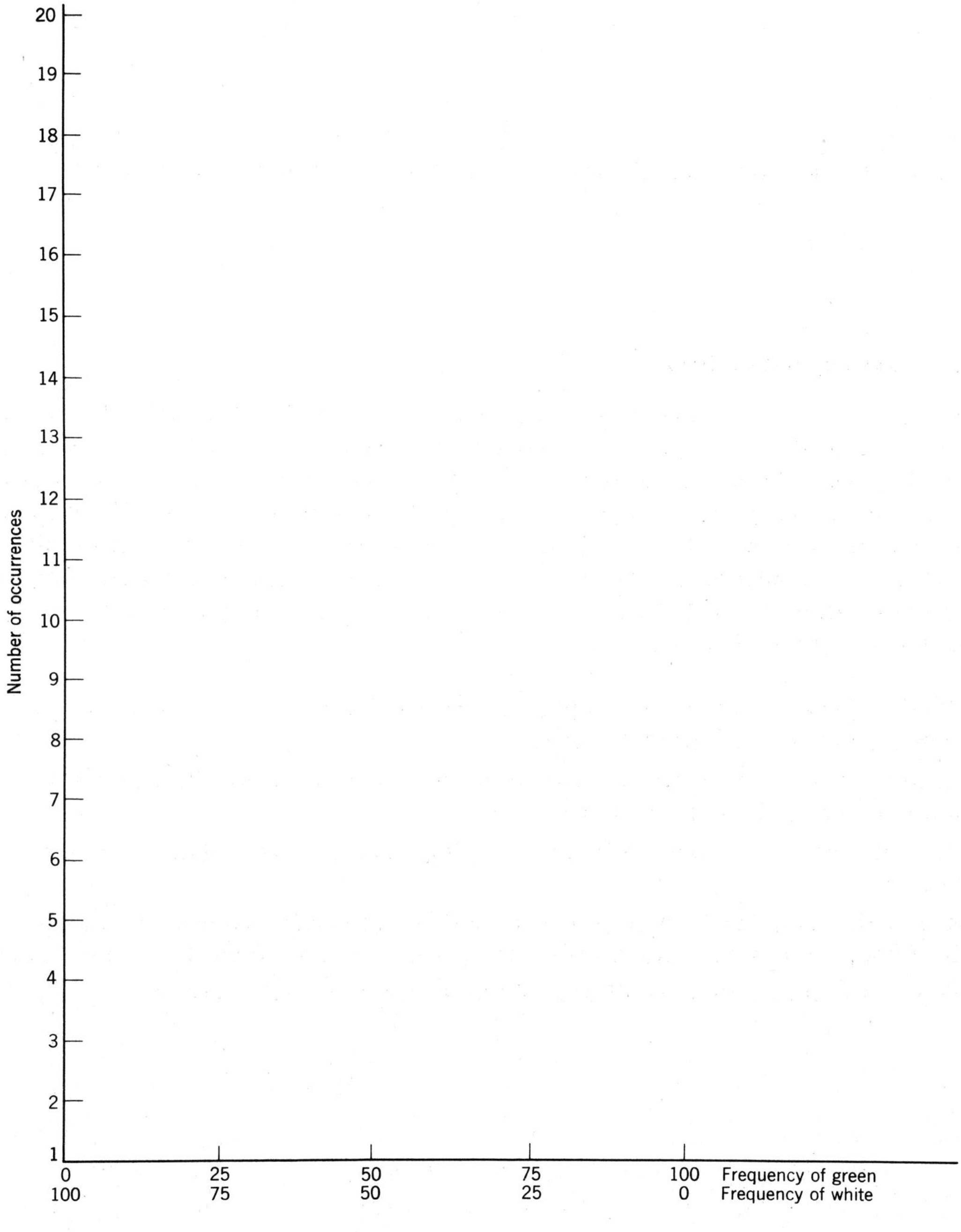
Number of occurrences
20
19
18
17
16
15
14
13
12
11
10
9
8
7
6
5
4
3
2
1
0 25 50 75 100 Frequency of green
100 75 50 25 0 Frequency of white

Figure 15.2

6. What effect does size of migrating sample have on gene frequency?

General Considerations of Hardy-Weinberg

It is not possible that only one assumption of the Hardy-Weinberg law can be violated. Severe selection, such as that found in the streptomycin experiment, reduces population size, negating two assumptions of the law. Migration also affects population size. Population size affects random mating. In nature, none of the assumptions on which the law depends exist.

A large, random change in the gene frequency of the alleles in the gene pool is called genetic drift.

1. How might migration lead to genetic drift?

2. Was the streptomycin experiment an example of genetic drift?

3. Why is the example of black people in West Africa and black people in the United States not an example of genetic drift?

4. Why does your calculation of gene frequencies for attached or free ear lobes, or curling or noncurling of the tongue in your classmates represent genetic drift?

ADAPTATION

The term adaptation is used to describe some organismal feature that fits some aspect of the environment. Since every feature of every organism is subjected to some selective pressure, every feature of every organism can be considered an adaptation.

Adaptations arise through variation and selective pressures. For example, let us say that organism *A* is the predator of organism *B*, which is, in turn, dependent on a precise ionic balance in an aquatic environment. In the population or organism *B* are some members that could survive in a slightly more acid environment and some members that could survive in a slightly more alkaline environment. If the environment becomes slightly more acid, those members of the population that possess the acid variation will survive and reproduce, while the others will perish; an altered form of type *B* will exist.

In the population of organism *A* there may be some members that can feed only on the original form of organism *B*, and others that can feed on both the original and the altered form. Therefore, those members that possess the variation for feeding on the altered form will survive and those that do not, will perish unless they possess a variation that permits them to feed on another organism instead, for example, organism *C*.

In this example, selection pressures have been such that those members of the population of organism *B* that were adapted to a slightly more acid environment survived, and those members of population of organism *A* that were *adapted* to feed on the altered form of organism *B* or *adapted* to accept organism *C* as a substitute food survived. Hence, the capacity to exist in areas with different ionic balances, the capacity to accept a slightly different food source, and the capacity to accept a substitute food source all became features of the respective organism.

The fact that mutation and the various sources of variation are detected in all investigated forms indicates that the process of adaptation is constantly occurring in all forms of life. The environment is constantly changing and organism variations must keep abreast of environment changes. An example of recent adaptive reactions may be seen in the Scotch heather, which in northern Europe in early and mid-summer is pollinated by thrips (a type of insect). Toward the end of summer, the flower is wind pollinated. In southern Europe, the plant is pollinated by bees, which are not native to the north. The heather in the northern regions has adapted in such a way as to attract the thrips. However, thrips are limited in pollinating the heather because they have a relatively short life span. The flower has, in a sense, compensated for the short life span of the thrips and increased its own pollination period by adapting to wind pollination for late summer, when there are no thrips. There remains, therefore, a close association between flower types and animal life, or between predator and prey, but these relationships are always shifting. Nature is dynamic; the balance remains, but the participants change.

Gene mutations, recombinations, and other phenomena occurring in blind fashion are the bases of these changes. Life is built to contain a rich source of variation, permitting it to cope with the most unpredictable shifts in the environment.

ADAPTATION OF PELVIC BONE STRUCTURE

As the vertebrate organisms (those with backbones) have evolved, the bone structure and organization have undergone considerable adaptation to fit the needs of the organisms. This is not meant from a teleological approach. Certain bone structures have been of evolutionary advantage to the organism possessing the change and this organism has survived to pass the genes for this bone structure to its offspring. Many of the bones found at the different levels of organization in the vertebrates are *homologous,* having the same embryological origin although if differing in their final position or function in the more complex form. These homologous structures will be examined from the standpoint of adaptive results of evolution.

1. Examine the bones comprising the pelvic region of the frog, cat, and human skeleton, using Fig. 15.3 as a guide.
2. If the skeletons are available, locate these bones and examine their relative size, fusion of bones, and thickness.
3. Using your observations, fill in Chart 15.2.

CHART 15.2

Organism	Distribution of Body Weight	Adaptations of Pelvic Bones	Primary Function of Pelvic Girdle
Frog			
Cat			
Human			

4. How is the pelvic region adapted to enclose the female reproductive organs in the human? In the cat?

5. Examine the relative size of the pectoral girdle (the girdle of bones that supports the upper limbs) of the cat and the human. Do their sizes confirm your conclusions regarding the distribution of body weight in both organisms? Why?

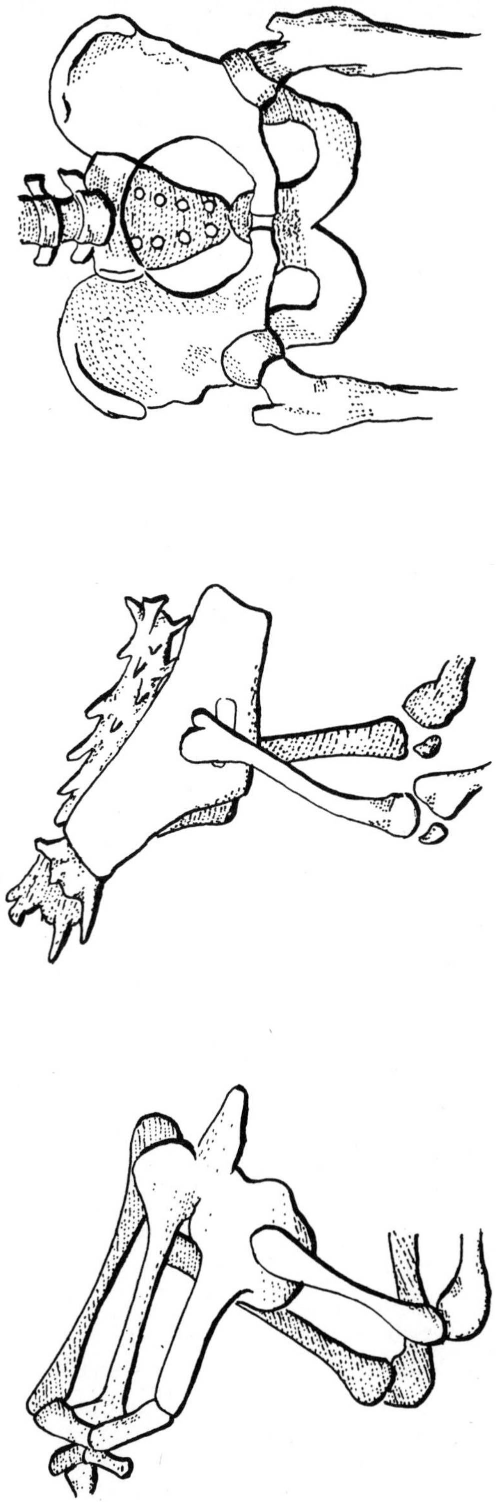

Figure 15.3

6. How has the structure of the pelvic region become adapted for the human to walk upright?

7. What type of pelvic girdle would you expect to find in the dog? The ape?

ADAPTATIONS OF THE HEART

In the course of evolution, the heart has undergone a complex series of changes. These changes are closely associated with the adaptation of the vertebrates to a terrestrial environment. However, the primary concern of this exercise is the muscularity of the heart, not the means of blood passage.

1. Examine the heart of a dogfish shark, a frog, and a sheep.
2. Note the degree of muscularity each possesses and enter on Chart 15.3, using the following notations:

   – = no muscularity
   + = some muscularity
   ++ = greater muscularity, etc.

3. Dry the hearts as thoroughly as possible to remove any excess preserving fluid and weigh each on the gram scale.
4. Enter the data on Chart 15.3.
5. Obtain an undissected dogfish shark of average size and an undissected bullfrog.
6. Dry each of these thoroughly and weigh on the scale.
7. Replace the hearts and the specimens in the preserving fluid immediately after weighing to prevent their decay.
8. Enter the data on Chart 15.2.

CHART 15.3

Organism	Muscularity	Body Weight (grams)	Heart Weight (grams)	Heart Weight % of Body Weight
Shark				
Frog				
Sheep		114 kg.		

9. The percentage of body weight of the heart is correlated with the physical activity of the animal. From the adaptations of the hearts you have examined, can you conclude which of these animals is most active?

10. The size and degree of muscularity of the heart are directly correlated with the demand placed on the heart by the system. A terrestrial environment demands a greater pressure in pumping the blood than an aquatic environment. How is this strain exhibited in the adaptation of the sheep heart?

ADAPTATIONS OF FLOWERS

The flower has evolved in harmony with its environment, adapting in such a way as to utilize factors within the environment to its own end, pollination. This process involves the transmission of the male gametophyte (pollen) to the female plant organ, the stigma. The means by which the pollen is carried varies, employing many agents. These means range from the relatively simple process of self-pollination to that used by a tropical species that involves the nuzzling of an elephant's trunk to effect the transfer. In each case, however, the process of pollination is very specific, often based on naturally occurring animals in the same area. Four means of pollination will be examined to demonstrate how organisms adapt to specific conditions for accomplishing important phases of their life cycle.

The Pea (or Violet)

1. Obtain a flower of the garden pea and notice that it has no opening at maturity for external agents to enter.
2. Examine the flower to see if there is any pollen on the stigmatic surface.

3. Draw the essential features of a self-pollinating flower in the space provided.

4. Why is the pea flower so useful in genetic research?

5. What might be an evolutionary disadvantage to a self-fertilizing system? (Would pea plants tend to be homozygous or heterozygous?)

The Snapdragon Flower

Some flowers attract small insects by their color and scent. As the insect collects the nectar, which is produced by organs called nectaries, it haphazardly transmits the pollen to all parts of the flowers it visits, including the stigma.

1. Observe the floral arrangement of the snapdragon flower. Notice the small entrance and the amount of pollen produced on the anther.

2. In the space provided, draw the external parts of the flower, including the position of the stamens and pistil. As the bumblebee alights on the lip of the corolla, its weight opens the flower and the bee moves to the bottom, dusting the pollen off the stamen.

3. Does the flower offer any advantage to the bee? What?

4. Why are bee-pollinated flowers open only during the day? What is the adaptive advantage?

5. Bee-pollinated flowers have both stamen and pistils. Why are both present if it is not for self-fertilization since both do not mature simultaneously?

The Wild Rose (Magnolia or Water Lily)

Some flowers are completely open and contain relatively large quantities of pollen. This is characteristic of the wild rose. This condition permits less adapted animals to be the pollinating agents. Beetles visit such flowers, consuming the pollen as food, and, in the process, transmit some to the stigmatic surface of the same and other flowers.

1. Examine a flower of wild rose.
2. Draw a rose flower in the space provided, indicating the floral characteristics important for pollination.

3. Can you give another example of a beetle-pollinated flower?

The Grass Flower

Most grasses, including the domestic cereals, produce copious quantities of pollen which are carried by the wind for cross pollination. The pollen is more dusty than sticky as in the cases of pea, snapdragon, and rose, in order to permit easy dissociation from the anther.

1. Observe a grass flower (wheat, etc.) with a hand lens or a dissecting microscope.
2. Draw the grass flower in the space provided, indicating the size, color, and scent. Notice the stamens protruding above the flower.

3. Shake the flower to see if any pollen is dislodged.
4. List four differences between insect- and wind-pollinated flowers.

5. Certain human ailments (allergies) are caused by pollen in the air. Can you add wind-pollinated plants other than the grasses to the list?

6. Why do wind-pollinated flowers usually produce one seed whereas animal-pollinated flowers produce many seeds?

ADAPTATIONS OF LEAVES

The main activity of the leaf in seed plants is carbon fixation through the photosynthetic mechanisms. Carbon dioxide is one requirement and oxygen is one by-product, indicating that in the course of the photosynthetic process there is a considerable exchange of gases necessary for optimal activity. If the leaf is to permit gas flow, there will also be, unfortunately, an outward movement of water from the moisture-laden cells of the internal tissue. The problem is, then, how the leaf will allow sufficient CO_2–O_2 exchange without excessive loss of water. By the nature of the problem some water will escape, but plants have found by adaptive mechanisms, ways to lower the rate of water depletion. The leaves of several different plants will be studied to discover how the problem has been partially solved. Keep in mind once again that these means are the results of gene mutations and recombinations until an efficient system was evolved which had a selective advantage.

Ligustrum leaf (Privet)

1. Obtain a cross section of privet leaf. Observe the typical tissues using your textbook as a guide. Find the stomatal apparatus. The stomatal apparatus opens during periods of photosynthesis and closes during periods when no photosynthesis occurs.
2. Count the number of stomatal apparatus on the top and bottom surfaces of the entire leaf.

   top surface =
   bottom surface =

3. Explain why this distribution is beneficial in preventing excess loss of water from the leaf. (Hint: Water vapor evaporates upward.)

4. What other structure on the epidermis suppresses water loss?

Corn *(Zea maize)*

1. Obtain a prepared slide of a corn leaf in cross section and observe the outside and inside layers of epidermis.
2. Notice the bulliform cells on the outer epidermis. They are large water-holding cells that occur in groups.
3. During periods of drought the corn leaf curls; during periods of sufficient water the leaf is expanded. The bulliform cells control this curling of the leaf. Explain how the bulliform cells control leaf curling.

4. How does the condition of parallel veination complement the bulliform cells' activity? (Observe a demonstration of a corn plant to see parallel veination.)

Pine (or *Oleander)*

1. Observe a cross section of a pine leaf (needle).
2. Notice that the stomatal apparatus is found at the bottom of a depressed area, the stomatal cavity.
3. Explain how the stomatal cavity contributes to the prevention of excess water loss.

4. Why cannot the pine leaf employ the mechanisms used either by the *Ligustrum* or the corn leaf?

5. Can you think of a plant that has stomatal apparatus only on the top surface?

6. Of what advantage is the shape of the pine needle?

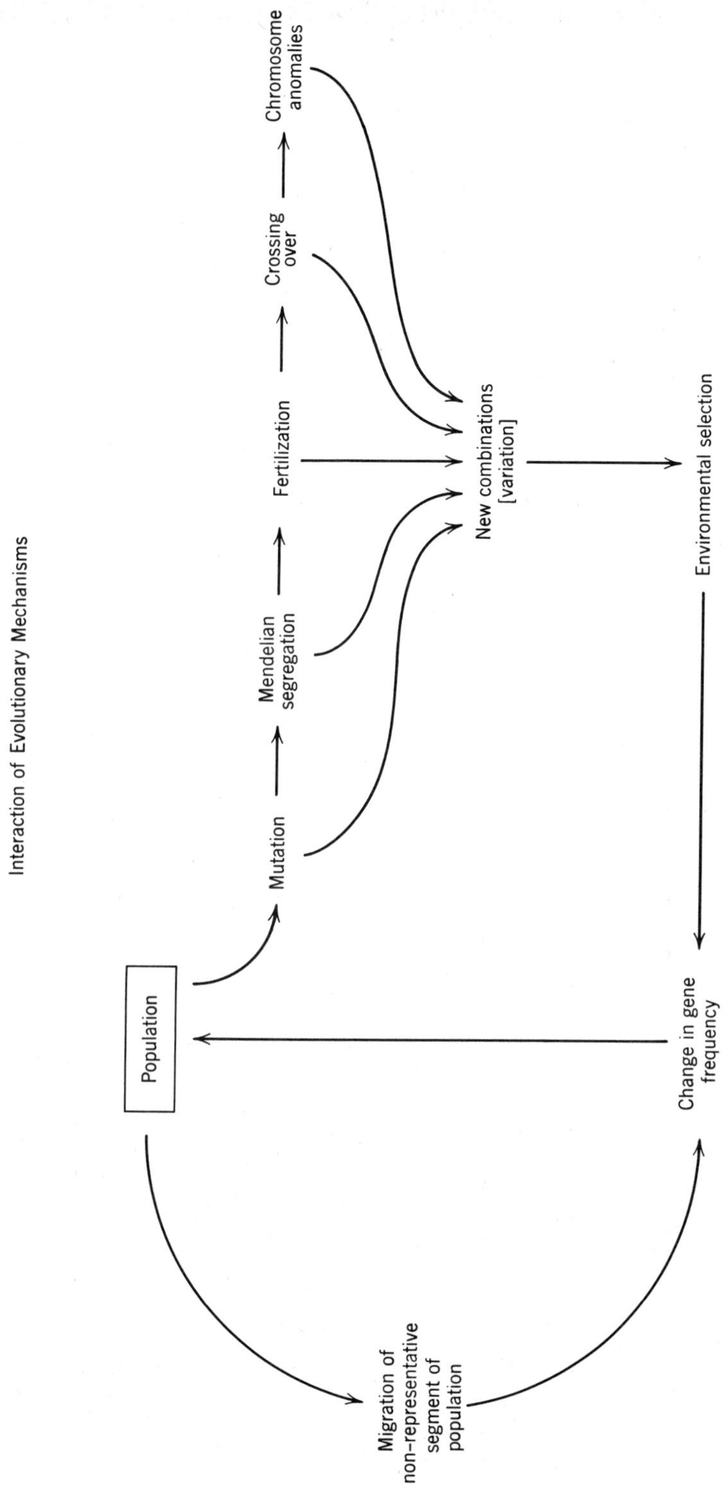
Interaction of Evolutionary Mechanisms
Population
Mutation
Mendelian segregation
Fertilization
Crossing over
Chromosome anomalies
New combinations [variation]
Environmental selection
Change in gene frequency
Migration of non-representative segment of population

Figure 15.4

16

The Origin of Life

MATERIALS

CLASS STOCK

chemicals
- 3.0% ammonium solution
- 1.0% formaldehyde solution
- amino acid solution (1.0% glycine)
- 1.0% glucose solution
- chromatography solvent
- ninhydrin spray
- 20 amino acids in crystalline form
- conc. KOH
- 0.5% copper sulfate

yeast suspension
radish seeds

STUDENT STATION

chromatography paper
large beaker
glass rod
plastic wrap
crucible
oven
Pasteur pipette
test tubes
slides
cover slips
microscope
microscope lamp
lens paper
cardboard
large nail
hammer
dissecting pan
wax paper cups
scissors
test tube rack

INTRODUCTION

Scientists once felt that life could arise *de novo*, that is, by itself, without any external intervention. This ancient idea of spontaneous generation as articulated by scientists up to the middle of the nineteenth century has been discarded, but the essential framework of the concept has survived in a more modernistic interpretation. In the present day concept, it is not flies, beasts, or even bacteria that generate spontaneously, but molecular complexes. The time for the emergence of life is now extended beyond the anticipation of the early scientists, to a perspective of millions of years. And causation of biogenesis has shifted from a mysterious "vital force" to the concept that chemical and physical phenomena, operating in accordance with their respective laws, can and have formed the bases for the origin of life.

Life has not yet been produced in the laboratory, nor have any experiments even come close to this goal. However, there has been an elucidation of some mechanisms and conditions that apparently had major roles in this process. The investigation of biopoesis is unique in biology in that the individual steps are not important; the long sequence of small, repeating events are important.

FORMATION OF ORGANIC MOLECULES

Organic molecules are the structural basis of living things. The first step toward the creation of life had to be a creation of organic molecules from available material. Spectroscopic analysis of certain galactic regions of space reveal that four molecules are present. These have been considered the raw materials on the early earth—hydrogen (H_2), ammonia (NH_4^+), formaldehyde (H_2CO) and water (H_2O). (Formaldehyde, like carbon dioxide, is not considered an organic molecule).

The following exercise is to demonstrate how more complex molecules might have been formed under conditions of the early Earth. BECAUSE HYDROGEN CYANIDE, A GAS WHICH IS LETHAL IN EXTREMELY SMALL QUANTITIES, (10-20 ppm) IS FORMED IN THE EXPERIMENT DESCRIBED BELOW, YOU CANNOT PERFORM THIS EXPERIMENT. THE EXPERIMENT IS DESCRIBED AND THE RESULTS ARE GIVEN ON A CHROMATOGRAPH SO THAT YOU MAY COMPARE EXISTING BIOLOGICAL COMPOUNDS WITH THOSE PRODUCED AS A RESULT OF THE EXPERIMENT.

Ammonia solution, formaldehyde, and water were mixed in a beaker and transferred to a quartz or a glass tube. A rubber stopper was placed in the tube as illustrated in Fig. 16.1. Hydrogen gas was bubbled into the tube to flush out all oxygen. When the oxygen was flushed out, both outlets were clamped. The tube was placed below an ultraviolet (UV) lamp for one week, or near a window. Enough UV rays penetrated the window pane and the glass tube to bring about reaction. After a few days the surface of the liquid became yellow in color. Some of the liquid from the tube incubated for one week was placed on a strip of chromatography paper, dried, and chromatographed using 4 parts butanol, 1 part glacial acetic acid, and 5 parts water as a solvent. The chromatograph was sprayed with ninhydrin, after which it had the appearance of the chromatograph of Fig. 16.2.

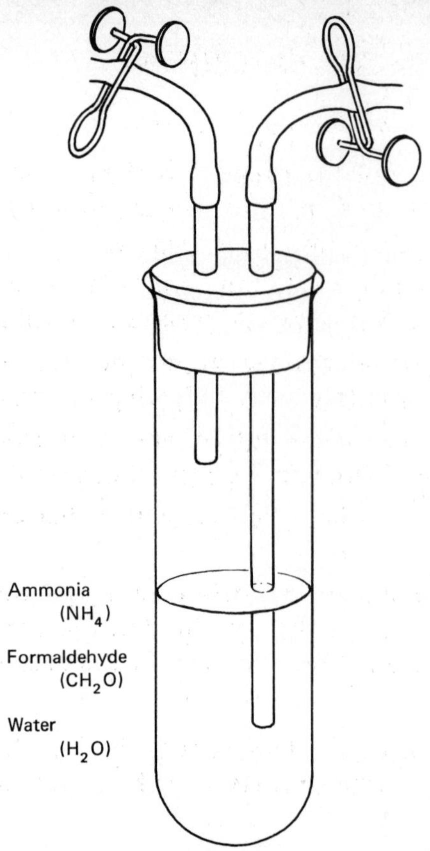

Figure 16.1

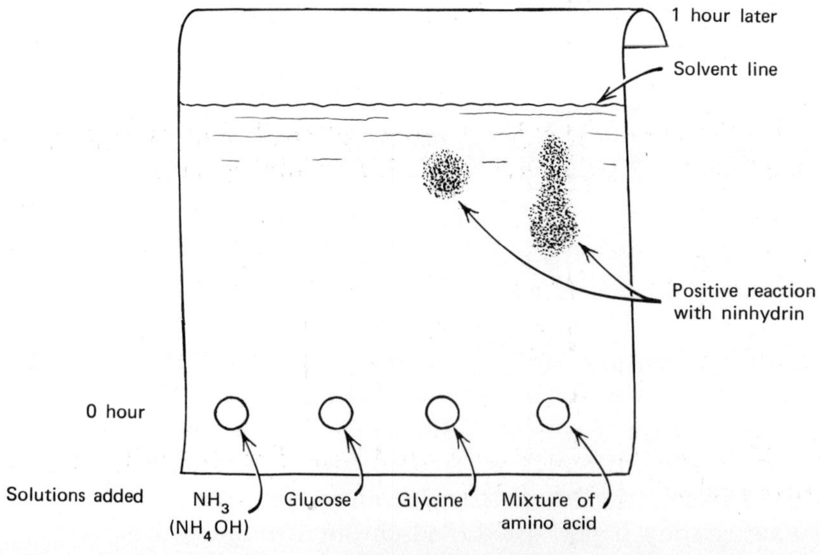

Figure 16.2

PERFORM THIS PORTION OF THE EXPERIMENT.

1. Obtain a strip of chromatography paper 3 - 4 in. wide.
2. About ¾ in. from the edge of the paper, and ½ in. from the bottom, place about five drops of ammonia solution, permitting each drop to dry before putting on another, and keeping all drops within a pinhead area.
3. About ½ in. from the ammonia solution, and ½ in. from the bottom, place five drops of formaldehyde solution, following the procedure outlined in 2.
4. About ½ in. from the formaldehyde solution, and ½ in. from the bottom, place ten drops of amino acid solution (glycine), following the procedure outlined in 2.
5. About ½ in. from the amino acid solution, and ½ in. from the bottom, place ten drops of glucose solution, following the procedures outlined in 2.
6. Place 10 ml of solvent into a large beaker, with a diameter at least as large as the chromatography strip.
7. Hold the top of the chromatography strip over a glass rod and insert the strip so that the bottom edge is immersed in the solvent, but not the dried drops. The glass rod sits on the lip of the beaker. Cover the top of the beaker with plastic wrap.
8. Leave for 20 min.
9. Remove the strip of paper dry and spray it with ninhydrin. Heat in oven (60°C) for one hour. Above which of the solutions on the paper did you get a positive ninhydrin test?

10. For what does ninhydrin test?

11. The result can be produced under a variety of conditions as long as a reducing environment is present. What creates the reducing atmosphere?

12. What type of organic molecules have been formed by H_2, NH_4^+, H_2CO, and H_2O in the experiment described on p. 255?

13. The formation of organic molecules is an energy requiring process. Where did the energy come from in the experiment described on p. 255?

14. What other forms of energy were probably available on the early Earth?

15. Defend the statement "Organic molecules can arise by a spontaneous generation under certain conditions."

PROTEIN SYNTHESIS

As you have demonstrated in the previous exercise, organic amines can be spontaneously generated under certain circumstances. In fact, they can be generated under circumstances that in many ways resemble those present on the early Earth. Some of these organic amines can have the same structure as one or more of the twenty amino acids.

Biological specificity is dependent on proteins or a conglomeration of the amino acids of which proteins are composed. Can the peptide bond between amino acids be spontaneously generated to form proteins?

1. Add a small amount of each crystalline amino acid (about the size of a match head) to a small crucible.
2. Mix the dry 20 amino acids crystals together with a glass rod.
3. Place the crucible on a hot plate at highest temperature in a fume hood until the material smoulders (approximately one-half hour).
4. Remove the crucible and add a few (3 to 6) drops of water with a Pasteur pipette. Notice how the mixture sizzles. Add 3 ml water until you have a brown solution.
5. Permit the crucible to cool. Stir while cooling.

6. Place five drops of the suspension in a test tube with 1 ml of H_2O and perform the biuret test.
7. Does the mixture give a positive test for protein or for polypeptide?

8. Can peptide linkages between amino acids be spontaneously generated?

9. Are the proteins formed here specific for any particular function? Explain.

10. Place a drop or two of the suspension from the crucible on a slide and prepare a wet mount.
11. Observe under high power (or oil immersion magnification if available).
12. Notice the numerous small spherical microspheres. Draw several in the space provided.

13. Prepare a water mount from a yeast suspension and notice the similarities to the microspheres. Did any of the microspheres have "buds" like those of yeast cells?

14. Are the microspheres equivalent to the yeast cells in shape? In chemical content?

15. Are the microspheres alive?

16. Why is it believed that some entity similar to the microsphere, with specific shape and make-up of protein, was a necessary intermediary of biopoesis (origin of life)?

SELF-ASSEMBLY

The shape of the microspheres was generated by a self-assembly of many proteins. Self-assembly would involve a coming-together, a creation of order out of disorder. This seems to operate in violation of the second law of thermodynamics, which predicts that all systems experience increasing degrees of disorganization or entropy. Actually, biological systems operate in accordance with this law, and have, curiously, explored it in numerous ways. Entropy must have been involved in the origin of life, and although it would appear to deter evolution, the tendency toward entropy actually may have contributed positively to the elaboration of molecular complexity.

Entropy can be stated in two ways: (1) a tendency toward *maximum* randomness, or (2) a tendency toward *minimum* free energy (usable energy). The following exercises illustrate certain potentialities of matter that could have contributed to biopoesis.

1. Mark a piece of cardboard into 100 squares of 1 cm^2 (10 cm × 10 cm). Press or pound the head of a large nail into each square thereby forming depressions.
2. Count out 100 small (radish) seeds, and pile them up in center of the checkerboard. Place cardboard on a wire test tube rack.

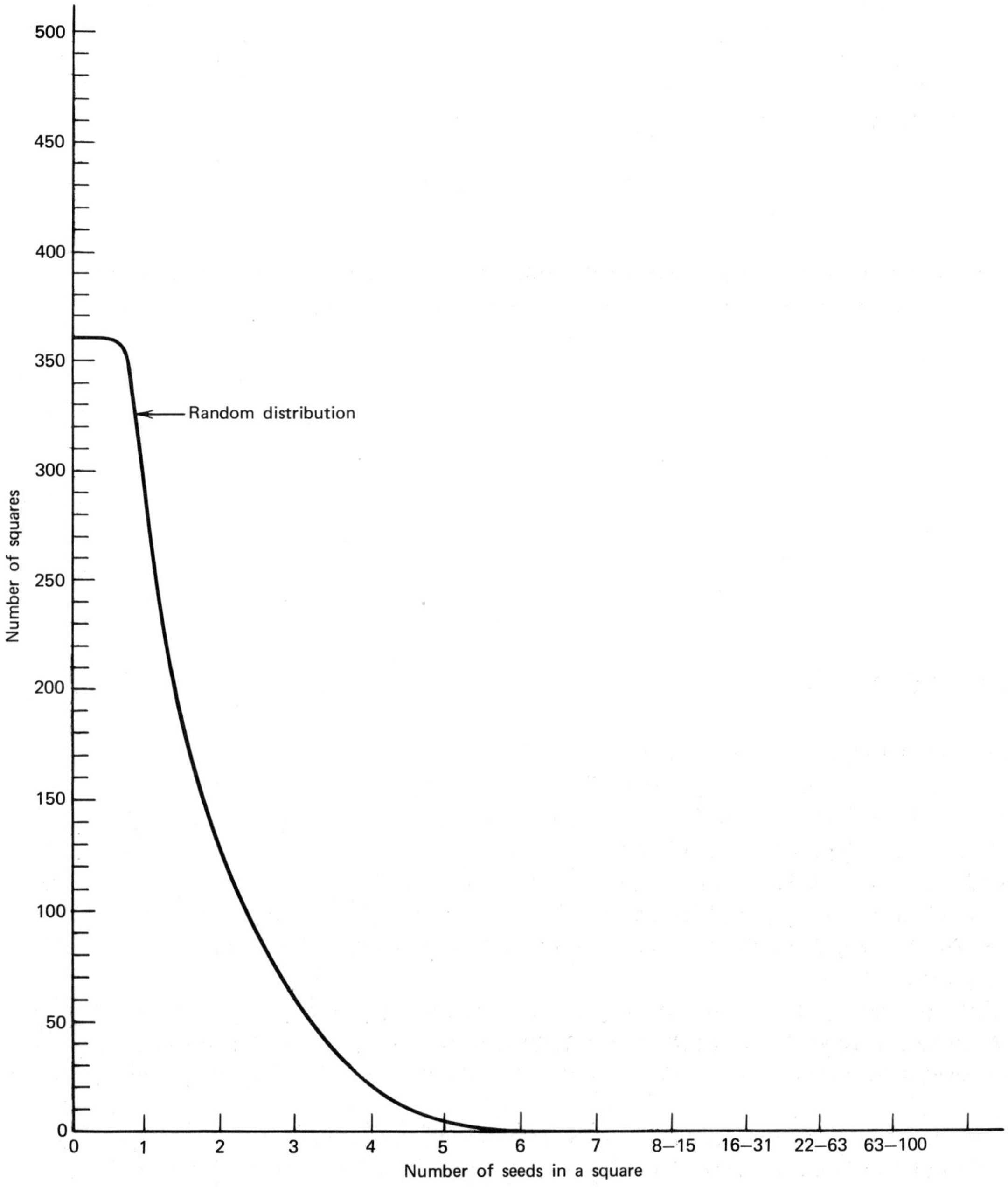

Figure 16.3 Distribution of seeds after 100 taps.

3. Tap the cardboard gently with a pencil at different sides 5 times so that some seeds move.

4. Count how many squares have 0 seeds, 1 seed, 2 seeds, and so on. Plot this distribution on the graph of Fig. 16.3.

5. Tap the board 5 more times, and determine the distribution. Record on Fig. 16.3.

6. Tap the board 10 more times (a total of 20), and determine the seed distribution. Plot your data on the graph of Fig. 16.3.

7. The line marked on the graph indicates the expected pattern for randomly distributed seeds. What happens to the distribution of seeds after 5 taps? 10 taps? 20 taps?

8. Why does the population of seeds approach the most stable state (unchanged pattern)?

9. What can you say about predicting the position of a particular seed at different times?

10. In which perspective of entropy would you categorize this phenomenon—a tendency toward randomness or a tendency toward minimum free energy? Explain.

11. Fill a small dissecting pan with water.
12. Cut a wax paper cup into about 25 crosses, as illustrated in Fig. 16.4.

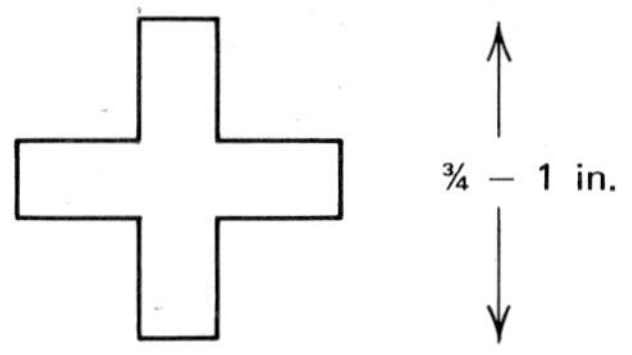

Figure 16.4

13. Drop the strips one by one onto the surface of the water in the dissecting pan, and wait until the system reaches an equilibrium.
14. The free energy involved in the repulsion of the wax on the cardboard from the water (or vice versa) is the most dominant energy in the system. Would this surface free energy increase or decrease as the cardboard pieces interact?

15. In which perspective of entropy would you categorize this phenomenon—a tendency toward maximum randomness or a tendency toward minimum free energy? Explain.

16. Is the randomness of particles and changes in minimum free energy always synonymous? Why?

17. Blow gently on the surface and observe what happens.
18. Drawing from the information gained in your observation of the waxed cardboard and water model, explain how protein molecules can self assemble to generate order (as in the microspheres) although the amount of free energy decreases.

19. Is it possible for a pattern to arise through chance associations alone?

20. Can you predict the pattern that will be achieved? In the waxed cardboard and water model, what factors determined the pattern? Could those factors be repeated?

The Concept of Order

Patternization in the previous experiment was achieved because of the *high* surface free energy in the repulsion of the wax for water that decreases as the pieces of waxed cardboard fit together. The waxed cardboard pieces floated apart in groups when kinetic energy was added to the surface by blowing heavily. Thus, patternization is due to the relative amounts of kinetic and free energies involved in the system, in this case, high free energy coupled with low kinetic energy leads to an organized, highly stable pattern, while high free energy plus high kinetic energy brings about ordered units that are random with respect to one another. In other words, if the free energy is initially high, but decreases when the attractive particles are in proximity, and the kinetic energy is initially low so that the pattern is undisturbed, the pattern will persist. If, however, the other energy conditions exist, the pattern will disintergrate in accordance with those other conditions, ultimately reaching a stable state.

Assume that all populations of units began as pattern *d* in Fig. 6.5 and in accordance with their respective kinetic and free energies, moved to the positions indicated in 16.5 *(a), b, c,* and *d*. For these movements, the initial kinetic and/or free energies could be either low or high.

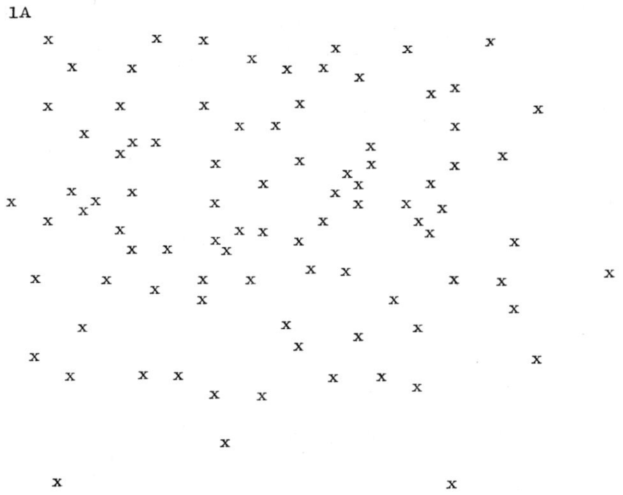

Figure 16.5 (a)

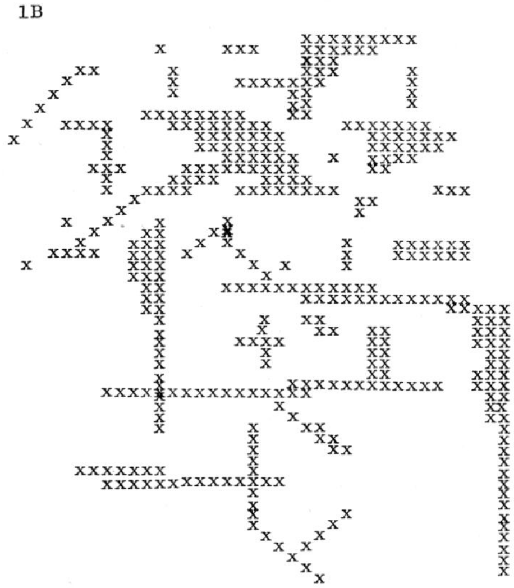

Figure 16.5 (b)

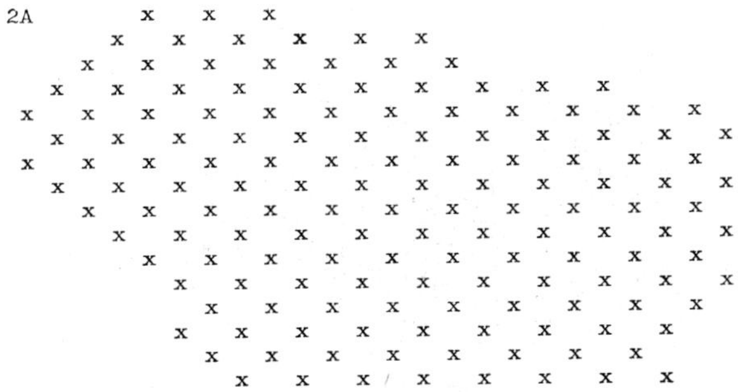

Figure 16.5 (c)

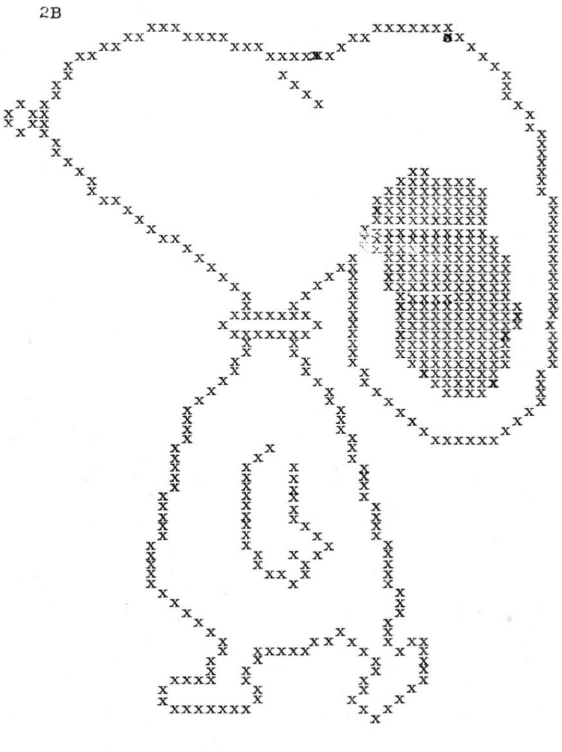

Figure 16.5 (d)

Initial kinetic energy	Initial free energy
Low	Low
High	Low
Low	High
High	High

1. Examine pattern *a*.
2. Is the pattern random or ordered?
3. What final state of kinetic energy would produce this pattern?
4. Do the *x*'s have much attraction for one another?
5. Do the *x*'s contain high or low levels of free energy?
6. What state of kinetic energy and free energy would have to exist for pattern *d* to move to a pattern such as *a*?
7. Examine pattern *b*.
8. Is the pattern random or ordered?
9. What state of kinetic energy would produce this pattern?

10. Do the *x*'s have much attraction for one another?

11. Do the *x*'s contain high or low levels of free energy?

12. What state of kinetic energy and free energy would have to exist for pattern *d* to move to a pattern such as *b*?

13. Examine pattern *c*.
14. Is the pattern random or ordered?

15. What state of kinetic energy would produce this pattern?

16. Do the *x*'s have much attraction for one another?

17. Do the *x*'s contain high or low levels of free energy?

18. What state of kinetic energy and free energy would have to exist for pattern *d* to move to a pattern such as *c*?

19. Examine pattern *d*.
20. Is the pattern random or ordered?

21. Do the x's have much attraction for one another?

22. Do the x's eventually contain relatively high or low levels of free energy?

23. What state of kinetic energy and free energy would have to exist for pattern 2*b* to remain?

24. Why did the microspheres remain in a constant shape?

25. Why is the three-dimensional protein in its folded state considered to be in its most stable thermodynamic state?

26. A molecule of DNA is linear and is also a double helix. Although there are no cross linkages, as in protein, to give it stability, why does it assume this characteristic configuration?

Part V

ECOLOGY

17

Ecology

MATERIALS

CLASS STOCK	STUDENT STATION
Pond samples frozen or thawed.	slides
from surface (100 ml)	cover slips
from 2-10 cm level (100 ml)	pipettes
from 10-20 cm (100 ml)	Stender dish
from bottom level (100 ml)	dissecting microscope
from subbottom level (100 ml)	microscope
	microscope lamp
	lens paper
	petri plates

INTRODUCTION

Ecology is the study of organismal-environmental interaction and the consequences of these interactions. Organisms are not simply spread evenly over the surface of the earth. They occur in groups belonging to the same species, that is, as *populations;* they are limited to specific combinations of environmental factors, by living in certain *habitats;* they exist as members of complex groups that include many species, *(communities);* and they share common materials within a small region *(ecosystems).*

The interrelationships between organisms involve such complex balances that only recently have biologists begun to comprehend the spectrum of factors involved. The task of placing quantitative values on ecological relationships is often side-tracked by gathering any data that appears to be available, whether or not it is of value for ascertaining fundamental relationships. The goal of the following set of exercises is to discern some order within an ecological community and to express this order in some quantitative fashion.

COMPOSITION OF A POND

The first task in an ecological study is to identify the types of entitites within a habitat. A pond is perhaps the easiest ecological situation to analyze because it is bounded by a terrestrial margin which prevents emigration, contains some uniformity of organismal types throughout, and representative samples can easily taken.

A pond is an *ecosystem* and as such, contains many habitats, including those at different distances from shore, and those at different depths. For this study you will analyze only those at different depths.

1. Obtain a pond sample taken from the bottom. Swirl the liquid before sampling.
2. Prepare a wet mount.
3. Identify *all* entities belonging to one of the following classes, following the drawings of Fig. 17.1. Record the presence of an entity on Chart 17.1, making a small sketch of the organisms.

   Algae: green, brown, or yellow organisms with cell walls.

   Protozoans: usually colorless, motile, unicells without cell walls.

   Animals: multicellular organisms capable of motion.

   Alien entities: pollen grains, spiral thickenings of xylem cells, seeds, etc.

   Debris: inorganic–sand grains, other crystals; organic–decomposing material from organisms.
4. Obtain a five-ml sample from the same source, remembering to swirl the sample first, and place in a shallow petri dish.
5. Observe under a dissecting microscope.
6. Identify the larger objects according to the list in step 3, and record on Chart 17.1.
7. Pour the material back into the sample jar when you are finished.
8. Are there any objects you have found which do not fit into any of the categories? How would you augment this listing?

CHART 17.1

Entities	Pond Bottom Sketches of entities found
Algae	
Protozoans	
Animals	
Alien entities	
Debris	

IDENTIFICATION OF TROPHIC LEVELS

Each species in the pond is dependent on some other species in some way. The most common dependence is as a food source. So each type of organism can be further categorized according to their source of food, and, in turn, the organisms that use them as a source of food. The collective organisms used as food for another collective group of organisms is called a *trophic level.* The following is a listing of trophic levels in a pond.

Producers: those organisms that make their own food by photosynthesis (algae).
Herbivores: those organisms that consume producers for food (copopod, some protozoans).
Primary carnivores: those organisms that eat herbivores (small crustaceans, insect larvae).
Secondary carnivores: those organisms that eat primary carnivores (beetles, insects, crustaceans, fish).
Scavengers: those organisms that consume either solid or dissolved organic debris (bacteria, worms, some beetles).

1. Prepare another wet mount of the sample from the bottom of the pond.
2. Observe and look for organisms representing each trophic level.
3. Draw several organisms for the trophic levels of producers, herbivores, primary carnivores, and secondary carnivores.

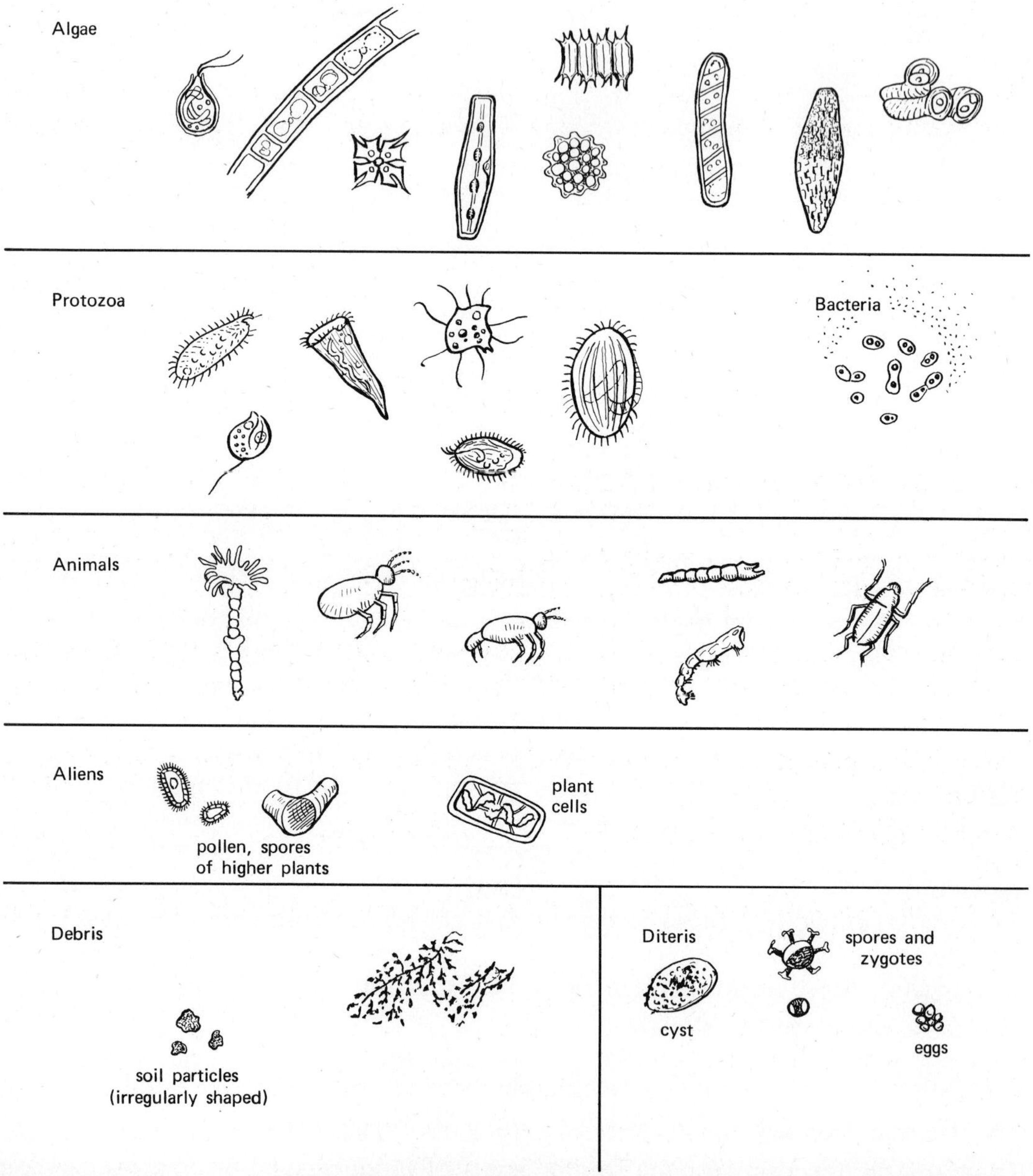
Algae
Protozoa
Bacteria
Animals
Aliens
pollen, spores
of higher plants
plant
cells
Debris
soil particles
(irregularly shaped)
Diteris
cyst
spores and
zygotes
eggs

Figure 17.1

4. Would some of the organisms you identified in the pond sample serve for more than one trophic level (for example, as primary *and* secondary carnivores)? If so, which ones?

STRATIFICATION OF THE POND

A more realistic picture of the habitats of a pond can be obtained by examining the different strata of water depths. This exercise is best done in groups of four.

1. Prepare a wet mount for the surface of the pond, 2-10 cm below the surface, 10-20 cm from the surface, and 2-10 cm below the bottom of the pond. Each student will study one of the slides.
2. Observe and indicate the presence of each type of entity by drawings and indentifications.
3. After all observations have been made, take a 5 ml sample and place it in a petri dish.
4. Examine under a dissecting microscope, and continue drawings and identifications. Repeat until all samples are analyzed.

CHART 17.2 Pond Surface

Entities	Sketches of entities found
Algae	
Protozoans	
Animals	
Alien entities	
Debris	

CHART 17.3 2-10 cm from Pond Surface

Entities	Sketches of entities found
Algae	
Protozoans	
Animals	
Alien entities	
Debris	

CHART 17.4 10-20 cm from Pond Surface

Entities	Sketches of entities found
Algae	
Protozoans	
Animals	
Alien entities	
Debris	

CHART 17.5 2-10 cm below Pond Bottom

Entities	Sketches of entities found
Algae	
Protozoans	
Animals	
Alien entities	
Debris	

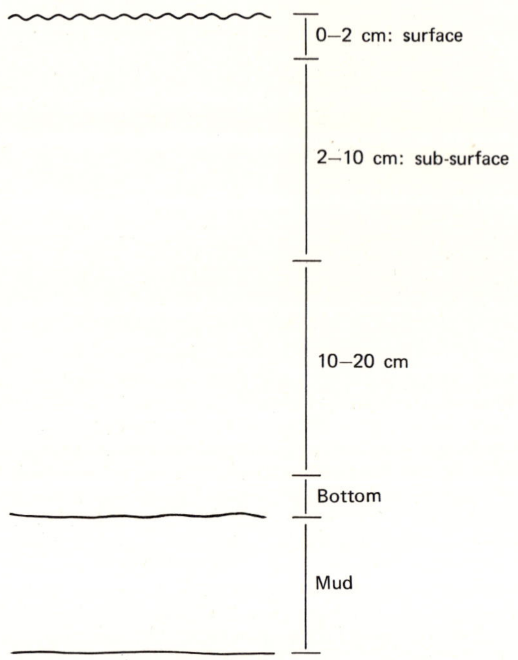

Figure 17.2

CONSTRUCTION OF A BIO-PYRAMID

The final step in your analysis of this ecosystem with its community of organisms is to gather some quantitative data on the composition of the pond. Since organisms can be classified according to the trophic level they belong to, and this classification indicates some fundamental relationship, quantitative knowledge of trophic levels permit a bio-pyramid to be constructed. This exercise is best done in groups of four.

1. Prepare a wet mount by taking a drop from a sample of the substratum level (2-10 cm below the pond bottom). One student will count the number of producers and herbivores on the slide, and record it on Chart 17.6.
2. Take a 5-ml sample from the substratum, and examine it under a dissecting scope. One student will count the number of herbivores present in the sample.
3. Take a 5-ml sample from the substratum, and examine it under the dissecting scope. One student will count the number of primary and secondary carnivores present in the sample.

4. Take a 5-ml sample from the substratum, and examine it under the dissecting scope. One student will count the number of scavengers in the sample.
5. Record all data on Chart 17.6.

CHART 17.6 Trophic Levels of Substratum

Trophic Level	Number of organisms
Producers[a]	
Herbivores[b]	
Herbivores[c]	
Primary carnivores	
Secondary carnivores	
Scavengers	

[a]One drop was used for this calculation, whereas 5 ml was used for the other calculations. To reach a comparable volume, the number of organisms found must be multiplied by 140.

[b]The herbivores found in one drop must be adjusted to a volume of 5 ml by multiplying by 140.

[c]Herbivores from 5-ml sample.

6. Repeat the procedure for the other strata of the pond.
7. Record results on Charts 17.7, 17.8, 17.9, and 17.10.

CHART 17.7 Trophic Levels of Pond Bottom

Trophic level	Number of organisms
Producers[a]	
Herbivores[b]	
Herbivores[c]	
Primary carnivores	
Secondary carnivores	
Scavengers	

See foot notes for Chart 17.6.

CHART 17.8 Trophic Levels of 10-20 cm Below Pond Surface

Trophic level	Number of organisms
Producers[a]	
Herbivores[b]	
Herbivores[c]	
Primary carnivores	
Secondary carnivores	
Scavengers	

See foot notes for Chart 17.6.

CHART 17.9 Trophic Levels of 2-10 cm Below Pond Surface

Trophic level	Number of organisms
Producers[a]	
Herbivores[b]	
Herbivores[c]	
Primary carnivores	
Secondary carnivores	
Scavengers	

See foot notes for Chart 17.6.

CHART 17.10 Trophic Levels at Pond Surface

Trophic level	Number of organisms
Producers[a]	
Herbivores[b]	
Herbivores[c]	
Primary carnivores	
Secondary carnivores	
Scavengers	

See foot notes for Chart 17.6.

8. Construct a pyramid for each strata in the following way: Let 1 mm equal one organism. Focusing attention on the producer trophic level, draw a line in which the length corresponds to the number of organisms. For example, if there were 200 organisms, then the line would be 200 mm long. Label the line "producers." Above that line draw another to represent the number of herbivores, and label accordingly. Above the herbivore line draw one to correspond to the number of primary carnivores, and so on.

9. Where would you put the scavengers with respect to the pyramid?

Part VI

PHYLOGENY

HISTORY OF EVOLUTION

Adaptation and speciation are the steps through which evolution is expressed. But single steps do not indicate the overall direction in which change occurred. An examination of extinct and extant organisms on the other hand, provides a broad historical account of evolution; an account of the directions, in some cases, the specific changes that occurred. These histories of evolution provide the basis for taxonomic classification of organisms.

In a discussion of the origin of life, one is apt to overgeneralize and create a unidirectional picture as a progression toward more complex molecular aggregates that finally have the collective qualities of life. The discussion of the evolutionary events after the first living form arose, however, must be governed by the knowledge of various extant and extinct organisms. Evolution did not produce a single lineage of forms, each of which replaced some prior one, but rather a series of *adaptive radiations* at successively more complex levels of biological organization. A major contribution of taxonomy has been the identification of significant adaptive changes and the establishment of criteria by which different levels of evolutionary progression can be judged.

Before embarking on the exercise for history of evolution, it is necessary to establish some ground rules. The first pertains to the separation of fact and speculation in regard to the order in which various organisms appeared on the Earth. That organism of different levels of biological organization exist is fact. The evidence that a progression of organismal types of increasing complexity sequentially inhabited the Earth is so overwhelming that it too can be regarded as fact. In some cases there is sufficient evidence that a given group gave rise to another group to regard as a particular evolutionary sequence as fact. In most cases, however, the specific origins of groups are obscure, and their taxonomic classification is made on incomplete evidence. Here evolutionists can only say that the most probable evolutionary origin of the group was . . . etc. The combination of facts, semi-facts, and educated guesses gives a more meaningful picture of biological diversity and provides guidelines for further study. Yet it is important to remain aware of the combination of sources, and to recognize the factual and speculative statements in a discussion of evolutionary origins.

The second ground rule pertains to the speculations with respect to adaptive values of morphological and physiological traits. We know from the principles of evolution that the form that an organism takes is functionally complementary to its ecological niche. Neutral and negative traits are subject to selection, so that virtually every physiological and morphological trait must have an adaptive value. But assigning a specific adaptive value to each and every trait is impossible. Often, the knowledge about an organism is too meager to make definite statements about their traits. Also, adaptations could have occurred in response to an environmental factor that has long since passed from the scene, and these adaptations have been retained because of their physiological complementarity to other traits directly involved in organismal survival. Hence it would be incorrect to directly relate the trait itself to some positive feature which contributes to survival. Nevertheless, it is man's nature to give meaning to life by speculating on the advantages of organismal form and function.

18

Monera and Viruses

MATERIALS

CLASS STOCK

E. coli 4 × 10^8 cells/ml
"Spectronic 20"
water baths: 37°C, boiling, 45°C
chemicals
- crystal violet
- gram iodine
- 95% ethyl alcohol
- aqueous safranin solution
- 1.0% versene and 2.0% lysozyme solution
- xylol
- giemsa stain

immersion oil
B. subtilis culture, fresh and old
Staphlococcus aureus
plates of *E. coli*
plates of *B. megatherium*
filamentous blue-green alga slides or culture
plating *E. coli* culture (10^9 ml)
T-2 phage
37° incubator
Petroff-Hauser counter

STUDENT STATION

flask with 100 ml nutrient broth
flask with 80 ml nutrient broth
sterile pipettes
Bunsen burner
colorimeter tubes
slides
microscope
microscope lamp
lens paper
test tubes
paper towels or cheesecloth
coverslip
ice bucket with ice
14 test tubes with 1.0% agar
18 dilution tubes 1/100
14 dilution tubes 1/10
sterile 0.1 ml pipettes
sterile 1.0 ml pipettes
14 petri plates with 1.5% agar
wax pencil

INTRODUCTION

Cytologists recognize two basic patterns in the internal fine structure of cells. One is the *procaryotic* pattern in which enzymes and heredity material are either bound to membranes or are free in the cytoplasmic soup. The other is the *eucaryotic* pattern in which various cellular functions are localized in specific cell organelles. The eucaryotic cell types were discussed in Chapters 2 and 8.

The first cells to inhabit the Earth apparently had a procaryotic organization and were probably similar to some of the bacteria present on the Earth today. Procaryotic cells do not have a true nucleus, and contain only one or a few centrally located chromosomes.* These cells do not contain mitochondria, but have enzymes of similar function bound to the cell membrane. There are no chloroplasts in photosynthetic forms, only vesicles containing pigments or extensive regions of chlorophyllous lamellae (layers). They exhibit no mitotic figures or spindle formation during cell division. There is no endoplasmic reticulum, golgi body, or centriole.

In the course of their adaptive radiation, procaryotic cells most likely exploited many different environmental conditions. Organisms resulted from the adaptive radiation are grouped together as the *Moneran Kingdom*. Undoubtedly many organismal types that arose have, for a variety of reasons, become extinct. It is nevertheless possible to gain some understanding of the physiological and structural patterns that were explored in the procaryotic radiation from those froms present today.

Viruses have been placed in a separate category which cannot truly be called a kingdom. Viruses are more chemical entities, possessing a nucleic acid core and a protein coat. They are wholly parasitic, having no metabolism of their own, but depending on the host cell for biosynthesis of viral material.

MONERAN ADAPTATIONS

As with any organism, all features of the members of the Moneran Kingdom may be considered adaptations. Therefore, these features must have an adaptive value. In many cases, the adaptative values may be difficult to ascertain, because the changes may represent an adaptation to a past environment and has been retained because of the physiological balance necessary in the cell.

Mass Effect

There is no true multicellularity among procaryotic forms. Despite this, there are some of the characteristics of multicellularity in the Monera. Certain bacteria do not separate after cytokinesis and groups of cells are formed. Others share monomers, producing a shortened

*Bacterial geneticists standardly use the term chromosome to refer to the heredity material of bacteria that have been genetically studied. While other bacteria do not exhibit similar strands, bacteriologists have been reluctant to identify these as chromosomes.

lag period of growth by what is known as the mass effect.

1. Obtain two culture flasks with medium from the instructor.
2. Mark the flask containing 100-ml nutrient broth *A*, and the flask containing 80-ml nutrient broth *B*.
3. *E. coli* culture contains approximately 4×10^8 cells (four hundred million bacteria/ml). It has been refrigerated until needed.
4. Carry both flasks of medium to where the *E. coli* is.
5. Remove the cotton plug from the culture, by grasping and pulling with the small finger of your right hand (if you are right-handed; the reverse situation if you are left-handed). Pass the mouth of the culture flask over the flame. With your thumb and forefinger remove a 2-ml pipette from the pipette tube. Insert into the culture tube and draw up 2 ml of suspension. After removing the pipette from the culture flask, pass the mouth of the flask over the flame, and reinsert the cotton plug.

   Remove the cotton plug from flask *A*, and pass the mouth of flask *A* over the flame. Dispense the bacteria into the flask, pass the mouth of the flask over the flame again, and reinsert the cotton plug. Place the pipette into the box provided for used pipettes.
6. Repeat the procedure for flask *B* using a 10-ml pipette twice, so that 20 ml of bacterial culture are placed in the flask.
7. There are approximately 4×10^8 bacteria/ml in the culture flask. By placing 2 ml into 100 ml, you have made approximately a 1/50 dilution. How many bacteria are there at present in flask *A*? (Hint: $50 = 5 \times 10^1$, $4 \times 10^8 = 40 \times 10^7$).

8. By placing 20 ml of the bacterial culture into a flask containing 80 ml of nutrient broth, you have made approximately a 1/5 dilution. How many bacteria are there present in flask *B*? (Hint: $5 = 5 \times 10^0$).

9. Using the sterile procedure outlined in **5**, remove 5 ml from flask *A*, and transfer into a colorimeter tube.
10. Use the colorimeter tube with sterile nutrient broth as a standard, and taking a reading of the medium in flask *A* on the "Spectronic 20."
11. Enter your data on the graph in Fig. 18.1 as the zero-time reading.

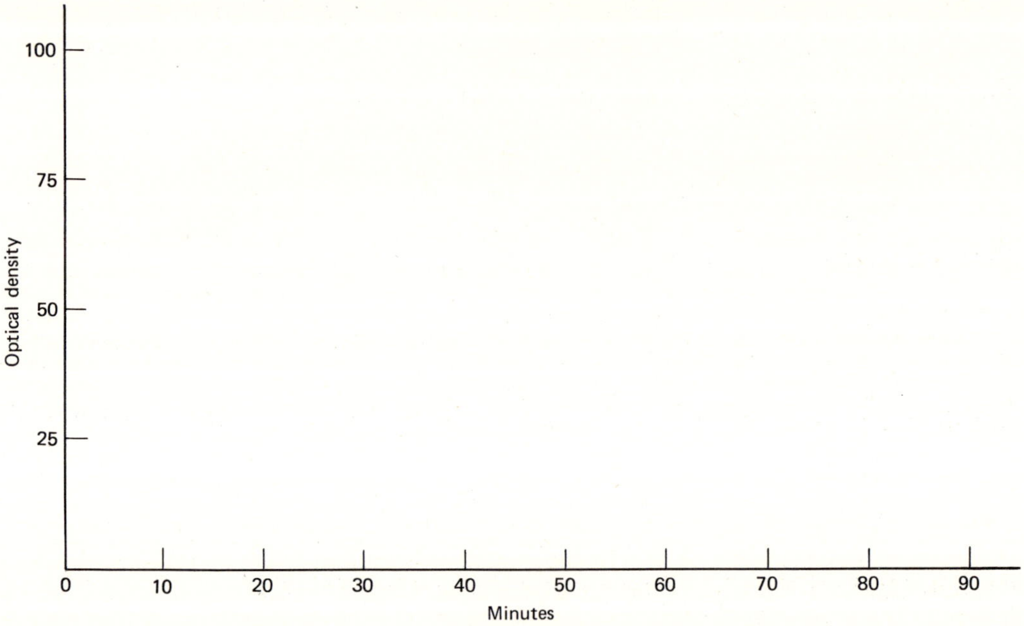

Figure 18.1

12. Using the sterile procedure outlined in **5**, remove 5 ml from flask *B*, and transfer to a *clean* colorimeter tube.
13. Read the absorption of light of the material on flask *B*, using the nutrient broth as a standard.
14. Enter your data on the graph in Fig. 18.1 as the zero-time reading for flask *B*.
15. Discard the material in the colorimeter tubes, and wash and dry thoroughly.
16. Place both flasks into a 37° water bath, preferably on a shaker. If no shaker is available, swirl the culture every few minutes.
17. At ten-minute intervals, take 5 ml from each culture, transfer each to a clean colorimeter tube, and again take a reading with the nutrient broth as a standard.
18. Enter the data as point on the graph in Fig. 18.1.
19. When you obtain readings that are two to three times higher than the initial readings, remove the flask from the water bath. Keep it on your desk; you will need the bacteria later for another exercise.
20. Continue taking readings on the other flask until you obtain readings that are two to three times higher than the initial readings. Remove the flask from the water bath and discard its contents.

21. Connect the points representing the readings from flask *A*. Do the same for the readings from flask *B*.
22. Draw a vertical line from the point where the slope starts to increase to the *x*-axis. How many minutes does this represent in each case?

23. What part of the growth curve have you been observing?

24. Why is turbidity used as an indicator for growth?

25. Did the flasks with different innoculations have different durations of lag phase?

26. Explain mass effect on the basis of sharing a monomer pool.

Gram Strain and Bacterial Structure

Bacteria are very difficult to see under the microscope unless they are stained. There are several simple stains designed for simple observation of bacteria. This exercise, however, will use a more complicated staining procedure to illustrate both the structure of bacterial cells, and the diagnostic gram stain.

1. Place a small drop of the *E. coli* from the flask on your desk on one end of a slide, using the sterile procedure when you remove the cotton plug from the flask.
2. With another slide, smear the drop across the slide.
3. Permit the slide to dry completely and pass it through a flame to heat fix the material. Do not hold the slide over the flame too long.
4. Flood the slide with crystal violet and leave for one minute.

5. Rince the slide with water.
6. Flood with gram iodine and leave for one minute.
7. Holding the slide over the sink, drip 95% ethyl alcohol on the slide for 10 sec.
8. Rinse the slide with water.
9. Flood the slide with aqueous safranin. Leave for one minute.
10. Rinse with water and permit the slide to dry.
11. Examine your slide under high power. If you feel you cannot see the bacteria clearly enough, place a drop of immersion oil on the slide, and use the oil immersion lens on the microscope (if available).
12. Gram positive bacteria retain the original purple stain. Gram negative bacteria lose the purple stain during the alcohol wash, and take on the red counterstain. Is *E. coli* gram positive or negative? Enter your information on Chart 18.1.

CHART 18.1

Bacteria	Gram Positive	Gram Negative
E. coli		
B. subtilis in culture		
B. subtilis in versene and lysozyme		
Staphlococcus aureus		

13. Obtain about 1/3 test tube full of *Bacillus subtilis* culture. Use the sterile technique when removing the bacteria from the culture flask, but make no attempt to keep your sample sterile.
14. Make a gram stain of *B. subtilis,* following the procedures outlined in # 1 through 11 above.
15. Is *B. subtilis* gram positive or gram negative? Enter your data on Chart 18.1.
16. Mark a test tube at the 4-mm level from the bottom.
17. Place the versene and lysozyme solution up to this mark.
18. Pipette 1 ml of the *B. subtilis* sample into the test tube with versene and lysozyme. Draw the fluid in the test tube up and down the pipette for 1 min.
19. Prepare a gram stain with *B. subtilis* in the versene and lysozyme solution, following the procedures outlined in 1 through 11 above.
20. Are the shapes of these bacteria different from those in the culture medium?

21. What seems to be missing in these *B. subtilis* cells?

22. Are these *B. subtilis* gram positive or gram negative? Enter your answer on Chart 18.1.

23. When *B. subtilis* is treated with versene and lysozyme, the cell wall disintegrates, leaving only the membrane bound cell. With this information and your results, explain why it is believed that the cell wall composition determines whether a species is gram positive of gram negative.

24. Obtain a drop from a *Staphlococcus aureus* culture using the sterile technique. Prepare a gram stain using procedures outlined above.

25. *Staphlo* in Greek means grapes. Why were these bacteria called Staphlococci?

26. Are Staphlococci gram positive or gram negative?

Physicological Variation

There is a wide range of physiological variation in the Moneran kingdom, ranging from motility to varied nutritional requirements. This exercise deals only with the capacity of some bacteria to form spores.

1. Obtain a loopful of *subtilis* culture and add to a clean slide, spreading a small circle area with the loop.
2. Permit the slide to dry.
3. Add a drop of Giemsa stain and after setting for a few seconds, wash the slide by dipping it into a finger bowl of water.
4. Wipe the water from the slide with a piece of paper toweling or cheesecloth, being careful not to wipe over the area coated with bacteria.
5. Allow the slide to dry.
6. Add a drop of xylol and place a cover slip over the area as in a wet mount. Examine under high power.

7. Draw the bacteria and colorless spores in the space provided.

8. What evidence is there that the bacterial spore is resistant to environmental shocks?

9. What percentage of the cells have spores? Why is it that they all do not have spores?

10. Why would these cells form spores when *E. coli* does not?

Colonial Growth

Some bacteria form colonies on semisolid surfaces such as agar, while others do not.

1. Examine the prepared petri plates of *E. coli* and *Bacillus megatherium.*
2. Which bacterium forms discrete colonies?
3. Each colony is indicative of a single cell originally in that place on the agar surface. Which bacterium would you use of you wanted to see the effect of radiation on cells?
4. Which bacterium would you use if you wanted to study motility on a semisolid surface?
5. Give two other experiments for which you might use either of these bacteria as research tools.

THE BLUE-GREEN ALGAE

The ancestral stock of higher plants and animals is believed to have been either the blue-green algae or some common ancestor of the blue-green algae. The procaryotic cells in the blue-green algae are generally larger than those found in bacteria and show some indication of compartmentalization that is found in later eucaryotic cells.

1. Obtain a prepared slide of a filamentous blue-green algae or prepare a water mount of fresh material.

2. Draw the alga in the space provided, labeling the following parts:

   chromatin material – in the central nongreen area are a number of small graules that give a positive test for DNA, the chromosomal material. Notice that there is no nucleolus or nuclear membrane.
   chlorophyll material – the outer regions of the cytoplasm are colored with blue-green pigments. Notice that there are no distinct bodies of chlorophyll (chloroplasts) present.

3. Would a physical breakage of the filament, such as in rough waters, be advantageous or disadvantageous to the survival of a filamentous species? Why?

VIRUSES

A growth curve for viruses is very different from that of other organisms. Viruses grow within cells, and it is not until there are from 10 to over 100 viruses that the host cell undergoes lysis. During the multiplication of viruses, then, the growth is masked by the apparent normal outward appearance of the cell.

Considerable work has been done on bacteriophage, viruses that attack bacteria.

1. Before starting your experiment, read the exercise throughly. You will have to work quickly—there will be no time for reading, checking, and preparing appropriate tubes later.
   (a) Fill a bucket with crushed ice.
   (b) Melt the 1% agar in nutrient broth in fourteen test tubes in a boiling water bath. Then place them into a 45°C water bath. Let the tubes adjust to the temperature of the bath for at least one-half hour.
   (c) Place two drops of *plating bacteria* (10^9/ml) into each tube of 1% agar, using the sterile technique and leave the tubes in the 45°C bath.
   (d) Place 18-1/100 dilution tubes and 14-1/10 dilution tubes in your rack. Arrange them in rows according to the following program and mark as indicated:

Front of Rack

			15 (1/10)	20 (1/10)	25 (1/10)	30 (1/10)
0 (1/10)	5 (1/10)	10 (1/10)	15 (1/10)	20 (1/10)	25 (1/10)	30 (1/10)
0 (1/10)	5 (1/10)	10 (1/10)	15 1/100	20 1/100	25 1/100	30 1/100
0 1/100	5 1/100	10 1/100	15 1/100	20 1/100	25 1/100	30 1/100
0 1/100	5 1/100	10 1/100	15 1/100	20 1/100	25 1/100	30 1/100

Back of Rack

   (e) Place a case of sterile 0.1 ml pipettes and 1.0 ml pipettes within reach.
   (f) Obtain 14 petri plates containing 1.5% agar. You are now ready to start.
2. Your instructor has been growing cultures of *E. coli* in aerated flasks, in several water baths. Choose one flask and put your name or initials on it.
3. Remove the cotton plug and aerating tube slightly, and draw some of the culture into a pipette.
4. Count the number of bacteria per milliliter either with a counting chamber or a Petroff-Hauser counter. Enter your results on Chart 18.2.

CHART 18.2

Petroff-Hauser Count	Dilution Factor	No. of Bacteria in Original Culture
	2×10^7	

*Circled dilutions are to be plated.

5. You should use three times more viruses than bacteria. Multiply the number of bacteria per milliliter and calculate the dilution you must make of the viral culture in order to have 3:1 ratio of viruses. Note that in placing the viral dilution into the culture, there will be at least a 1/100 dilution. For example: viral culture = 6×10^8 viruses. If 1 ml of 6×10^{10} viruses are placed in the flask, the final number of viruses per milliliter of culture would be 6×10^8; or if a 0.1 ml of 6×10^{11} were put in the flask, the final concentration would be 6×10^8 viruses per milliliter.

   Calculate how much you must dilute the virus culture, and do so. Add the final dilution to the *E. coli* culture.
6. As soon as you have added the phage to the culture, obtain a 0.1-ml pipette, and withdraw 0.1 ml aliquot from the culture. This is your zero-time sample. Place this 0.1 ml into the first tube on the left in your rack of dilution tubes. Swirl the tube after pipetting the 0.1 ml from the culture into it. Using a fresh sterile pipette, withdraw 0.1 ml from the tube, and place it in the one behind it in the rack. Swirl again. This time, take a 1.0 ml pipette, withdraw solution from the tube and place it in the next tube on the rack. From this tube take 1.0 ml and place it in the next tube. These last two are the source of material you will plate.
7. Mark these tubes 0-1 and 0-2, and place them in the crushed ice. The cold condition will stop all bacterial action, and hence, all viral activity.
8. Repeat the procedures outlined in **6** to **7** at 5-min, 10-min, 15-min, 20-min, 25-min, and 30-min intervals. Label the last two dilution tubes of each set appropriately
9. By the end of the experiment, you should have 14 dilution tubes in the crushed ice.
10. With a clean, sterile 0.1-ml pipette, remove 0.1 ml from tube 0-1, and place it in the tube with melted agar and plating bacteria.
11. Work very quickly. Swirl the melted agar, and pour it into a petri plate on top of the 1.5% agar. Move the plate around quickly to distribute the top agar layer evenly over the surface. Put the top back on the plate.
12. Wait until the agar has set. Turn the plate upside down and mark the bottom of the plate with the same label as the tube the material was taken from (0-1).
13. Repeat the procedure outlined in **11** and **12** for each dilution tube.
14. Place all plates in a 37°C incubator. Your instructor will remove them tomorrow and place them in the refrigerator until the next laboratory period.

The plating bacteria that you put into the melted agar will grow evenly on the plate, forming a thin film or meadow of bacteria. The viruses that were still in the bacterial cells when they were plated, will lyse the cell when it is on a semisolid medium. The only bacteria available for phage attack on the plate are the plating bacteria. The phage will atack these and they will lyse also. If there are about 0 to 5×10^2 phage or infected bacteria on the plate, countable uniform clear areas representing lysed plating bacteria will be seen. These are the plaques. Hence, a plaque represents the place where either there initially was one phage or where there initially was one infected cell containing many phages in the process of multiplication.

15. During the next laboratory period, retrieve your plates. Looking at them from the bottom, count the plaques. Mark each plaque with a wax pencil on the bottom of the petri plate to indicate you have counted it. If there are too many to count, make an eighth pie section of the plate. Count the number within the pie section and multiply by eight.
16. Enter your data on Chart 18.3.

CHART 18.3

Time	No. of Plaques	Dilution Factor	No. of Viruses or Infected Bacteria
0-1		10^6[a]	
0-2		10^7[b]	
5-1			
5-2			
10-1			
10-2			
15-1			
15-2			
20-1			
20-2			
25-1			
25-2			
30-1			
30-2			

[a]dilutions for 0-1 sample were as follows: 1/100, 1/100, 1/10, and since 0.1 ml was used in plating, 1/10 again. Converting these values to powers of 10: $1/10^2 \times 1/10^2 \times 1/10^1 \times 1/10^1 = 1/10^6$ (multiply by adding exponents).

Total dilutions = 10^{-6}, hence the number of plaques represent $1/10^6$ of the viruses or infected bacteria on the plate, and must be multiplied by 10^6 to give the number in the original sample.

[b]dilutions for 0-2 sample were as follows: 1/100, 1/100, 1/10, 1/10, and 1/10 for plating. $1/10^2 \times 1/10^2 \times 1/10^1 \times 1/10^1 = 1/10^7$

Number of plaques counted must be multiplied by 10^7 to give the number of viruses or infected bacteria in the original sample.

17. Multiply by the dilution factors to obtain the number of viruses or infected bacteria per milliliter.
18. Plot the values obtained for the number of viruses or infected bacteria at the different times on the graph of Fig. 18.2.
19. Why is it said that viruses exhibit a one-step growth curve?

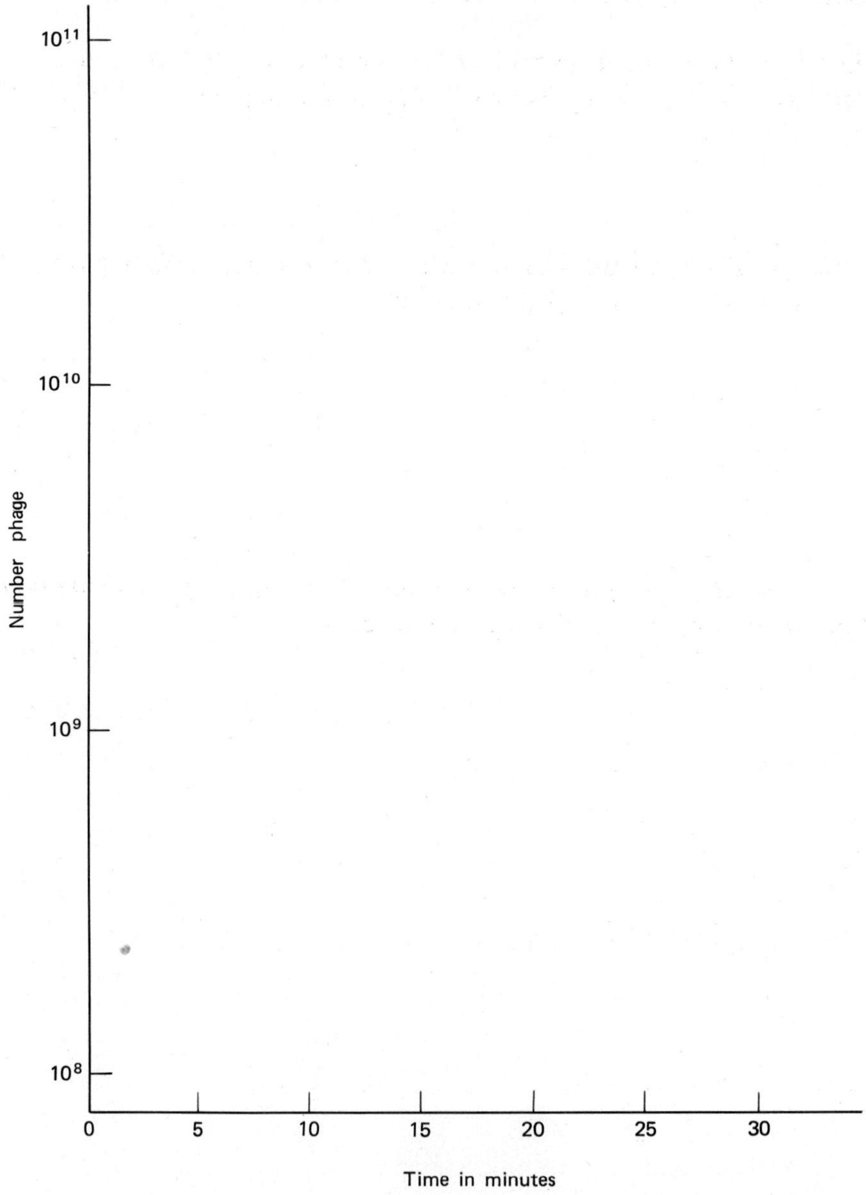

Figure 18.2

The sequence of events in bacteriophage infection is as follows: The bacteriophase attach to the villi on the bacterial cell wall and inject their DNA through the pore. Once inside the bacterial cell, the DNA takes over the metabolism of the bacterial cell, primarily the biosynthetic aspects of metabolism. Instead of producing bacterial RNA, the nucleotide pool in the bacteria is used for transcribing viral RNA. Ultimately, viral DNA is replicated and viral protein is formed. The DNA helices in some move into the newly formed protein shells, and when lysis of the cell occurs, viruses are released.

20. Add a line to your graph to indicate DNA increase during infection.
21. What criterion did you use to draw the line on the graph?

22. Considering your line on the graph, is viral growth much different from bacterial or for that matter, any organismal growth? Why?

23. Why was it said at the beginning of this section that the means of detection for viral growth produces a different type of growth curve?

19

Protista

MATERIALS

CLASS STOCK

culture of mixed desmids
culture of *Spirogyra*
culture of fungi
prepared slide of *Trypanosoma*
culture of *Amoeba*
culture of *Tetrahymena*
culture of *Paramecium*
culture of *Dinoflagellates*
10 petri plates containing double innoculation of either or *Phycomyces*
culture of *Chamydomonas moewuesii*
culture or slide of *Oedogonium*
Neurospora plates
  (1) Single mating type, 5 days old
  (2) Both mating types, 10 days old
  (3) Both mating types, 14 days old

STUDENT STATION

slides
cover slips
microscope
microscope lamp
lens paper
beaker
petri plates with nutrient agar

INTRODUCTION

At least one type of unicellular organism among the products of the procaryotic radiation—had acquired the genetic capacity for limited compartmentalization of cellular functions (probably an ancestral blue-green alga). Over the several million years that followed, organisms with such genetic capacities were selected, until a true eucaryotic cell type evolved.

Because it is extremely difficult to ascertain evolutionary relationships in unicellular organisms, it is not known in which lineage the first eucaryotic cell arose. Nor for that matter, is it known if there was only a single evolutionary conversion from procaryotic to eucaryotic cell types. One guess put the advent of eucaryotic cells in either the *Rhodophyta,* red algal lineage, or the *Chrysophyta,* the brown and golden algal lineage. There is sufficient evidence to show that the protistans, and ultimately animals and plants, trace their lineage back to one of these two groups.

The Protistan kingdom is comprised of the products of the eucaryotic radiation; the organisms that are classified as protistans resulted from adaptive changes in the compartmentalized eucaryotic cell.

EUCARYOTIC VARIATIONS IN PROTISTANS

Procaryotic cells appear to have a limited potential for differentiation. Bacteria are either spherical, rod shaped, or corkscrew shaped. Rickettsia are small spherical cells. Only the blue-green algae exhibit variance in cell shape, but the range of variation is nothing compared to that in protistan forms.

It would be impossible to view all the existing variations of the protistan kingdom. A few types will be examined to give some idea of the genetic potential for differentiation through elaboration of organelles and macromolecular complexes in the protists.

Desmids

1. Make a wet mount of a mixed culture of desmids.
2. Observe several different types of desmids. Notice that each cell is composed of two halves or semicells connected by an isthmus. The spherical light region in the isthmus is the nucleus.
3. Draw several desmid cells of different taxonomic species in the space provided.

4. Explain beneath each drawing how differentiation through different rates and directions of growth could result in the evaluation of the different types.
5. Evidence indicates that the nucleus controls the deposition of cell wall material, and hence the shape of the cell. If the desmids are similar to other plant cells, the chloroplasts control their own shape and number. Assuming that desmid evolution began with two spherical semicells, describe the genetic events that would lead to a *Closterium,* the half-moon shaped desmid.

Filamentous algae

1. Obtain a prepared slide or make a wet mount of *Spirogyra.*
2. Spirogyra is a filamentous green alga. Draw a section of the filament in the space provided, labeling the following:

   vegetative cells — contain helical chloroplasts.
   zygotes — darkly pigmented football-shaped protoplasts in the old vegetative cells. These cells achieve dormancy by producing a resistant cell wall.

3. Are the chloroplasts of adjacent cells connected to one another?

4. In what way would *Spirogyra* be considered a unicellular organism, and in what way would it be considered a type of primitive multicellular organism?

5. What adaptation permits *Spirogyra* to survive in adverse environments?

6. Why does this adaptation include the formation of spores somewhere along the length of the filament even in acceptable environments?

7. How is the chloroplast adapted to capture light for photosynthesis?

8. Is this chloroplastic arrangement the most efficient of all the algae? Why or why not?

Fungal Filaments

1. Place a drop of water on a slide.
2. Flame a loop, permit it to cool, and take a loopful of fungal hyphae from the petri plate provided by your instructor.
3. Place the loopful of fungal hyphae in the water drop on the slide and mix.
4. Prepare a water mount.
5. Observe the fungal hyphae under low and high power.
6. Draw a filament in the space provided.

7. Compare and contrast the fungal filament with the algal filament.

Mastigophoran Protozoans

The class Mastigophora includes what are thought to be the most primitive of the protozoans. Many of the members of this class are parasitic, and all are identified by their dominant feature—a single, long whiplash cilium.

1. Obtain a prepared slide of *Trypanosoma. Trypanosoma* is the protozoan responsible for a type of encephalitis known as sleeping sickness.

2. Draw the *Trypanosoma* in the space provided, labeling the following parts:

   undulating membrane—extends from the body and runs almost the full length of the cell.
   cilium—found at the free edge of the undulating membrane, it continues as a free cilium posteriorly.
   kinetoplast—base of the flagellum, found at the anterior portion of the cell.
   nucleus—dark-straining spherical body centrally located.

3. From viewing the morphological features or adaptations of the trypanosome, would you deduce that it is motile? Why?

4. From examining the trypanosome, what route in the body would this organism take to reach brain tissue?

5. How would you design an experiment to check your answer to question 4?

The Sarcodinian Protozoans

The sarcodinian protozoans are believed to have been derived from the mastogophoran protozoans through adaptations that involved loss of the flagellum and modification of the already elastic cell body to form projections or pseudopodium (false feet).

1. Prepare a wet mount of *Amoeba* and *Tetrahymena* by placing drops from each culture on top of each other. The Amoebae have been starved for 24 hrs and use *Tetrahymena* as a food source.
2. Observe the movement of the *Amoeba* and its engulfing of the food material.
3. Draw the *Amoeba* in the space provided, and label the following parts:

   plasma membrane—membrane surrounding the cell
   pseudopodium—false foot formed by the extension of cellular material
   food vacuole—round bubble like vacuole that contains solid material.
   nucleus—circular, densely granular, usually centrally located in the cell.

4. Why is the *Amoeba* considered to be a terminal result of evolution instead of an intermediate form?

5. There are forms in the class Sarcodina that are parasitic. Which organelles would you suppose would be absent or nonfunctional in these forms? (Check your answer by looking up *Endoamoeba histolytica* in text or reference material.)

Ciliated Protozoans

Another group of protozoans which may have arisen from the mastogophoran group are those grouped in the class Ciliata.

1. Prepare a wet mount of a culture of *Paramecium,* using some cotton fibers over the drop to trap the organisms so that they may be observed.
2. Draw the *Paramecium* in the space provided, labeling the following parts:

   pellicle—membrane surround the cell.
   oral groove—extension of the pellicle into the cell on one side. *Paramecium* resembles a slipper in its shape. The oral groove is found in the instep of the slipper.
   macronucleus—large circular body found directly adjacent to the oral groove. This nucleus controls the metabolism of the cell.
   micronucleus—smaller densely granulated body associated with the macronucleus. This nucleus controls the heredity of the cell.
   contractile vacuoles—two clear bubble like structures, one at the anterior end of the cell, the other at the posterior end of the cell. These appear and disappear alternately.
   food vacuole—usually found near the pellicle. This vacuole contains solid cell debris from the material that is undergoing digestion.
   cilia—small hairlike projections that surround the organism.

3. Why is *Paramecium* considered to be a terminal result of evolution instead of an intermediate form?

A Dinoflagellate

The dinoflagellates are organisms classified as Chrysophyceae. It is believed that one of the protozoan types and the dinoflagellates have a common ancestor.

1. If available, prepare a wet mount of a culture of dinoflagellates. If not available, then obtain a prepared slide.
2. If live material is available, observe the movement of the dinoflagellates. Of which protozoan type is this movement characteristic?

3. Compare the internal structure of a dinoflagellate with your drawings of protozoan types. Which protozoan could have been derived from the same ancestor as the dinoflagellate?

4. What alterations would be necessary in the dinoflagellate cell to produce that protozoan type?

General Considerations of Eucaryotic Variation

Many of the variations you have observed are actually examples of convergent evolution—similar solutions to similar selective pressures at different evolutionary levels—in this case, the procaryotic and eucaryotic radiation levels.

PROTISTAN REPRODUCTION

The particular features involved in protistan reproduction are as varied as the protists themselves. Some are asexual and reproduce only by vegetative reproduction. Others are asex-

CHART 19.1

Feature	Procaryote	Eucaryote
Cellular constituents		
Cell wall		
Filamentous Forms		
Flagella		
Chloroplast		
Differentiation potential		
Spore-forming capacity		
Pseudopodium		

ual, but are capable of producing spores, which, on germination, give rise to the vegetative cell or body. Some are sexual and haploid, and produce gametes that fuse to form a diploid zygote, which undergoes meiosis to restore the haploid condition. Others are diploid and undergo meiosis in the production of gametes, the fusion of which restores the diploid condition.

Sexual reproduction does not necessarily imply the presence of a distinct male and female form. Some protists are homothallic, having cells that are indistinguishable from one another and yet, produce gametes that combine to form zygotes in any and all combinations. Other protists are heterothallic and produce gametes that will fuse only with gametes from the complementary cell. Several variations are among the heterothallic types. When the gametes from two complementary cells are indistinguishable from one another cytologically the species is called isogamous. When the gametes produced include a large egg and smaller sperm, the species is considered oogamous. In the course of protistan radiation, many forms of reproduction were found successful—all are based on the processes of mitotic and meiotic division, and some based on the process of fertilization.

The Mating Type Concept

Ten petri plates have been prepared for your examination. Each of these plates contains a double innoculation of either *Rhizopus* or *Phycomyces*. The innoculations were made on opposite sides of the plate and the plates were incubated for one week at room temperature. During this time, the fungus has reproduced asexually. When each has grown sufficiently, the growths will confront one another at the center line across the plate. If the two innoculations are of the same mating type, the growth will exhibit a uniform pattern. If, on the other hand, the two innoculations are of different mating types, sexual reproduction ensues and zygotes are produced along the central line of confrontation. These zygotes are large dark bodies and may be seen quite easily on the agar surface.

1. Examine each plate and determine whether mating has occurred.
2. Chart 19.2 corresponds to the numbers found on the back of the plates. Enter the occurrence of mating as *Z*, the absence of mating as –.
3. Given that strain 1 is arbitrarily mating type +, determine the mating types of the remaining three strains.

CHART 19.2

	1	2	3	4
1				
2				
3				
4				

Mating Types
(+) 1 _ _ _
(–) _ _ _ _

4. From the nature of sexual reproduction, why is there a tendency for a population to be 50 percent male, 50 percent female? (Hint: What does a zygote contain?)

5. Using the sterile procedure, place a loopful of each mating type (from a plate on which mating has occurred) in a drop of water on two slides.
6. Prepare wet mounts of both drops, and observe under low and high power.
7. Is there a morphological difference between the vegetative phases of the mating types?

8. Does the fact that only the filaments that produce gametes express morphological differences confirm your answer?

9. Why are mating type designations used instead of the terms maleness and femaleness?

Isogamous Reproduction

In many primitive plants, the gametes undergoing fusion are morphologically similar and are referred to as isogametes. Reproduction in these forms is called isogamy.

1. On a clean slide, place a drop of cell suspension of *Chlamydomonas moewuesii* of the (+) mating type at one end, a drop of the (–) mating type at the other, and a drop of each in the center.
2. Do not use a cover slip.
3. Notice in which drop the pairing of cells occurs.
4. Draw a pair of cells indicating the isogamous condition.

5. Defend the statement: "A cell of *Chlamydomonas* can be both vegetative and gametic."

6. Is *Chlamydomonas moeweusii* homothallic or heterothallic? Defend your answer.

Oogamous Reproduction

Fertilization involving gametes of different morphologies is termed oogamous reproduction. One gamete is motile and is called a sperm or antherozoid; the other is larger and immobile and is called the egg or ovum.

1. Obtain a prepared slide of *Oedogonium* or make a wet mount from living material.
2. Observe the following cell types, and draw each in the space provided.

   vegetative cells—cylindrical and containing an elaborate chloroplast.
   oogone—a large, spherical, green cell with a fertilization pore through which the sperm enters. The oogone is a one-celled sex cell or gametangium.
   egg—the protoplast of the oogone. The oogone is merely a cell wall.
   zygote—the fertilized egg, which has become brown and enclosed in a thick wall.
   antherids—small cells, usually empty; formerly contained the sperm. This is the male sex organ or gametangium; some antherids may still contain two sperm.

3. Which is the active (motile) gamete? Which is the passive?

4. Which stage would be prevalent during winter? Summer?

LIFE CYCLE OF NEUROSPORA CRASSA

All organisms pass through cyclical patterns. Life is a continuum; each individual is a parent of some and an offspring of others. Each organism passes through the same stages of development as other members of its species. This repeating pattern of events is called the life cycle and serves as a representation or model of an organism useful in comparison with other species, and useful in organizing facts.

1. The instructor has prepared three petri plates for your examination.
   (a) Single mating type (+ or –) 5 days old
   (b) Both mating types (+ and –) 10 days old
   (c) Both mating types (+ and –) 14 days old

2. Using the sterile technique, take a loopful of hyphae from plate (a) and prepare a wet mount.
3. Draw the hyphae in the appropriate place on Fig. 19.1.
4. Using the sterile technique, brush the loop over the surface of the hyphae to collect conidia.
5. Prepare a wet mount of conidia.
6. Draw the conidia in the appropriate place on Fig. 19.1.
7. The vegetative cells of *Neurospora crassa* are haploid. The conidia were formed in the plate with only one mating type. Are the conidia haploid or diploid?

8. On the basis of your answer, indicate ploidy next to your drawings (*N* or 2*N*).
9. Place about 5 ml of water into petri plate (a) and swirl it. The water will dislodge the conidia, but will leave the hyphae in place.
10. Pour the water from the petri plate into a beaker.
11. Take a few drops from the beaker, and place them in a sterile petri plate.
12. Mark the petri plate *C* for conidia, and incubate it for one week at room temperature.
13. Using the sterile technique, prepare a wet mount of the perithecia, dark reproductive bodies in plate (b).
14. The perithecia are products of dikaryotic hyphae, hyphae that contain both (+) and (−) nuclei. Press down on the cover slip to crack the perithecia open so that their contents can be seen.
15. Draw a perithecium on Fig. 19.1.
16. The spore containers in the perithecium are called asci. How many spores does an ascus contain?

17. The ascospores are haploid. Through what type of division process were they formed?

18. How many products are usually associated with this division process?

19. How many divisions occurred after the typical division process?

20. Place plate (c) in a 45°C incubator for 2 hr. Plate (c) contains ripe perithecia that have broken and released ascospores on the top of the plate. These ascospores require heat for germination.
21. After 2 hr, use the sterile technique to transfer a few spores from the top cover of plate (c) to a sterile petri plate.
22. Mark the plate *S* for spores, and incubate for one week at room temperature.
23. After one week examine plates marked *C* and *S*.
24. Is there any difference in the gross morphological features of the products of the conidia and the spore?

25. Make a water mount of each.
26. Is there any difference in the microscopic morphology of the products of the conidia and the spore?

27. Why do your answers indicate that you have examined each phase of the life cycle?
28. On the basis of the information given and your observations, fill in the ploidy level of each structure involved in the life cycle, and draw arrows connecting sequential phases.
29. Label the diagram in Fig. 19.1 according to the place in the life cycle that mitosis and meiosis occur.
30. Why does meiosis instead of fertilization restore the proper ploidy for the vegetative phase of *Neurospora crassa?*

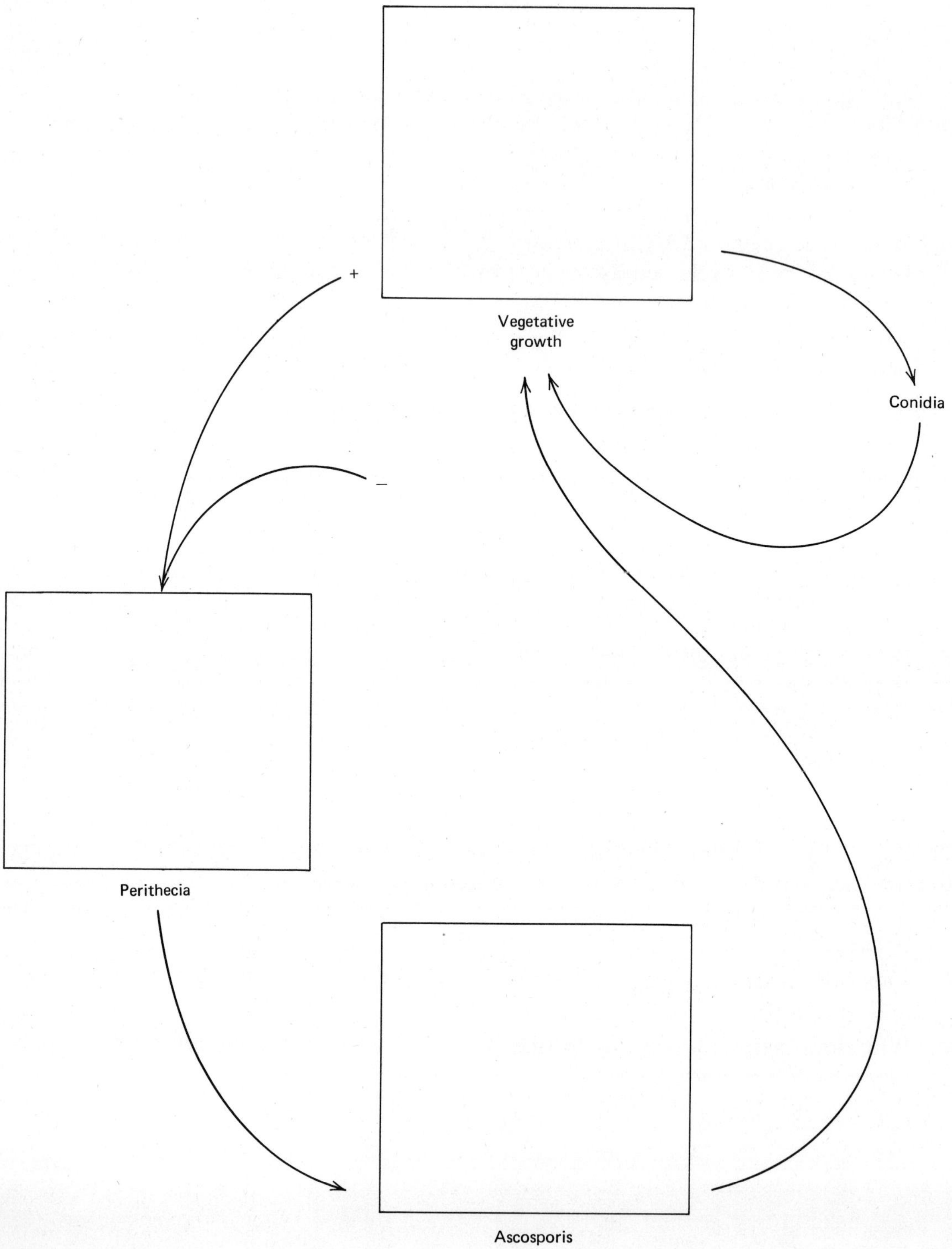

Figure 19.1

20

Land Invasion by Plants

MATERIALS

CLASS STOCK

moss plants *(Polytrichum)* with gametophyte and sporophyte
prepared slides
- longitudinal section of moss gametophore, male and female
- cross section of *Psilotum* stem
- cross section of fern rhizome
- fern prothallia showing archegone
- fern prothallia showing antheridia

fern spore germination of 10 days
fern spore germination of 14 days
fern mature prothallia
fern developing sporophyte
preserved specimen of *Psilotum*
germinating moss spores

STUDENT STATION

microscope
microscope lamp
lens paper
slides
cover slips
glass rod

INTRODUCTION

Life originated and underwent the procaryotic and eucaryotic radiations in the sea. Hence, from the beginning, life has been based on a water-mediated chemistry. The land represents a hostile environment to living systems, and in order to exploit the opportunities of this environment (no predators, no competition, etc.) drastic changes in structural and physiological features were mandatory. The move from an aquatic environment to a terrestrial one required (1) a means for obtaining and retaining water, and (2) the replacement of some water-based phenomena on which the life of the species depend, such as spore and gamete dispersal and other biological innovations that capitalize on atmospheric or other conditions. The appropriate changes in these directions appear to have formed the framework that generated major adaptive radiations; both plants and animals evolved to their greatest complexity when they invaded the land.

Plants have truly conquered the land on two occassions, first by the *Tracheophyta,* simple vascular plants, and second by the *Bryophytha,* or moss and mosslike plants. In both cases the successful adaptation was through multicellularity. These groups represent divisions of the multicellular plant kingdom, the *Metaphyta.*

The evolutionary changes in plants may be divided into two categories: (1) changes related to more elaborate reproductive processes, and (2) changes involved in more efficient means of photosynthesis. It must be stressed that this approach is artificial in that adaptations that affect reproductive systems also affect offspring dispersal, and are therefore related to photosynthetic efficiency. Nevertheless, the approach does serve to initially organize concepts and to relate numerous facts.

All plants have two morphologically distinct phases in their life cycle—a haploid gametophyte, which produces gametes, and a diploid sporophyte, which produces haploid spores through meiosis. Fusion of the gametes (fertilization) in the gametophyte gives rise to a zygote from which the sporophyte develops. Germination of the spores (meiospores) from the sporophyte gives rise to the gametophyte.

Both the gametophytic phase, which is dominant in the bryophytes (such as mosses and liverworts) and the sporophytic phase, which is dominant in the tracheophytes (such as ferns and seed plants) were tested and modified in the course of plant evolution. For various reasons, the adult sporophyte of the tracheophytes became the elaborate form, and is the form usually associated with the flowering plants.

Both bryophytes and tracheophytes were among the earliest plants to become established on the land. By interpreting their features as terrestrial adaptations, the meaning of plant evolution becomes clear.

BRYOPHYTES

1. Obtain a moss plant containing the gametophyte and the sporophyte.
2. Draw and label the following structures in the space provided.
   gametophyte—lower leafy portion containing "leaves" and "stem" and basally located hair-like rhizoids.
   sporophyte—upper portion of the plant containing the long-thin stalk or seta and the enlarged top or capsule, (the sporangium).

3. Observe and note the sizes of the sporophyte and gametophyte, the presence or absence of chlorophyll in each, and the relative dependency of each on the plant. Fill in this information on Chart 20.1.

CHART 20.1

		Moss	Fern
	size (mm)		
Gametophyte	chlorophyll		
	dependency		
	size (mm)		
Sporophyte	chlorophyll		
	dependency		

4. Give two functions of the rhizoids.

5. When you think of moss, do you envision the gametophyte or the sporophyte?

6. What is the selective advantage of the long, thin form of the seta?

7. The fact that the gametophyte did not grow taller would indicate either a lack of genetic opportunity or a disadvantage associated with a taller gametophyte. In what ways would it have been advantageous and disadvantageous for the gametophyte to attain greater height?

8. Observe a mat of moss plants. Notice that some gametophores (leafy stalks) have a cluster of larger terminal "leaves" or *rosettes.*
9. Draw several plants in a habitat sketch in the space provided.
10. If the differences stated above are sexual characteristics, how would you distinguish between male and female plants? (Hint: What type of plant are the sporophytes growing from?)

11. How is the growth habit of moss condusive to water retention?

12. What evidence is there that mosses thrive in shallow soils?

13. Obtain a prepared slide of a longitudinal section of a female moss gametophore.

14. Draw the vase-shaped archegone, labeling the following structures:
neck—long tapering tip of the archegone.
fertilization canal—a passageway running down through the neck from the top to the egg.
venter—the central, enlarged region of cells.
egg—a single large cell embedded in the venter. This is the female gamete.
stalk—the somewhat narrow base of the archegone attached to the moss plant.

15. Obtain a prepared slide of the longitudinal section of a male gametophore.

16. Draw the *anterid* or male sex organ and label the following:
antherozoids—thousands of small cells in patches of different sizes and color; the male gametes.
jacket—epidermal layer of cells surrounding the antherozoids.
stalk—attaches antherid to gametophore.

17. The male gametes of the moss are flagellated. Considering the structure of the archegone and the above information, how do the antherozoids reach the archegone?

18. What is the source of the medium for this movement?

19. Why would taller female gametophytes probably be selected against?

20. How would the advantage that might be gained from taller growth be offset by the disadvantage? Which adaptation was apparently more demanding for moss survival?

21. Where on the moss plant is the zygote formed?

22. Into what does the zygote develop?

23. The sporophyte is diploid and produces haploid spores. Prepare a wet mount from a culture of germinated spores.
24. Notice the colorless rhizoids and photosynthetic cells.
25. Draw the moss protonema in the space provided, labeling the rhizoids and the photosynthetic cells.

26. Is the prothallus haploid or diploid?

27. The prothallus will develop into one of two things. Name the two types of plants.

EARLY TRACHEOPHYTES

The earliest traceophytes, known only from fossil evidence, seem to have been similar to the only extant plant remaining in the Psilopsid group, *Psilotum*. *Psilotum* is characterized by a dominant sporophyte generation. This dominance is due mostly to the fact that the sporophyte, instead of the gametophyte possess chlorophyll. Since the sporophyte does not contribute to the fertilization process, but instead undergoes meiosis to form spores and functions in spore dispersal, it has no restrictions to a prostrate growth pattern.

Increase in height of the plant creates transport problems. Water must move up against gravity, and products of photosynthesis must move further to supply all regions of the plant with organic material. Concommitant with the increase in height, therefore, was the differentiation of special cells, which formed the vascular system of the plant.

1. Obtain a prepared cross-section of *Psilotum* stem.
2. Draw a pie-shaped section of the stem in the space provided, labeling the following tissues:

   xylem—located in the center of the stem, these cells are stained dark red. Notice that these cells give the same appearance as though you are looking through a cluster of red straws.
   phloem—found around the xylem in patches. These cells stain blue-green, and appear to be thinner than the xylem.
   cortex—large empty appearing cells surrounding the phloem.
   epidermis—narrow rectangular cells at the periphery of the stem.
   guard cells—paired cells interrupting the epidermis. These cells control the rate of evaporation of water from the epidermis by changing the space between them.
   stomate—space between the guard cells.

3. The xylem serves the plant in the transport of water; the phloem serves in the transport of photosynthetic products. From their cytological features and their positions in the plant, what other function do you think they serve?

4. Observe a preserved specimen of *Psilotum.*

5. What system usually associated with plants is not present in *Psilotum?*

6. What cells in the plant are responsible for photosynthesis?

The plant you have been observing is the sporophyte. You may be able to discern small sporangia on the tips of some branches.

7. What do the sporangia contain?

8. Are the contents of the sporangia haploid or diploid?

9. What will the contents of the sporangia develop into?

The gametophyte of *Psilotum* is nonphotosynthetic, small, and grows beneath the soil. As it grows, it develops both archegonia (the female organ) and antheridia, (the male organ). Fertilization takes place in the archegonia.

10. What is the ploidy of the zygote?

The zygote develops and germinates *in* the archegone of the gametophyte, and grows out of it.

11. What develops from the zygote?

12. In *Psilotum* is the sporophyte directly dependent on the gametophyte, or is the gametophyte directly dependent on the sporophyte?

13. Is this a similar dependency to what was seen in the mosses? How?

14. Why are tracheophytes sometimes referred to as embryophytes?

Ferns

The gametophyte in many of the primitive tracheophytes are difficult to find, and there are not enough available for student laboratories. The ferns, a later tracheophyte development, have a more prominent gametophytic stage. The life cycle of the fern will be examined by observation of the various stages.

1. Obtain a sporophyte of fern. You will note that this is the typical fern frond with which you are familiar.
2. Turn the leaf over and examine the sporangia on the underside of the frond.
3. Is the fern frond haploid or diploid?

4. Are the spores in the sporangia haploid or diploid?

5. Your instructor has prepared four stages of spore germination for your observation.
   (a) Spore germination and 10-days growth.
   (b) Spore germination and 14-days growth.
   (c) Mature prothallia.
   (d) Developing sporophyte.
6. Prepare a wet mount of the material in the flask of spore germination and 10-days growth, by inserting a glass rod into the flask, and washing off the material that has adhered to the surface of the rod on to a slide with a drop or two of water.
7. Observe and draw the developing gametophyte on Fig. 20.2.
8. What type of cell pattern is formed in the initial growing stages of the gametophyte?

9. Make a sketch indicating the direction of the spindle of mitotic division that produces this growth.

10. Prepare a wet mount of the material in the flask marked spore germination and 14-days growth.
11. Observe and draw the developing gametophyte on Fig. 20.2.
12. Sketch a group of cells indicating the direction of the spindle of mitotic division that produces this growth.

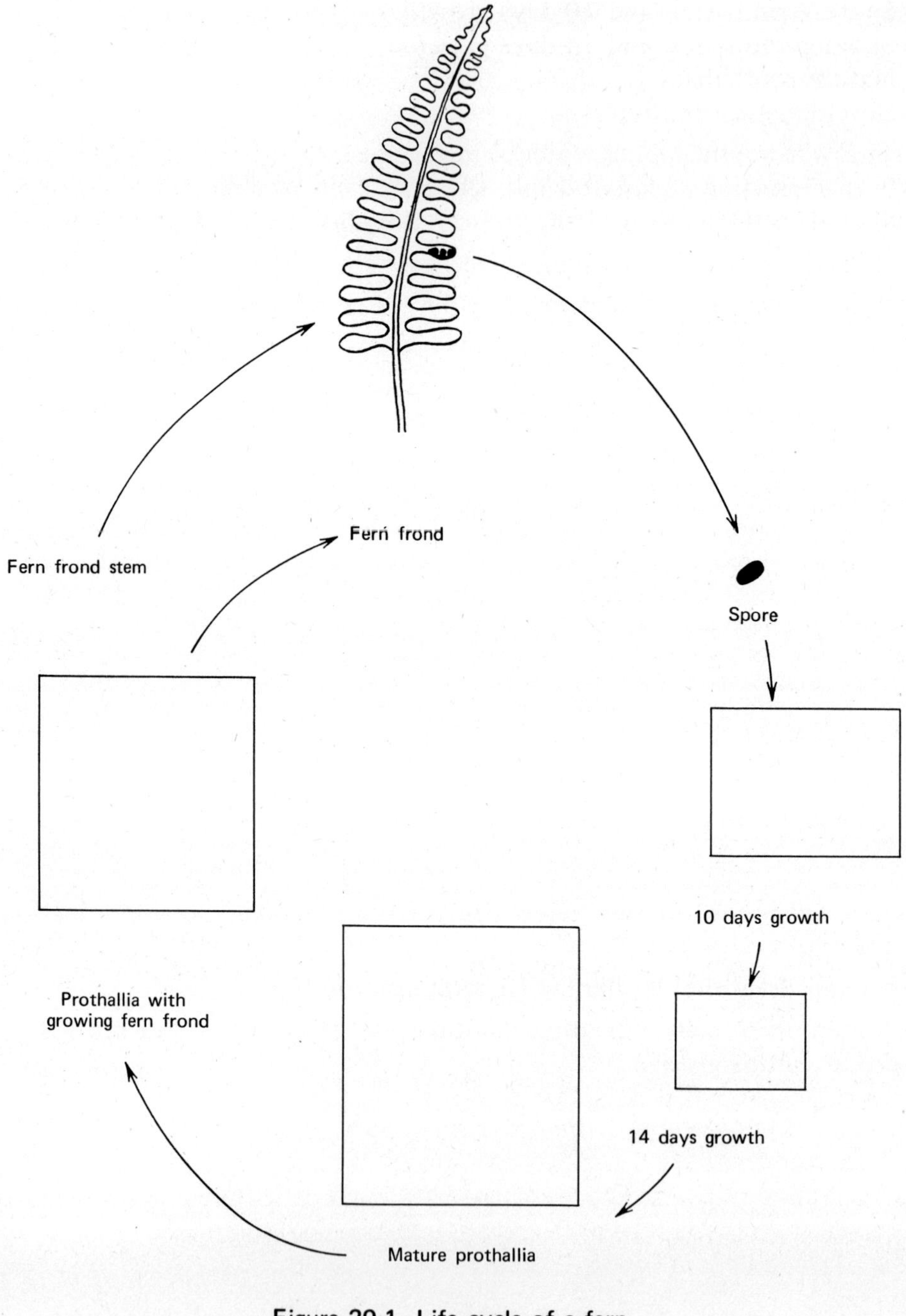

Figure 20.1 Life cycle of a fern.

13. From the information obtained during the experiment performed in Chapter 13, what causes the alteration of the plane of division?

14. The fern spore germinates underground, and produces a filament that grows upward. How has the growth of the prothallia adapted to (a) germinate in available light; (b) get to a full light source before alternating its plane of division?

15. What is the ploidy of the developing gametophyte?

16. Observe the fern gametophyte, called a prothallia, growing in the pot marked mature prothallia.

17. Draw the mature prothallia on Fig. 20.1.

18. What shape is the fern prothallum?

19. Obtain a slide of fern prothallia showing the archegones, and another slide showing the antheridia. Although both grow on the same prothalia, it is difficult to prepare a single slide that shows both.

20. The archegone is on a stalk and has a venter and one large egg. The antheridium is a stalk in which many antherozoids may be seen.

21. Draw an archegone and an antheridium on the drawing of the mature prothallia you have made on Fig. 20.1.

22. From observing the growth of the fern prothallia, and where this growth takes place, in what medium do the antherozoids move to the archegonium?

23. The fern prothallia is obviously photosynthetic. Why would selection favor only ground level growth for this phase?

24. Observe the prothallia with the growing sporophyte.
25. Draw this stage of development on Fig. 20.1.
26. Is the sporophyte physically dependent on the gametophyte or is the gametophyte physically dependent on the sporophyte?

27. Why are ferns considered embryophytes in the alternate classification scheme?

28. Obtain and examine a cross section of fern rhizome. The rhizome is the fern stem.
29. Draw this cross section in the appropriate place on Fig. 20.1.
30. Notice the arrangment of xylem and phloem into units called vascular bundles. How are these vascular bundles shaped?

31. Would the surface area/volume relationship of these vascular bundles be large or small?

32. The fern fronds come out from the rhizome at many levels, ranging from the base near the roots to the very top. Would the function of the vascular bundles and their shape be consistent with the arrangement of fronds? Why?

33. On Fig. 20.1, label the sporophytic and gametophytic generations, the ploidy of each stage, and where meiosis and fertilization occur.

21

The Seed Plants

MATERIALS

CLASS STOCK

Prepared slide
- cross section of complete gymnosperms stem
- cross section of a woody stem
- pine leaves
- cross section of pine leaves
- (carpellate strobulus of pine)
- bean seeds
- germinating bean or pumpkin seed (5 days old)
- 2-week-old bean seedling
- stem of horse chestnut
- potato tubers
- sweet pea flowers
- cross section of lily anther-1st division
- pollen grains
- double fertilization in lily-demonstration
- pea pod
- corn grain
- IKI solution

STUDENT STATION

- microscope
- microscope lamp
- lens paper
- slides
- cover slips

INTRODUCTION

The sporophyte of the primitive tracheophytes never achieved total nutritional independence from the gametophyte. In broth *Psilotum* and the fern, the sporophyte embryo developed within the gametophyte and continued directly into the development of the mature plant.

The independence of sporophyte maturation from the gametophyte appears to have first occurred in a fern lineage. The fossils of the *Pterisperms* (seed ferns) seem to indicate that embryological development occurred in a pod-like structure of the gametophyte, which dropped from the plant and matured or germinated when separated from the parent gametophyte. Because of this process, these fossils are called seed ferns.

The two major plant groups that presently dominate the Earth are the gymnosperms and angiosperms. In both groups not only is the sporophyte the dominate form, but the gametophyte is reduced in size and is totally dependent on the sporophyte.

THE GYMNOSPERMS

The familiar pine tree is an example of a sporophyte in the gymnosperm group

Stem Adaptations in Gymnosperms

1. Obtain a prepared slide of gymnosperm stem.
2. Observe and draw a sector of the stem in the space provided, labeling the following parts:

   xylem (wood)—large empty cells almost in the center.
   rays—strips of xylem cells that raditate from the center to near the margin (cells are radially elongated).
   cambium—dense, protoplasm-filled cells at the periphery of the xylem.
   phloem—small cells with thick walls to the outside of the cambium.
   pith—large cortical (thin-walled) cells in the center.
   cork—dead cells at the margin of the stem.
   bark—cork plus phloem.

3. The cambium consists of cells that are constantly dividing. Some of the division products are siphoned off from the inner surface and differentiate into xylem while others from the outer surface differentiate into phloem. What does this activity do to the diameter of the stem? To preexisting phloem cells?

4. The phloem cells are eventually crushed and die. Where would you expect to find the cambium in a cross section of a large woody stem?

5. The xylem sells take on thick cell walls as they become older. How does this contribute to the height that can be attained by pine trees?

6. Obtain a cross section of a woody stem.

7. In temperate zones, the cambium divides rapidly and produces many large xylem cells during the growing season (spring), and then slows during the subsequent months. This is reflected in the pattern of xylem deposition in the stem. Which area or areas in the stem developed during the growing season? What brought you to this conclusion?

8. Which area or areas in the stem developed during the colder months? What brought you to this conclusion?

9. What does one of each of the areas mentioned above represent?

10. How old was the tree that the cross section was taken from?

11. Do you think you could tell the age of a tree grown in the tropics?

Leaf Adaptations in Gymnosperms

1. Obtain some pine leaves.
2. Examine and draw the pine leaves in the space provided, making sure to represent the flattened surface of a group of leaves, and their position with respect to one another.

3. How would you characterize the surface area/volume relationship of a pine leaf?

4. What is the primary function of a pine leaf?

5. Pine, as an evergreen, retains its leaves in the winter months. How are the leaves adapted for this period?

6. How is the shape of the pine leaf a balance between its function and the conditions under which it exists?

7. Obtain a prepared slide of a cross section of a pine leaf.
8. Draw the cross section of the pine leaf in the space provided, labeling the following parts:

   epidermis—regularly shaped cells that surround the periphery.
   guard cells—kidney-shaped cells that interrupt the epidermis.
   stomate—opening between two guard cells.
   mesophyll—loosely organized cells central to the epidermis.
   vascular tissue—xylem and phloem in the center of the leaf.

9. Which cells are photosynthetic?

10. Is the epidermis frequently interrupted by guard cells?

11. What peculiar location of the guard cells in the epidermis is an adaption to prevent excessive water loss?

12. Before the continents drifted to their present positions, many broad-leaved gymnosperms lived on the North American continent. Why do narrow leaved-gymnosperms now thrive in this region?

Reproduction on Gymnosperms

The cone, a collection of sporangia, which originated in the lower plants *Lycopodium* (princess pine) or *Equisetum* (horsetails) achieved great complexity in the gymnosperms. The pine cone is formed by the fusion of leaf and stem elements, combined with sporangium.

1. Obtain a carpellate strobulus ("female" cone) of pine or some other genus and label the following parts on Fig. 21.1.

   ovuliferous scale—the "woody" leaves of the cone, which are actually fusions of stems and leaves.
   seeds—may or may not be present. Two are found on the top surface of each ovuliferous scale. If the seeds have dropped off, two scars are present.
   main axis—the stem on which the ovuliferous scales are born.
   spiral arrangement of parts
   multiplicity of parts
   fusion of parts (ovuliferous scale)

The sporangia were present at a very early stage. The spores from them produce gametophytes, which remain on the ovuliferous scale. These gametophytes (male and female) produce gametes and fertilization then occurs also on the ovuliferous scale. The resulting zygote produces the embryo, which developes along with the gametophyte and sporangium. These three together produce the seed, which eventually falls off the ovuliferous scale.

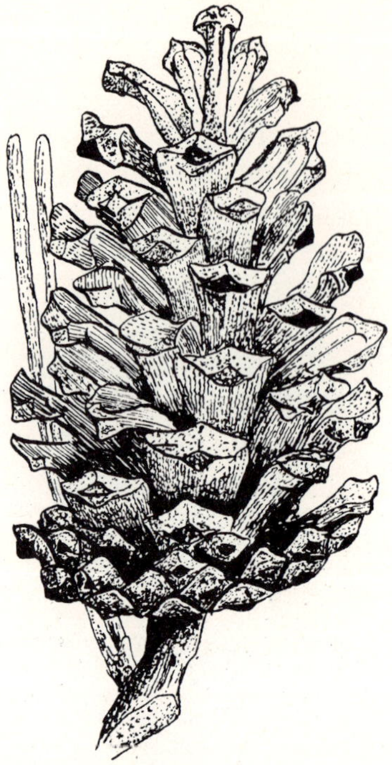

Figure 21.1

2. The staminate cone produces an extremely large number of pollen grains. What is the survival advantage of this to the species? Be specific.

3. The pollen grains are airborne; but once they are on the carpellate cone, they move in a fluid exuded by the cone itself. How does this illustrate a persisting commitment on the part of the gametophyte for an aquatic environment?

4. Why are gymnosperms considered seed plants?

THE ANGIOSPERMS

In order to familiarize you with the variation of plants grouped as angiosperms and the salient features of the life cycle, a life cycle will be made on the angiosperm using different plants to illustrate different phases. The diagram of a typical angiosperm life cycle is shown in Fig. 21.2.

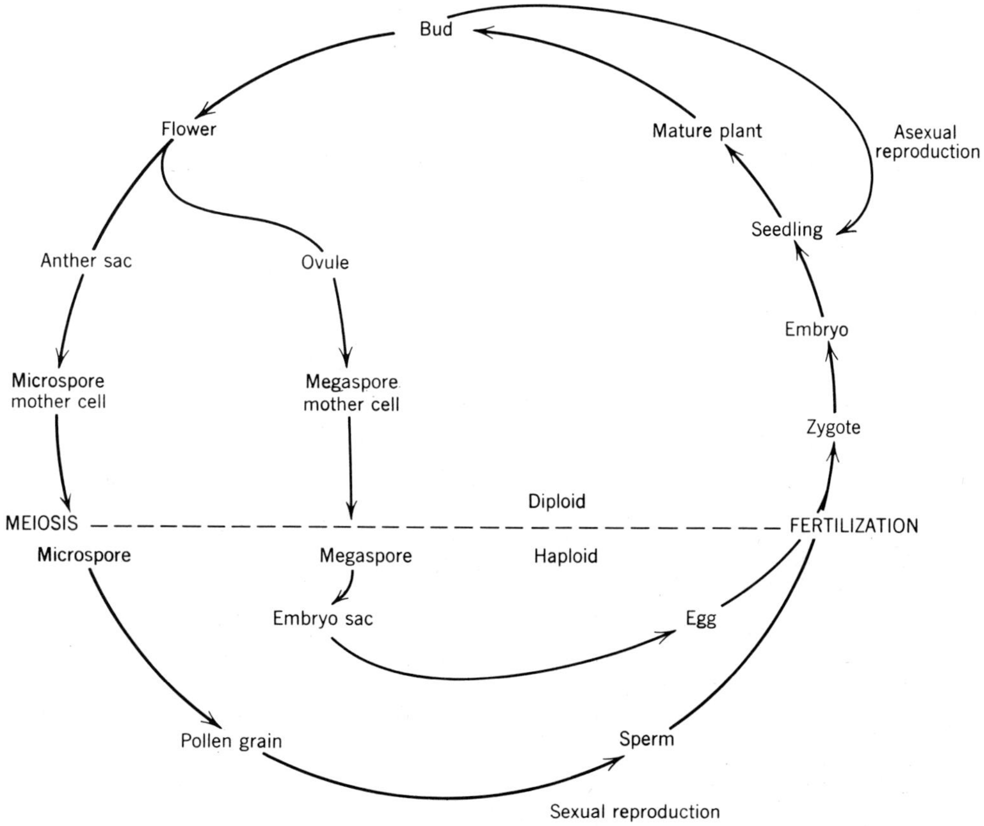

Figure 21.2

Note especially (1) where meiosis and fertilization occur, (2) which are the haploid and diploid phases, and (3) the asexual and sexual reproductive phases. The diploid phase is called the sporophyte and the haploid cells or structures are the gametophyte.

The Seed

For most flowering plants, the seed serves as a dormancy stage to overcome unfavorable conditions. It is a compact unit for dissemination and a storage structure for food for the young embryo within it.

1. Examine a bean seed.
2. In the space provided draw the bean seed, labeling the following parts:

   seed coat or testa—outer covering of the seed.
   hilum—small scar on the concave surface of the seed coat. Evidence that it was once attached to the parental plant.
   micropyle—small hole near the hilum used as a passage for exchange of gases.

3. Peel off the seed coat and separate the two cotyledons or seed leaves. Attached to the inner surface of one is the embryo.
4. Draw the cotyledon with the embryo in the space provided, labeling the following parts:

   cotyledon—seed leaves containing nutritional material for the embryo.
   embryo—composed of two white feather-like plumules and a finger-like hypocotyl.
   hypocotyl—finger-like structure of the embryo.

5. Why should an entrance be available for the passage of gases?

6. What else must enter the seed before germination will take place? Why?

THE GERMINATING SEED

When the period of dormancy ends, certain parts of the seed become active and grow, while other parts are broken down.

1. Examine a germinating bean and pumpkin seed.
2. In the space provided, draw both seeds, labeling the following parts:

   seed coat
   hypocotyl or root
   cotyledons—seed leaves.
   stem—region from cotyledons upward (if present).
   peg—in pumpkin seed only, the attachment between the seed coat and the hypocotyl.

3. Why does the root system first develop?

4. How does the peg aid in germination?

THE SEEDLING

When the full potential of the embryo has been reached, the original structures have been transformed into new parts. Some of these are matured and incapable of further development while other parts remain reproductively active (meristems) and are nourished by the mature tissues.

1. Obtain a 2-week-old bean seedling.
2. Draw the seedling labeling the following parts:

   root system—tends to have one main root with many secondary roots coming from it, and is, therefore, called a tap root system.
   stem—extends from the groundline upward.
   terminal bud—found at the top of the stem.
   cotyledons—quite small and shows signs of withering. Notice that these are green and therefore serve in a photosynthetic role as well as storage of food.
   leaf arrangement—both the cotyledons and the first two true leaves exhibit an opposite arrangement.
   node—region of the stem at which the leaf is attached.
   internode—the region between two consecutive nodes, for example, between the two cotyledons and the true leaves.
   blade—flat green area of the leaf.
   veins or vascular bundles—run across the blade in a pattern of net venation.
   petiole—part of the leaf that attaches the leaf to the stem.

lateral bud—found in the angle between the petiole and the stem.
apical meristems—located at the tips of the plant, namely, the bud and root tips.

3. Select one structure and show how its form is adapted to its function(s).

4. From where did the first two true leaves originate?

THE STEM

The stem is an organized group of tissues which can continue to produce more stem material by further growth upward at the tip and laterally by branch formation, Theoretically, this process can proceed indefinitely. The sharp distinction between mature and embryonic areas is unlike that of the animal body and so leaves a record of past developmental events.

1. Obtain and examine the stem of the horse chestnut or some similar stem.
2. Draw the stem in the space provided, labeling the following parts:

   terminal bud—at the tip.
   bud scales—modified leaves which cover the bud. They are brown in color and are covered with a protective sticky material during dormant periods.
   bud scale scars—occur in clusters around the stem at different levels, indicating the end of growth for past years. In the spring the bud scales fall off as the bud elongates into more stem.
   leaf scars—shield-shaped scars on the stem which indicate where leaves have fallen off in the autumn.
   bundle scars—numerous marks in the leaf scar which serve as evidence that the vascular bundles run from the stem into the leaf through the petiole.
   lateral bud—found above each leaf scar. Capable of forming a branch.
   bark—protective layer of dead cells.
   lenticles—small pores on the bark which serve for the exchange of gases.

3. Determine the age of the twig by the number of bud scale scar clusters. Do the same by counting the number of annual rings observed where the stem was cut.

4. What is the leaf arrangement?

ASEXUAL REPRODUCTION: SPROUTING OF POTATO

Although the lateral buds can produce branches, these stems are not new individuals. In plants, special lateral stems can become separated from the parental plant and develop into new individuals. This occurence of asexual reproduction is an efficient means of propagation in potato, tulip, onion, and other plants.

The potato is a tuber, which is a type of modified stem, and hence the tuber should have the same structures noted in the stem of the previous exercise.

1. Examine the potato tuber.
2. Draw the tuber in the space provided, labeling the following structures:

   lateral bud—commonly called the "eye" of the potato.
   leaf scar—curved side of the eye, or the "eyebrow."

Notice that the sprouts arise from the eyes of the potato and give rise to branches which are active growing regions. These branches are capable of giving rise to a new plant. Here is an example of asexual reproduction by the production of a tuber and this tuber's behaving like a seed: undergoing dormancy, storing food for forthcoming sprouts, and effecting a separation from the parent plant.

3. Why are tubers produced underground?

4. If time and labor were no problem, how would you ensure that potatoes would never sprout?

SEXUAL REPRODUCTION: THE FLOWER

At certain predictable times, the apical meristem will switch from making vegetative shoots (stems and leaves) to forming floral structures, thus changing from a vegetative stage to a reproductive stage. The meristem loses its identity by forming various floral parts, but within the flower itself, new meristematic cells will arise.

1. Obtain a flower of the sweet pea or some similar plant.
2. Draw the flower in the space provided, labeling the following parts:

peduncle—main stalk which supports the flower clusters.
pedicel—small stalk which attaches the flower to the peduncle.
receptacle—swelling in the pedicel where it attaches to the flower.
sepals—outer row of green floral leaves.
calyx—all the sepals.
petals—five colored, divided into two rear ones and three forward ones, the lip.
corolla—all the petals.
perianth—calyx and corolla collectively.
stamens—many long green stalklike structures in the center of flower.
filament—green stalk of the stamen.
anther—swollen tip of the stamen.
pollen grains—yellow powderlike substance on the anther.
pistil—slender structure found in the center of the stamens.
stigma—sticky, split top of the pistil.
style—long white stalk of the pistil.
ovary—base of the pistil.

3. Break open the ovary and find the small, smooth, beadlike ovules ("seeds").
4. Flowers are pollinated either by wind or by insects. Which is the case for sweet pea? (Hint: Notice any smell or color?)

5. Which parts of the flower appear to be modified leaves? Stem? Both combined?

MICROSPOROGENESIS: FIRST DIVISION

Within the anther certain cells, spore mother cells, become active by undergoing meiosis. In this way the meristematic continuity of the plant is preserved. It is during this nuclear division that the chromosome number is reduced and recombination occurs.

1. Obtain a slide of a cross section of an anther of lily and observe under low power.
2. Draw the cross section of the anther in the space provided, labeling the following parts:

   anther sac—four chambers.
   microspore mother cells—cells contained in the anther sacs.
   filament—in the center (seen in a cross section).
   tapetum cells—layer of cells around the anther sac which is thought to nourish the sporogenous cells.
   chromosomes—deeply stained within the microspore mother cells. These cells are in the process of meiosis and are passing from a diploid to a haploid condition.

POLLEN GRAINS

The microspore mother cells undergo meiosis to produce microspores. These cells then divide into two-celled pollen grains. The pollen grain is the only haploid structure besides the embryo sac cells in the higher plant and constitutes the male gametophyte.

1. Obtain a small amount of pollen material and make a water mount. These are the immature male gametophytes and contain only a few cells.
2. Draw the pollen grain in the space provided, labeling the following parts:

   cell wall or exine—special cell wall of the pollen grain.
   protoplasmic material

FERTILIZATION (DEMONSTRATION)

Once the pollen grain is formed it may be carried to the stigma to the pistil. The grain then germinates by the growth of the germ tube down the style and into the ovary. As it reaches the opening of the ovule (micropyle) the tip bursts, liberating two nuclei or male gametes. One male nucleus fuses with that of the egg to form a zygote and the other with two polar nuclei to form the first cell of the endosperm.

1. Observe in the demonstration microscope double fertilization in the lily.
2. Draw and label the following:

   embryo sac—of female gametophyte, which is composed of the following cells:
   three antipodal cells—located at the top or chalazal end of the embryo sac.
   one polar cell—a large central cell with two nuclei.
   two synergid cells—located at the micropylar end or opening of the ovule.
   one egg cell—between the two smaller synergids.
   sperm nuclei—worm-like nuclei, one associated with the egg nucleus and the other fusing with the two polar nuclei.

3. Which gamete has more cytoplasm associated with it, the sperm or the egg?

4. How many sets of chromosomes are there in the egg? The sperm? The zygote? The synergid cells? The endosperm? The spore mother cell?

THE FRUIT: THE PEA POD

After fertilization, the ovule develops into the seed and the ovary (sac containing ovules) enlarges into the fruit. Although the fruit eventually disintegrates, it serves an important role in the successful dispersal of the seed.

1. Obtain and examine a pea pod.
2. Open the pod in such a way that the seeds (peas) remain on one side of the pod.
3. In the space provided draw the open pod, labeling the following parts:

   <u>pericarp</u>—fruit wall.
   <u>carpel</u>—chamber within the pod.
   <u>seeds</u>—green spherical "peas."
   <u>funiclus</u>—attachment of seed to pericarp.
   <u>pedicel</u>—stalk at the base of the pod.
   <u>perianth</u>—withered, may remain attached to the pedicel.

4. What parts of the flower remain as functional parts of the fruit?

5. What must happen to the pericarp in order for the seeds to arrive at a seedling stage?

ANOTHER FRUIT: THE GRAIN

The grain is a specialized fruit type of great economic importance (corn, wheat, etc.). It is especially characterized by a thin pericarp, and the storage of food in a tissue called the endosperm.

1. Obtain a soaked grain of corn.
2. Observe the following parts and incorporate in the drawings of Fig. 21.3.

   pericarp—or fruit wall, is a very thin and can easily be peeled off. The cells which make up the pericarp are diploid.
   embryo—is in evidence as the white shield or face on one of the two broad surfaces. This is diploid.
   predicel—partial remains at the bottom as a point.
   style—remains at the top as a point. Look carefully, because this is not too obvious.

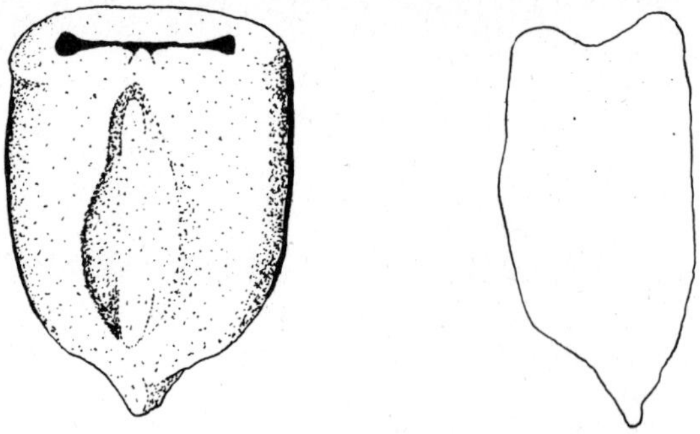

Figure 21.3

3. With a razor blade, cut through the face of the fruit and add a drop of IKI solution to one of the cut surfaces. Look for the following:

   endosperm—the storage region containing starch for the future seedling. This gives a positive starch reaction.
   embryo—does not stain with IKI.
   aleurone—a thin layer just inside the pericap giving the yellow coloration to the seed. This tissue is triploid (3N).

DIAGRAMMATIC SUMMARY OF THE LIFE CYCLE

Below are a number of structures specifically placed for you to indicate the origin and fate of each. Use symbols as shown to indicate the proper relationships.

origin ⟶
dead end ⊣

lateral bud
bark
terminal bud
flower meristem
seed coat ⊣
stem
leaf
embryo
root
pedicel
cotyledon
peg ⊣
branch root
petals
pericarp
ovary
fruit
ovule
stamens
seed
pollen
anther sac
tapetum

22

Physiological Adaptations in Plants

MATERIALS

CLASS STOCK

chemicals
- 1.0% safranin solution
- Benedict's solution
- lanolin
- lanolin + 10^{-3} M indole acetic acid
- gibberellic acid (10^{-7} M, 10^{-6} M, 10^{-5} M, 10^{-4} M)
- benzoadenine (10^{-7} M, 10^{-6} M, 10^{-5} M, 10^{-4} M)

celery set-up
carrot set-up
light source
lettuce seeds
bean seeds
soaked radish seeds
red cellophane
small brown bags

STUDENT STATION

beaker
ring stand
knife
4 pots with 10- to 14-day-old bean plants
3 pots with 10- to 14-day-old bean plants
light-tight box
petri plates
stapler
filter paper

INTRODUCTION

In the last two chapters you have examined the morphological adaptations of plants, some of which have developed in conjunction with certain physiological adaptations. For example, taller plants could not have evolved without some means by which water could be brought to the higher regions of the plant body. Although xylem tissue fulfilled the morphological aspect of bringing water up the stem of the plant, it could only have done so by operating in conjunction with physical and chemical properties of water. The same situation is true for all aspects of plant morphology. Hormonal control induces the growth of structures in a way that is typical of plants. Light induces seed germination in some species without, if there was no light there would be no plant. Thus physiological adaptations and morphological features had to develop together in the evolutionary scheme.

THE MOVEMENT OF WATER

As mentioned above, there are two aspects to the movement of water: (1) the tissue adaptation for water movement, and (2) the complementarity of the water moving mechanism and the physical and chemical properties of water.

Tissue Adaptation for Water Movement

1. Pour a 1% safranin solution into a 500-ml beaker until it is about 1/5 full.
2. Place the beaker on the base of a ring stand and adjust the ring to about five inches above the mouth of the beaker.
3. Obtain a stalk of fresh celery with leaves.
4. Cut off the base of the celery stalk with a sharp knife, and immediately place the celery into the beaker of safranin so that the stalk is held upright by the ring.
5. Observe the movement of the safranin solution up the celery stalk.
6. Counting the time you placed the celery stalk into the safranin solution as zero-time, how long did it take the solution to move up into the leaves?
7. After the safranin solution has moved to the leaves, remove the celery stalk from the solution, and cut it in half with a sharp knife.
8. Looking down at the cross section you have made, are all the tissues of the stalk stained red with safranin?

9. Through which tissue does the safranin solution pass?

10. Which tissues of the plant are adapted for movement of water up the stem?

Complementarity of Water Movement with Properties of Water

1. Examine the demonstration set up by the instructor in which a carrot has been cored, and the core has been partially replaced and extended by a glass tube after being filled with 2% glucose solution. The carrot is suspended in a solution of neutral red dye.
2. Test the solution around the carrot for the presence of glucose (Benedict's test).
3. What color is the solution moving up the tube? Why?

4. Indicate on Fig. 22.1 the relative concentrations of glucose and neutral red in the cells and in the core of the carrot. With arrows indicate the direction of the movement of the two molecular species.
5. Why is fluid moving up the tube?

6. Is the fluid moving up the tube because it is being pushed or pulled?

7. If a fan were placed next to the funnel at the top of the glass tube, would water move more or less rapidly through the carrot?

8. Would the evaporation of the water from the top of the funnel have the effect of pulling or pushing the water through the tube?

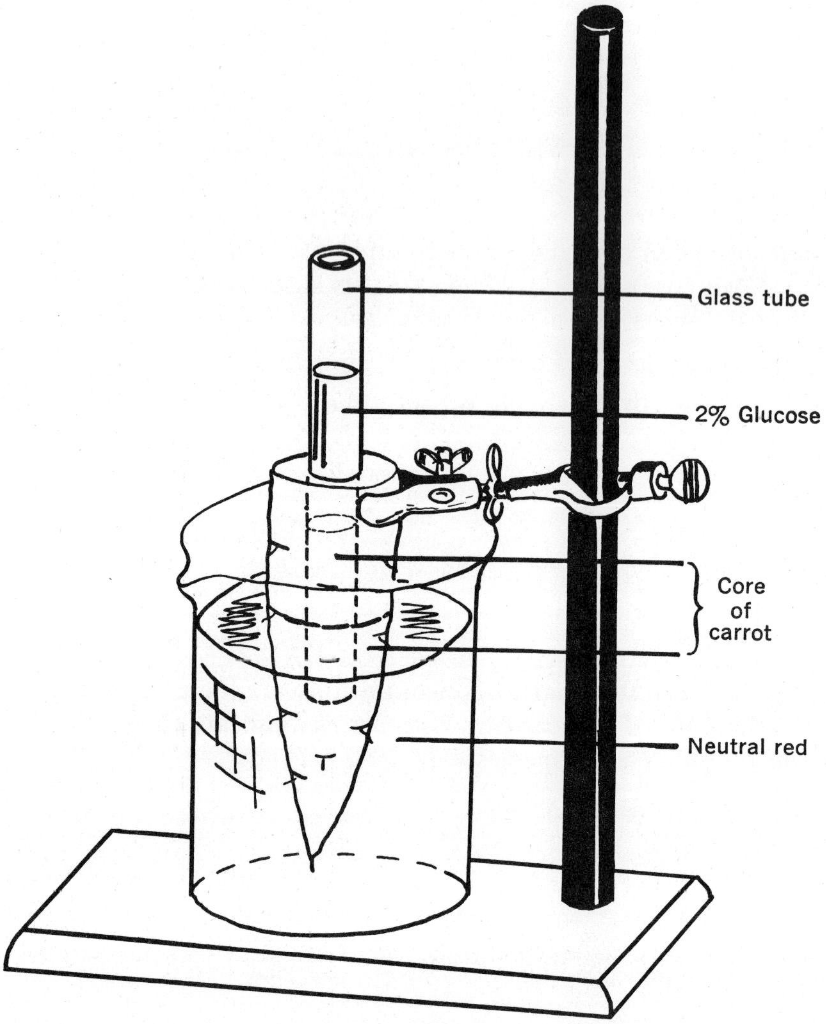

Figure 22.1

9. The carrot functions in a capacity similar to which organ of the plant?

10. What is the surface area/volume relationship of a leaf on a carrot, or for that matter, any plant?

11. What two factors are responsible for the movement of water up the stem of a plant?

12. In what way does evaporating water play a role in the movement of water up a stem?

13. How does the dipolar nature of water and the bondings between water molecules derived from the dipolar nature play a role in moving water up a stem?

PLANT HORMONES

Plant growth and differentiation are controlled by hormones—substances in small quantities produced and secreted by tissues, which affect either neighboring tissues or the tissues in which they are produced.

Auxins

Auxins, such as *indole acetic acid,* control growth and development in a variety of specific ways. Auxins are the basis of geotropic responses such as the downward growth of the roots (positive geotropism) and the upward orientation of the stems (negative geotropism), and for the apical dominance, which leads to the conical shape of pine trees. At a cellular level, auxins increase the plasticity of cell walls.

1. Obtain 4 pots containing 10- to 14-day old bean seedlings.
2. Mark and treat the pots as follows: (include your initials and the day of the laboratory)
   (a) Untreated.
   (b) Place a small dab of lanolin to the side of the plant stem just below the first pair of true leaves.
   (c) Place a small dab of lanolin + 10^{-3} μM indole acetic acid below the first pair of true leaves.
   (d) Place the pot on its side an illuminated location where it can be left for the next four days.

3. After four to six days, observe the seedlings, and draw them in the space provided.

4. Why was one set of seedlings treated with lanolin alone?

5. Cell wall plasticity enables cells to elongate. What effect does cell elongation have on the plants in pot (c).

6. Like any chemical entity, auxins tend to follow the laws of gravity. How can you explain the results in pot (d) on the basis of the above information, and the events in pot (c)?

Gibberellins

Gibberellins are another group of plant hormones, of which giberellic acid (GA) is the most common one investigated. Giberellic acid produces a set of characteristic effects on plates that are distinct from those produced by auxins.

1. Obtain 3 pots with 10- to 14-day-old bean seedlings.
2. Mark and treat each in the following way:
   (a) Untreated—this can be the same as pot (a) in the previous experiment.
   (b) Place in the dark under a large, light-tight box.
   (c) Place several small drops of gibberellic acid solution on the surface of one of the large true leaves and one or two drops in the axis of these leaves and the stem.
3. After four days observe the seedlings and draw them in the space provided.

4. Which hormone would a germinating seed be most likely to product first—auxin or gibberellic acid? Explain.

5. Suggest one factor that has an influence on the production of gibberellic acid (see answer to **4**).

Phytochrome

Phytochromes are among the most recently discovered plant hormones and have influence on many growth patterns, such as seed germination, flowering, folding of the leaves of a sensitive plant *(Mimosa)*, and even the orientation of the chloroplasts in filamentous algae.

1. Place 10 lettuce seeds in each of three petri plates.
2. Label and treat the petri plates in the following way:
   (a) Cover with red cellophane and staple.
   (b) Place in the dark (place in two small paper bags)
   (c) Leave exposed to white light.
3. After one week, determine the percentage of seeds that germinated in each dish and record your data on Chart 22.1.

CHART 22.1

Illumination	Percent Germination
Red light (a)	
Far red light (b)	
White light (c)	

4. Would seeds germinate better if placed 2 in. below the ground or on the surface of the soil?

5. Phytochrome exposed to far red light (P_r) is thought to affect cell permeability. How would this state of the pigment explain germination? (Hint: consider what is commonly done in order to germinate seeds.)

6. Consider the four factors—auxin, gibberellic acid, phytochrome, and water. List them in order of sequential effects on seed germination and seedling development.

Interaction of Plant Hormones

Plant hormones appear to operate in conjunction with each other. Some promote growth, others retard growth; by their relative antagonistic effects, the growth rate can be regulated. Also, some hormones enhance the effects of others, creating situations that are impossible with one hormone alone. For example, auxins increase cell size and cytokinins induce cell division. Therefore, their mutual operation makes plant growth possible.

Hormonal interactions can be identified. Suppose increasing concentrations of auxin increase growth accordingly. When gibberellic acid, for instance, is added, one of two results would be expected. Either the two curves obtained from plotting the concentration of each hormone against the increase in size would be parallel, indicating no interaction between the hormones, or they would not be parallel, demonstrating that interaction is present (see Fig. 22.2). In this exercise you will determine whether a cytokinin (benzoadenine –BA), and the gibberellin (gibberellic acid –GA) interact.

1. Obtain five petri plates. Line them with two layers of filter paper.
2. Add 5.0 ml of the appropriate concentrations of the cytokinin solutions to the petri plates.
   (a) 0 benzoadenine
   (b) 10^{-7} M benzoadenine
   (c) 10^{-6} M benzoadenine
   (d) 10^{-5} M benzoadenine
   (e) 10^{-4} M benzoadenine
3. Obtain another five petri plates.
4. Add 0.5 ml of the appropriate concentrations of gibberellic acid to the petri plates.
   (f) 0 gibberellic acid
   (g) 10^{-7} M gibberellic acid
   (h) 10^{-6} M gibberellic acid
   (i) 10^{-5} M gibberellic acid
   (j) 10^{-4} M gibberellic acid
5. Obtain another five petri plates.
6. Add 2.5 ml of each solution in appropriate concentrations to the petri plates.
   (k) 0 benzoadenine, 10^{-5} M gibberellic acid
   (l) 10^{-7} M benzoadenine, 10^{-5} M gibberellic acid
   (m) 10^{-6} M benzoadenine, 10^{-5} M gibberellic acid
   (n) 10^{-5} M benzoadenine, 10^{-5} M gibberellic acid
   (o) 10^{-3} M benzoadenine, 10^{-5} M gibberellic acid

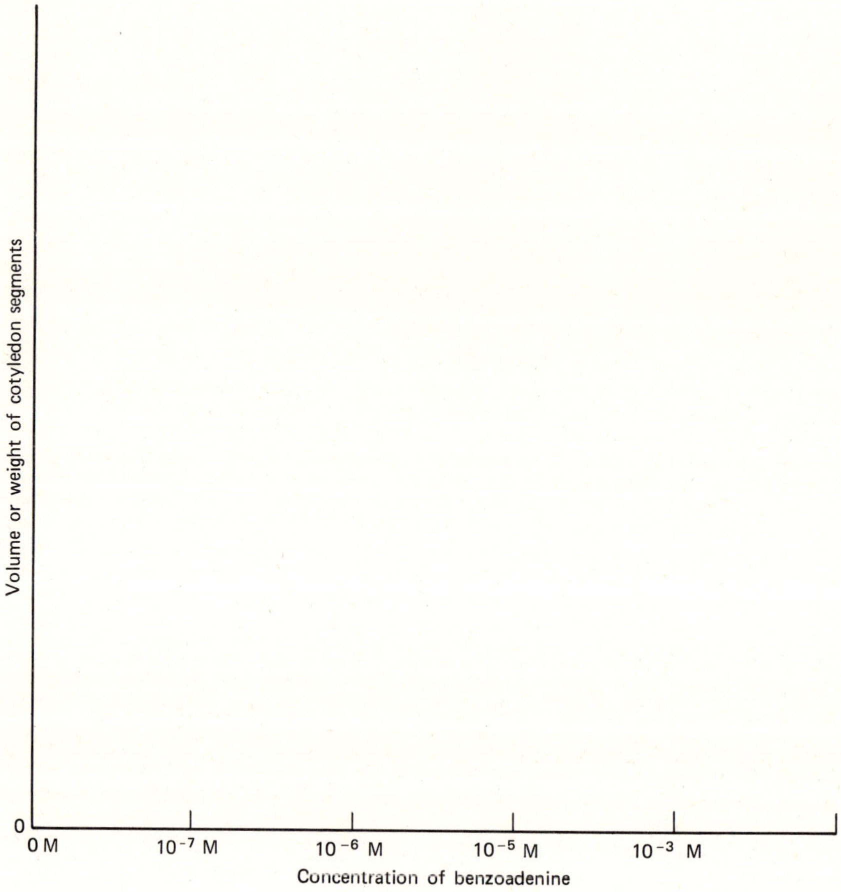

Figure 22.2

7. Remove the seed coats from presoaked radish seeds, and separate the cotyledons. Measure the largest diameter on each of five cotyledons; average for each group of five.
8. Place five in each of the fifteen dishes above.
9. After 1 day measure the size of the cotyledons in each dish. Average for each group in a plate and subtract appropriate original average.
10. Plot the volume difference of the cotyledons in the graph of Fig. 22.2.
11. Are these two hormones interacting in the growth of the cotyledon tissue?

23

Trends in Animal Evolution

MATERIALS

CLASS STOCK

Preserved specimens (or slides)
- *Scypha*
- *Spongilla*
- *Chalina*
- *Hydra*
- *Aurelia*
- *Metridium*
- *Pleurobrachia*
- *Planaria*
- *Taenia pisiformis*
- *Nereis* or *Lumbricus*
- *Helix* or *Venus*
- member of Phylum Arthropoda
- crayfish
- grasshoppers
- *Ascaris*

prepared slides
- *Fasciola*

STUDENT STATION

microscope
microscope lamp
lens paper
red pencil
dissecting pan
pins
scissors
hand lens

INTRODUCTION

Each form of life represents a successful solution to the problems presented by the environment. The particular solution depends on the genetic capability of the organism and the stage of evolutionary development.

The basic problems of survival have been identical throughout the evolutionary process—obtaining the energy needed for sustenance and continuance of life and utilizing this energy to procure food, protect the individual, and reproduce in order to preserve the species. Therefore, dividing the bodily systems of an organism into energy-obtaining and energy-utilization systems is not altogether artificial; it permits a greater integration than does the more standard means of approaching organismal study (for example, digestive system and circulatory system). The energy-obtaining systems include all those that contribute factors or functions to the cellular respiratory process, including the passage of foodstuffs into the body, the breakdown of these foodstuffs so that they can be handled on a cellular level, the transportation of the foodstuffs to the cellular level, the provision of the necessary oxygen as a final hydrogen acceptor in cellular resporation, and the excretion of toxic metabolic wastes from the body. The systems that perform these activities are the digestive system, certain aspects of the circulatory system, the respiratory system, and the excretory system. At low levels of Metazoan organization there may be no anatomical distinction between these systems.

These systems that are directly involved in obtaining energy also use some of this energy in order to function. In this sense they are also energy-utilization systems. However, their primary function is to obtain energy. The systems associated with energy utilization are those usually involved in the response or bodily adapation of the organism to the environment. This adaptation may take the form of organs actively involved in awareness and those capable of response, such as nervous systems which are usually involved in both awareness and response, and muscle systems which are primarily responsive. The reproductive system, which is also considered an energy-utilization system, utilizes energy for the maintenance of gonads (the gamete producing organs), the production of gametes, and in some animals, the nurturing of young, all of which contribute to the perpetuation of the species.

In this chapter we shall examine the solutions of organisms at various levels of organization to the problems of obtaining the necessary energy.

In order to comprehend the developments of these systems it is necessary first to understand the probable evolutionary events in the animal kingdom. Cytological evidence indicates that the animal lineage is traceable to the green algae, instead of to protozoans as had been assumed for many years. Somewhere in the multicellular stages of plant development, heterotrophy occurred, producing a metazoan branch. None of the intermediates have survived, although a few groups are thought to be offshoots of these intermediates, among which are the sponges.

Sponges are characterized by having two cell layers—endosperm and ectoderm. All other animals have three cell layers—ectoderm, endoderm, and a middle layer of mesoderm. Animals are generally classified according to the means by which the third middle layer of mesoderm is formed and the type of body cavity that results. The means of forming mesoderm appears to be intimately associated with the differentiation potential of the tissue. Highly differentiated organ systems are either derived from or initiated by mesoderm.

Mesoderm may be derived embryologically from the existing ectoderm, or endoderm, or both. Mesoderm that has been derived from ectoderm seems very limited in potential differentiating ability, and tends to form little more than filler tissue. This filler tissue consists of embryonic cells and contributes to the regeneration capacity of the organism. Mesoderm that has been derived from ectoderm and/or endoderm seems capable of wide differentiation patterns.

The classification of organisms into groups according to their mesoderm and body cavity formations is very real in terms of genetic potential. Certain groups of organisms cannot go beyond a particular level of organization because of limited genetic potential and, hence, differentiation potential.

> Acoelomates (a = without; coel = coelom = body cavity). These organisms show a solid cell mass instead of open spaces within the body. The mesoderm of this solid cell mass is derived from ectoderm. This group includes the Platyhelminthes, and it constitutes one line of evolutionary development.
>
> Pseudocoelomates (pseudo = false; coel = body cavity). The pseudocoelomates contain special areas around various organs, but these areas are lined with ectoderm instead of mesoderm in many cases. The pseudocoelomates include Rotifera and Nematoda, and constitute another line of evolutionary development.
>
> Schizocoelomates (schizo = formed by splitting, coel = body cavity). The schizocoel possesses a body cavity lined with mesoderm. The mesoderm may be of ectodermal and endodermal origin, or from endodermal origin alone, and is formed by a splitting off of cells from the layer of origin in the course of embryological development. This group includes the primitive nonsegmentated forms, the mollusks, as well as the segmented forms (those that have a body plan laid down in sections), annelids, and the most advanced invertebrate form, arthropods.
>
> Enterocoelomates (enter = gut; coel = body cavity). These organisms have a body cavity lined with cells of mesodermal origin. The cavity is formed by the outpocketings of endomesoderm. This group includes the echinoderms (starfish and sea urchin) and the chordates, (primitive fish → fish → amphibians → reptiles ↔ birds, → mammals).

Another feature used to classify animals is the eventual use of the second opening developed embryologically. During development, all animals develop a blastopore, an opening from the outside to the gut, duing gastrulation. All animals except acoelomates eventually develop another opening. These organisms, which have complete digestive systems, are either oriented in such a way that the first opening becomes the mouth, and the second, the anus, or that the second becomes the mouth, and the first the anus. Pseudocoels and schizocoels are called Protostomes, because the first opening, the blastopore, develops into the mouth of the adult organism. Enterocoelomates are called Deuterostomes because the second opening eventually develops into the mouth of the adult organism.

Phylum Porifera

Sponges have always been classified as animals, and although they do not fit into the above discussion regarding mesoderm and body cavity formations, they will be examined here.

General Characteristics

Two cell layers—diploblast; no organ systems; few cell types; cylindrical, globular, or irregular. Form maintained by "skeleton" of tiny $CaCO_3$ or silica spicules or a network of tough protein fibers.

Energy-Obtaining Systems

The circulatory, digestive, and respiratory systems are inseparable in these primitive organisms. The pore and canal system of the sponge permits the circulation of water from the environment throughout the body. Microorganisms in the circulating water are taken into the cells of the sponge in much the same way as the *Amoeba* engulfs organisms. Digestion occurs intracellularly, and the simplified nutrient material is distributed to surrounding cells by diffusion. Oxygen, which is dissolved in the water, is taken directly into the cells for aerobic respiration.

Energy-Utilization Systems

Excretion is accomplished by simple diffusion out of the cell into the surrounding and circulating fluid. Reproduction of a sponge may be either sexual or asexual (without sexual union of gametes). Asexually the sponge will start to grow a bud near the base. This bud will eventually develop all the characteristics of a mature sponge but will remain attached to the parent. Sexually, the sponge is hermaphroditic, capable of producing gametes of both sexual types. The sperm migrate to the egg, fertilizing it within the sponge ovary. A ciliated larva is then formed which is free-living and may swim until it settles down to form the mature sponge. There is virtually no organized nervous or muscular system of the sponge, except for the general ability of individual cells to respond to adverse conditions in their environment.

1. Examine the jars of *Scypha* (sometimes called *Grantia*), *Spongilla,* and *Chalina.*
   Scypha—found in shallow marine water. Possesses a skeleton of $CaCO_3$ in the form of calcareous spicules.
   Spongilla—found in fresh water, this is one of the few species of sponge that is adapted to a fresh-water environment. It contains a skeleton of silica, which is unusual for fresh-water types.
   Chalina—found in sea water and known as "finger sponge" because of its branching form. This sponge contains a skeleton of both protein fibers and silica spicules.
2. Draw one of these sponges in the space provided, labeling the following parts:

spicules—may not be seen in all specimens.
bodily orientation—anterior opening, if any, used as both mouth and anus; posterior point of attachment.
pores—for passage of water and nutrients. Scypha has a major central chamber which is called a spongocoel. The opening to this chamber is the anterior of the animal, the osculum.

3. Examine a prepared slide of a cross section of *Scypha* under low power.
4. Draw and label the following parts in the space provided.

spongocoel—which is now seen as the hollow center of the ring of cells.
epithelial cells—the cells on the outside of the body wall; they are characteristically flat.
collar cells—flagellate cells which are directly responsible for the ingestion of nutrients.
spicules—the spines of $CaCO_3$ found in the gelatinous substance between the epithelial layer and the collar cells.
mesenchyme cells—irregularly shaped and found in the gelatinous substance. These are embryonic cells that are capable of differentiation under certain conditions.
pore cells—hollow cells that permit the circulation of water. At the position of the pore cell, the animal is only one cell layer thick.

ACOELOMATES

Phylum Coelenterata

General Characteristics

Three cell layers—gastrodermis lining the digestive tract, epidermis covering the outside of the body, and inner "mesoderm" layer; few cell types; sac-within-a-sac body plan forming a central digestive cavity. Some forms have no skeleton, others have calcareous skeleton to maintain shape; radially symmetrical.

mesoglea—a gelatinous matrix with few cells found between the gastrodermis and epidermis.

nematocysts—coiled threads that contain poison to paralyze their prey.

Classification of the Coelenterates

The coelenterates are divided into classes according to the dominant embryonic state. In the course of development, the animal progresses from a hydroid to a medusa stage. The hydroid is sessile, the medusa free-living.

Class Hydrozoa—Both embryonic stages may be present and fully developed or the medusa stage may be reduced or lacking.

Class Scyphozoa—This class includes the jellyfishes. The medusa (free-living) predominates.

Class Anthozoa—This class includes the sea anemones and corals. The members of this class have no medusa stage at all.

Energy-Obtaining System

The circulatory, digestive, and respiratory systems are inseparable. In some forms, the water containing unicellular organisms that are food for the coelenterates is drawn into the central cavity by the tentacles which surround the opening; in others the tentacles trap organisms and bring them to the mouth. Digestion takes place extracellularly in the cavity into which cells secrete digestive enzymes. Respiration is by the simple diffusion of oxygen into the cells and CO_2 out.

Energy-Utilization Systems

The excretion of wastes in the coelenterates is accomplished without the aid of specialized cells, each cell eliminating the waste products of metabolism by simple diffusion. Patterns of reproduction among the coelenterates vary considerably. Some forms are hermaphroditic, whereas other species have separate sexes. The pattern of development that involves a medusa and hydroid type is reminiscent of the alteration of generations found in the plant kingdom. The analogy does not hold, however, because both forms of the coelenterates are diploid. The hydra does not have any organized form of nervous system. Jellyfish have a light-sensitive "sense organ," which is found near a nerve ring. At the end of each tentacle is a statocyst, which aids in maintaining the equilibrium of the individual. The movement of the medusoid jellyfish also implies the presence of muscle-type fibers which contract and relax in unison. Neither the muscular nor the nervous system is highly devleoped.

CLASS HYDROZOA

1. Examine the prepared specimen of *Hydra*. If available, examine prepared slides of *Hydra* under low power.
2. Draw the general form of the *Hydra* in the space provided, labeling the following parts:
   mouth—single opening at the anterior end of animal body.
   tentacles—surround the mouth and serve to bring food (organisms) into the cavity.
   holdfast—found at the base of the animal and is used to hold the hydra in place (sessile).

CLASS SCYPHOZOA

1. Examine the specimen of *Aurelia,* the common jellyfish.
2. Draw the specimen in the space provided, labeling the following parts:
   umbrella-shaped body
   tentacles—short projections around the margins of the umbrellalike body.
   mouth—found in the center of the underside surrounded by four arms.
   digestive system—long canals that may be seen through the body wall.

3. With a red pencil, follow the pathway of a unicellular organism being digested.
4. Where are solid wastes given off?

CLASS ANTHOZOA

1. Examine the specimen of *Metridium*, the sea anemone.
2. Draw the *Metridium* in the space provided, labeling the following parts:
   mouth—opening to the gullet.
   tentacles—surrounding the mouth.
   gullet—portion of the gastrovascular cavity directly inside the mouth.

3. With a red pencil, follow an organism being digested by the *Metridium,* using broken lines to represent the digestion taking place inside the organism.

Phylum Platyhelminthes

General Characteristics

Bilateral symmetry (right-left mirror images of one another); triploblast—endoderm, mesoderm, and ectoderm; flattened dorso-ventrally; definite organ systems, digestive, excretory, nervous, and muscular.

Classification

Class Turbellaria—free-living.
Class Tremetoda—parasitic, possessing a digestive cavity. This class includes the "flukes."
Class Cestoda—parasitic, not possessing a digestive cavity. This class includes the tapeworms.

Energy-Obtaining Systems

The circulatory, digestive, and excretory systems are complete and anatomically separate in the Platyhelminthes. The organ systems of the parasitic forms have become modified in accommodating to their mode of life.

The members of the class Turbellaria *(Planaria)* are free-living. Food in the form of microorganisms enters the mouth and proceeds to the intestine through the proboscis. Digestion and absorption of food materials take place in the intestine. No circulatory systems exist and the nutrient material diffuses directly through the body. The size and flattened dorso-ventral shape permit materials to leave and other materials of the surrounding media to bathe the tissues freely, lessen the need for an organized circulatory system. Respiration takes place by means of simple diffusion. The parasitic members of this phylum also respire by means of direct diffusion. The source of molecular oxygen, however, is the host's tissues. Food materials are taken in elemental form from the host, and therefore require little digestion. This system has been totally eliminated by the tapeworms. Excretion of metabolic wastes involves an elaborate excretory system made up of cells that collect the wastes, flame cells, and deposit them into the excretory canal from which they go into the surrounding environment through excretory pores.

Energy-Utilization Systems

Reproduction of the *Platyhelminthes* is varied. Planarians, although hermaphroditic, are incapable of self-fertilization. During copulation, the eggs of both participating planarians are fertilized. The zygotes are discharged in a sac or cocoon, which attaches to a rock or to debris on the bottom of the stream. The parasitic forms in this phylum are hemaphroditic and are self-fertilizing. The life cycle of these parasitic forms is highly complex, usually involving the invasion of several hosts before the adult form is achieved.

The nervous system is developed in *Planaria* to include a primitive brain and two major lateral nerve cords. These nerve cords are connected along the length of the body giving a ladder configuration.

CLASS TURBELLARIA

1. Examine the preserved *Planaria* and a whole mount of *Planaria* under low power.
2. Draw the animal in the space provided, labeling the following parts:

   mouth—opening found at the mid-section on the ventral side.
   pharynx—the tube which encloses the mouth. This tube may be extended in the course of ingesting foods.
   intestine—branches into a single anterior extension and two posterior extensions from the pharynx.
   diverticula—pocketings of the intestine which show marked convolutions.
   eyespots—anteriorly located, the eyespots give a cross-eyed appearance because of the position of the pigment, which is on the medial side of both.
   excretory canal—fine canal which can be seen most clearly in areas where there is no intestine (at the extreme anterior or posterior end).
   flame cells—(under high power) shaped like bulbs with several cilia oriented downward in the cell.
   reproductive system—both oviduct with the many enlarged yolk glands and the testes are found on the lateral sides of the *Planaria.* Since they are so difficult to separate visually, label as reproductive system.
   brain—found just posterior to the eyespots as two hemispheres of heavy tissue.

3. Why does the structure of the digestive system of *Planaria* demand that food the organism consumes be limited to bacteria, minute protozoans, decaying forms, and soluble proteins?

4. How do the diverticula aid in the digestive process? (Hint: Note the effect on surface area.)

5. Why can *Hydra,* a less organized organism, consume more structurally complex forms than *Planaria?*

6. What might this indicate concerning the relative time during which the *Hydra* and the *Planaria* in the forms we have today came into existence?

CLASS TREMATODA

1. Examine the preserved specimen and/or a prepared whole mount slide of *Fasciola.*
2. Draw *Fasciola* in the space provided, labeling the following parts:
   digestive tract—branched, originating at the anterior end of the organism and covering about half its length.
   anterior oral sucker—holding organ on the ventral side of the organism.
   reproductive system—found at the lateral sides of the organism, it is composed of enlarged yolk glands and testes. May be seen best in posterior half of organism. it is not concealed by the digestive system.
   mouth—center of anterior oral sucker.
   ventral sucker—holding organ found posterior to the anterior oral sucker on the ventral side.

CLASS CESTODA

1. Examine the preserved specimen or slides of selected segments of *Taenia pisiformis*, the dog tapeworm (the entire tapeworm is too large to fit on a single slide).
2. Draw the *Taenia* in the space provided, labeling the following parts:

   scolex—head region usually containing suckers with hooks for attachment to the intestinal wall.
   suckers and hooks
   mature proglottids—segments in the mid-region. These are just reaching a reproductive stage.
   ripe (gravid) proglottids—hindmost region, contain the zygotes.
   uterus—centrally located structures in the mature proglottids, seen as a thin line.
   testes—appears as grapelike structures in the mature proglottid surrounding the uterus.
   ovary—darkly stained grapelike cluster at the base of the uterus.
   yolk gland—posterior to the uterus at the base of the proglottid.
   Note the absence of a digestive system.

3. How does the tapeworm subsist without a digestive system?

4. Tapeworms are parasitic on many chordate forms. What does this indicate concerning the time of evolutionary appearance of the tapeworm in its present-day form?

5. What is the advantage to flukes and tapeworms of being hermaphrodites capable of self-fertilization?

PSEUDOCOELOMATES

Phylum Aschelminthes, Class Nematoda

General Characteristics

Approximately circular in cross section; size range—microscopic to more than 1 ft long; internal organs lie in a body cavity; digestive tract has both mouth and anus.

Energy-Obtaining Systems

The digestive system consists of a tube which runs the length of the organism, beginning at the mouth and terminating with the anus. In free-living members of this phylum, digestive juices are secreted into the intestine. These glands, in the parasitic form *Ascaris,* have degenerated probably as a result of a selection against them, since they are unnecessary in the worm living within muscle tissue.

The body cavity is filled with a fluid through which the nutrient materials diffuse. Although this fluid does not move, it serves as a forerunner to the circulatory system. Respiration for *Ascaris* is anaerobic; it is one of the few animals that only partially breaks down nutrient materials (anerobic respiration). Excretion is carried out by two lateral tubes that join anteriorly and exit by a pore shortly posterior to the mouth.

Energy-Utilization Systems

The nervous system is made up of two nerve cords originating from a nerve ring. These nerve cords run on the dorsal and ventral sides of the worm, and connecting circular rings provide sensory and motor receptors to all parts of the body. The sexes of *Ascaris* are separate, and the male and female meet and copulate in the intestine. The eggs are released with the feces and proceed through a complex life cycle before forming the adult.

1. Examine the preserved specimen of *Ascaris.*
2. Draw the *Ascaris* in the space provided, labeling the following parts:
   external chitinous cuticle
   mouth—found at the anteriormost end of the *Ascaris.*
   excretory pore—on the ventral side, just posterior to the mouth.
   anus—seen as a slit on the ventral side just before the anterior end.

3. Describe how digestive glands could degenerate through evolution.

4. How do pseudocoels and acoels differ from diploblasts in organization?

SCHIZOCOELOMATES

Phylum Annelida

General Characteristics

Segmented, repetition of units; definite organ systems; body cavity.

Energy-Obtaining Systems

The digestive system is complete, beginning with a mouth and terminating with an anus. This digestive tube which exhibits thickenings and narrowings provides specialized areas for digestive processes. Food, which consists primarily of dead vegetable material, enters the mouth and is ingested with the aid of the muscular pharynx. The food is passed down the esophagus by the rhythmic movements known as peristalsis. The food empties into the crop, the first expansion of the tube. This serves as a storage unit from which food is released in small amounts into the next expanded area, the gizzard. The food is ground in the gizzard with the aid of some sand grains; the food then passes into the intestine, where it is enzymatically digested. The elemental food material is absorbed into the blood vessels which supply the intestine. The intestine curves ventrally to form a loop known as the typhlosole.

The earthworm has a closed circulatory system. The blood is circulated by ten hearts which surround the esophagus. Two major blood vessels, one ventral and one dorsal, with a number of branches and capillaries, circulate the blood. The blood contains a form of hemoglobin capable of binding oxygen and carrying it to parts of the body.

Respiration is carried out by oxygen that dissolves in the moist covering of the skin. The oxygen is absorbed by the hemoglobin of the blood in the capillaries which come close to the surface. Carbon dioxide is given off through the skin. The excretion of metabolic wastes is taken care of by a pair of nephridia in each segment (except the first three and the last). Each nephridium consists of a funnel lined with cilia which connects to a coiled tube. The tube opens at a pore on the side of the worm.

Energy-Utilization Systems

Reproduction of the earthworm resembles that of *Planaria,* in that the individuals are hermaphroditic but incapable of self-fertilization, The ovaries and testes are in specific segments. The nervous system consists of an anterior dorsally located "brain" and a single

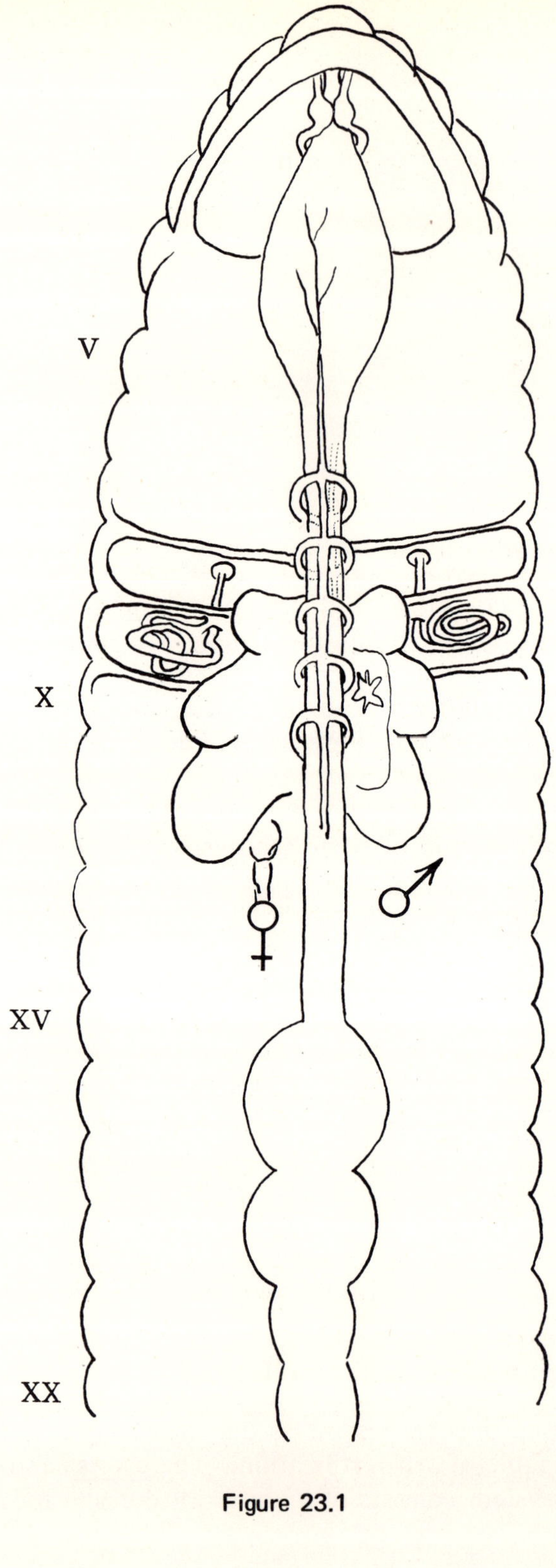

Figure 23.1

ventral nerve cord, which consists of ganglia (nerve cell bodies) in each segment except the first four. These ganglia send off branches around the worm body. There are both circular and longitudinal muscle fibers which control locomotion.

1. Examine the preserved specimen of either *Nereis,* the clam worm, or *Lumbricus,* the common earthworm.
2. If enough worms are available, students may dissect. If not, several dissecting specimens should be available for drawing.
   (a) Place the *Lumbricus* on a dissecting pan, dorsal side up. Pin both the anterior and posterior ends.
   (b) With fine scissors, cut through the clitellum, the swollen area.
   (c) Cut anteriorly along the mid-dorsal line, being careful to cut only the body wall.
   (d) Cut posteriorly a short distance.
   (e) Carefully clip the septa which separate the segments internally.
   (f) Pin the edges of the body wall to the pan.
   (g) Cover with a small amount of water to keep the specimen moist, and study the internal organs. Use a hand lens when necessary.
3. Label the systems associated with energy obtaining and energy utilization in Fig. 23.1 Use the above discussion of systems for locating them.

mouth	hearts	brain
pharynx	ventral blood vessel	ventral nerve cord
esophagus	dorsal blood vessel	ganglia
crop	nephridia (use hand lens)	
gizzard	testes (in seminal vessicles)	
intestine	ovaries	
typhlosole		

Phylum Mollusca

General Characteristics

Organ systems, including circulatory; no segmentation—body is not in definite zones, nor is it laid down segmentally from an embryological viewpoint; shell or mantle (fold of skin which covers the body—octopus or squid).

Classification

Class Gastropoda—"stomach-foot," this group includes the snails, slugs, conches, etc. This is the only class that includes terrestrial forms.

Class Pelecypoda—"hatchet-foot," this group includes the bivalve type. These forms have a shell consisting of two parts which completely enclose the body. The "half-shells" are held together by strong muscles.

Energy-Obtaining Systems

The organ systems of the phylum Mollusca are highly varied, for both aquatic and terrestial forms exist. Generally, the digestive system in both forms is developed into a mouth, an esophagus leading from the mouth to the stomach where digestion occurs, an intestine leading from the stomach which serves for the absorption of nutrients, and a terminating anus which removes solid wastes. The respiratory system is adapted to cope with the conditions of the environment. Aquatic molluscs respire by means of a series of gills. Water passing through siphons passes over the gills, and dissolved oxygen diffuses into the blood as CO_2 diffuses out. Some terrestrial molluscs dissolve atmospheric oxygen on the surface of the mantle of skin which surrounds the body and which is moistened by internal secretions. The circulatory system of both groups of molluscs is an open one, for the blood is not confined to the vessels at all times and circulates freely through some organs. A beating heart helps keep the blood circulating. Excretion is accomplished by the excretory organ, the kidney, a large bladder-like strucutre that collects wastes from the body and eliminates them through the dorsal siphon.

Energy-Utilization Systems

AQUATIC MOLLUSCS

The sexes are separate in these molluscs. The clam releases sperm through the excurrent siphon. Some of these are taken up by the incurrent siphon of the female clam, and the eggs are fertilized on the gills. The eggs are released and the small larvae must attach to fish and live a parasitic life until they become free-living. The nervous system of the aquatic molluscs has rather poorly developed sense organs, although the system itself is quite highly developed and consists of a brain and several pairs of ganglia. The muscular system is also highly developed. In the clam, two parts (valves) of the shell are closed by extremely strong muscles. Movement is achieved by extension and expansion of a muscular foot. This is highly efficient for burrowing into the sand.

TERRESTRIAL MOLLUSCS

A kidney, which functions in much the same manner as in the aquatic forms, collects the wastes from the pericardial cavity (which surrounds the heart) and discharges them into the mantle cavity. Snails tend to be hermaphroditic, containing both reproductive organs in the form of the ovo-testis which lies high in the shell. The products of this organ, egg and sperm, then proceed down separate ducts which join at the genital atrium. The nervous system is somewhat condensed compared with that of the aquatic molluscs, but the sense organs are more highly developed. The snail has two pairs of tentacles, the posterior ones containing eyes with primitive cornea, lens, and retina, and perhaps an olfactory or odor-sensitive organ. Movement is accomplished by extension of the foot over a "slime track" of mucus which is secreted by a gland beneath the mouth.

1. Examine preserved specimens of *Helix,* the land snail, and/or *Venus,* the clam.
2. Follow the instructions for the organism you are examining.

   Helix. Draw the land snail in the space provided, labeling the following parts:

   shell—hardened spiral structure which is protective in function. Indicate the position of the cavity between the mantle and the body directly under the shell which is used as a site for respiration. The primitive "lung" is like the vertebrate lung in function (analogous) but is totally unlike the vertebrate form in its development pattern (non-homologous).
   tentacles—two pair located on the dorsal side of the head.
   stalked eyes—atop the posterior pair of tentacles.
   radula—or mouth is found at the anterior portion of the head and contains small teeth.
   genital pore—small opening directly posterior to the anterior tentacles.
   anus—small opening in the mantle between the snail and the shell. This can only be seen if the snail is fully extended from the shell.
   respiratory pore—larger opening anterior to the anus between the snail and the shell. This opens into the mantle cavity and can be seen only when the snail is fully extended from the shell.
   foot—large muscular portion of the body extruding from the shell.

Venus. If enough clams are available, follow the instructions for dissection. If not, observe the demonstration specimens and follow the descriptive instructions.

1. Open the shell of the clam by prying the shells apart until the valves are exposed.
2. Note the mantle, a membranous tissue just inside the shell. Cut the mantle away from the left valve.
3. Cut the muscles and open the clam gently.
4. Note the structure of the mantle which thickens and comes together to form the siphons. Water enters through the ventral siphon and circulates over the gills where oxygen is taken up and CO_2 is given off into the water.

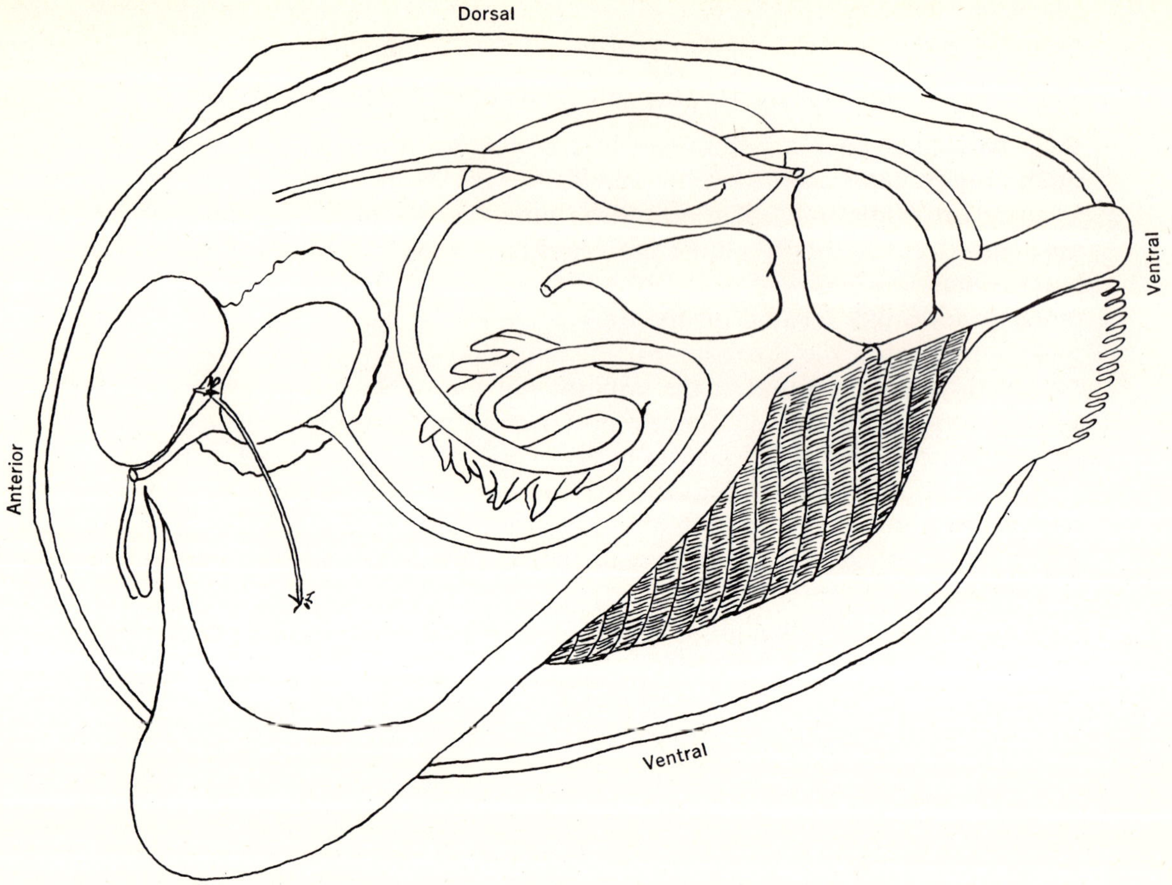

Figure 23.2

5. On the illustration of the clam, Fig. 23.2, label and give characteristic texture to the following organs associated with energy-obtaining and energy utilization systems.

mantle—membranous lining which surrounds the body of the clam.
anterior abductor muscles—located above the muscular foot, these muscles hold the bivalves together.
posterior abductor muscles—located on the other side of the umbo (hump in the shell), these muscles also serve to hold the bivalves together.
mouth—opening directly dorsal to the foot.
esophagus—tube leading from the mouth to the enlarged portion of the digestive tract, the stomach.
stomach—enlarged portion of the digestive tract.
liver—organ surrounding the stomach.
intestine—portion of the digestive tract leading posteriorly. This passes out of sight behind the heart but reappears posterior to the heart.
rectum—terminal portion of the intestine.

anus—opening of the digestive tract into the excurrent siphon.
incurrent siphon—located at the extreme posterior end of the clam, it is characterized by feathery projections which aid in bring water in.
gills—feather-like organ partially hidden by the muscular foot, this runs along the ventral side of the clam.
suprabranchial chamber—duct on the dorsal side of the gills.
excurrent siphon—posterior to the suprabranchial chamber. Water which has passed over the gills and solid wastes from the digestive tract exit from this opening.
heart—dorsal and anterior to the posterior abductor muscle, this is enclosed in a cavity.
ventricle—dorsal side of the heart.
auricle—ventral side of the heart.
pericardium—cavity in which the heart lies.
anterior aorta—blood vessel running anteriorly from the heart.
posterior aorta—blood vessel running posteriorly from the heart.
kidney—bladder-like organ directly ventral to the pericardium.
cerebral ganglia—knot of nervous tissue found on the dorsal side of the esophagus.
gonad—found near the loops of the intestine. The male and female gonads need not be distinguished.
foot—large muscular protrusion.

Phylum Arthropoda

General Characteristics

Segmented body plan—segments highly specilized in function. Body functionally divided into head, thorax, and abdomen. Exoskeleton, skeleton of chitin, a nitrogenous polysaccharide. Jointed appendages; highly developed systems; compound eye.

Classification

Class Crustacea—Lobsters, shrimp, crayfish
Class Diplopoda—millipedes
Class Chilopoda—centipedes
Class Arachnida—Spiders, scorpions, mites, ticks, king crab
Class Insecta—insects
Order Orthoptera—grasshoppers, cockroaches

Energy-Obtaining Systems

The arthropods are the most highly developed of the invertebrates. The fact that this phylum shows the greatest diversity as well as the greatest number of multicellular individuals within the animal kingdom attests to the success of its adaptability. The digestive system is highly developed in all of the arthropods. The food, which may range from minute sea organisms consumed by the crustacea to the vegetable matter eaten by some insects, is digested in the stomach and absorbed in the intestine. The respiratory systems vary according to the environment. A system of gills, not unlike those found in the molluscs, are present in the crustacea, which exist in an aquatic environment. Those anthropods which exist in a terrestrial environment are adapted in several ways. The millipedes and centipedes dissolve oxygen in the liquid phase on their skin, while insects have many passageways or tracheids which contain pools of hemoglobin-like substance, the "dead end" pockets or sinuses in which oxygen dissolves. The circulatory system is generally of the open type, where the blood flows into the sinuses of the body, leaving the blood vessels temporarily. In the closed type of circulatory system, the blood never leaves the blood vessels, and circulation through tissues is achieved by capillaries. The heart of the arthropod pumps blood into arteries. There are no veins, and the heart picks blood up directly from the surrounding pericardial sinus. Slight variations of this type exist throughout the arthropods. The excretory systems of the arthropods vary greatly from species to species. Some crustacea have a "green gland" found in the cephalothorax which extracts wastes from the system. Some forms have nephridia-like collecting tubules which collect wastes.

Energy-Utilization Systems

Reproductive patterns also vary greatly. Almost all the members have separate sexes, but the development of the organism into an adult may involve relatively simple transition or the highly complex patterns of metamorphosis. The nervous system generally follows the pattern of that laid down in the Annelid forms—a single ventral nerve cord with several ganglia. The muscle systems are highly complex and vary in action from slow strong movement in the appendages of the crustacea to the rapid movement necessary in wing movement of the insects.

1. Examine preserved specimens of members of the Phylum Arthropoda. Observe at least *one* from each of the noted classes.

Crustacea— *Balanus*—rock barnacle
crayfish—fresh water
lobster—salt water
Diplopoda— *Spirobolus*
Chilopoda— small centipede
Arachnida— *Argiope*—garden spider
Centrurus—scorpion
Dermacentor—wood tick
Limulus—king crab (horseshoe crab)

Insecta—		
	grasshopper	*Cicada*
	praying mantis	ground beetle
	may fly	monarch butterfly
	dragon fly	cabbage butterfly
	giant water bug	horse fly

2. Examine a dissected demonstration of crayfish and grasshopper and label Figs. 23.3 and 23.4, locating those organs associated with obtaining and utilizing energy.

Crayfish

green gland—associated with excretion of metabolic wastes. Found in the cephalothorax near the eyes.
mouth—directly posterior to the opening for the green gland. Large opening anteriorly located on ventral side.
esophagus—short tube leading from the mouth almost directly dorsal to the stomach.
stomach—enlarged part of the digestive tube where major part of digestion takes place.
intestine—straight tube leading from the stomach to the anus.
heart—dorsal to the beginning of the intestine. Appears as an elongated swelling in the arteries.
liver—anterior to this portion of the intestine is the spongy tissue of the liver.
brain—found in the dorsal head region.
ventral nerve cord—major nerve tract running along the ventral side of the crayfish.
gonads—male testes or female ovaries are found ventral to the heart.

Grasshopper

hemocoel—body cavity, space between body wall and internal organs. This is filled with colorless blood during life.
mouth—opens between two lateral jaws or mandibles.
esophagus—short tube leading from the mouth to the foregut.
foregut or crop—first enlarged area of the digestive tract.
midgut or stomach—enlarged area of the digestive tract following the crop. This is where digestion takes place.
caeca—pouches which extend from the stomach; secrete digestive juices.
hindgut—leads to the anus.
spiracles—small valved openings found laterally in the thin membranes between the thoracic and abdominal segments.
trachea—tubes entering the body from the spiracles through which air penetrates into the body.
heart—dorsally located in a sinus.
aorta—only blood vessel; runs toward the head on the dorsal side.
ovaries or testes—found lying above the posterior portion of the gut. If female, the ovaries will be seen as tubes containing the developing eggs; if male, narrow tubes in which the sperm develop will be noted.
ventral nerve cord—runs along the ventral side of the grasshopper.

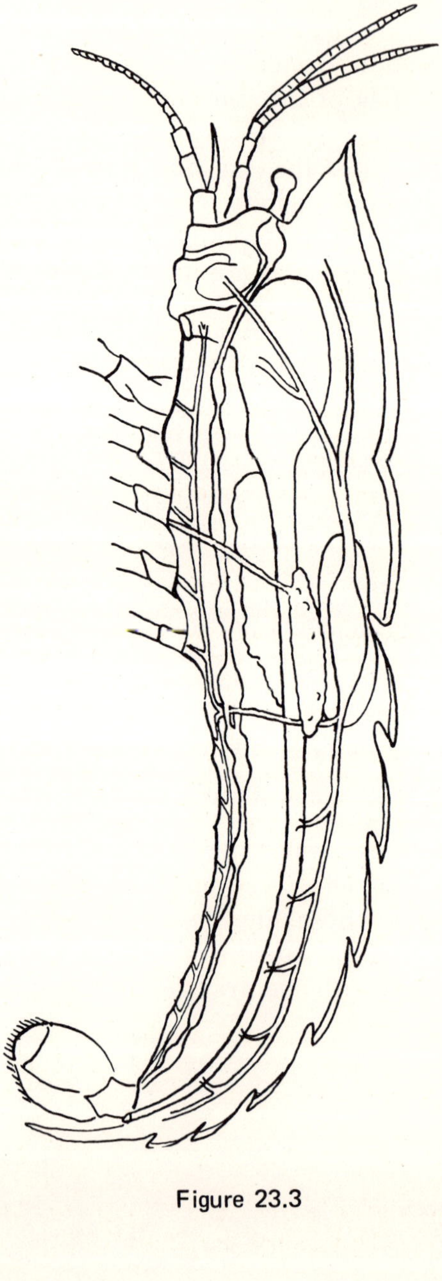

Figure 23.3

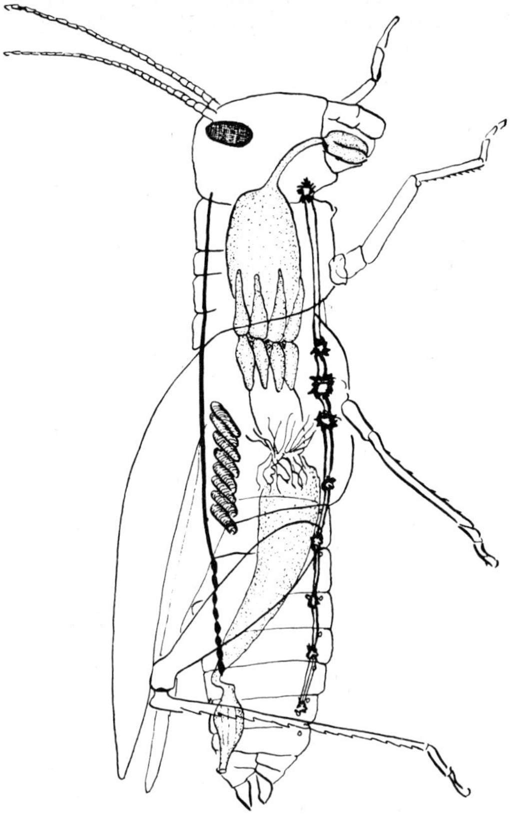

Figure 23.4

3. How have the different members of the phylum Arthropoda adapted to aquatic and terrestrial environments and which seem to be more advanced in terms of organization?

4. Why has there been a repeated tendency toward terrestrial life?

5. What factors of arthropod structure have limited the development so that they now form what is termed an "evolutionary dead end"?

24

Animal Evolution – the Deuterostomate Line

MATERIALS

CLASS STOCK

Preserved specimens
- *Arbacia*
- *Balanoglossus*
- *Molgula* or *Corella*
- *Amphioxus* (or slide)
- lamprey
- *Squalus* pup
- perch
- frogs and newts
- turtle
- lizard
- snake
- chick
- pig embryo
- mouse
- bat

Prepared slides
- tunicate *(Molgula, Corella, etc.)*
- *Amphioxis*

STUDENT STATION

- dissecting pan
- pins
- scissors
- microscope
- microscope lamp
- lens paper

INTRODUCTION

Unlike that of the preceding three-layered animals, the lining of the body cavity of the Enterocoelomates is mesoderm of endodermal origin formed by an outpocketing during embryological development. Another feature that distinguishes the enterocoelomates is the manner of formation of the complete digestive tract. In acoelomates there remains only a single orifice through which food material enters and waste material exits. In both Pseudocoelomates and Schizocolomates, the blastopore opening, which is developed early in embryological life, becomes the mouth, and a second opening to the digestive tract, which is formed in the course of development, becomes the anus. Because of this sequence of events, these groups are called Protostomes. The enterocoelomates differ in that the blastopore opening becomes the anus, and a new opening develops for the mouth. These are therefore known as Deuterostomes.

The fate of the second opening to the digestive tract, and other evidence involving the particulars of larval development, have caused the grouping of the seemingly unlikely phyla Echinodermate (starfish) and Chordata (vertebrate forms including man) as Enterocoelomates (see Fig. 24.1).

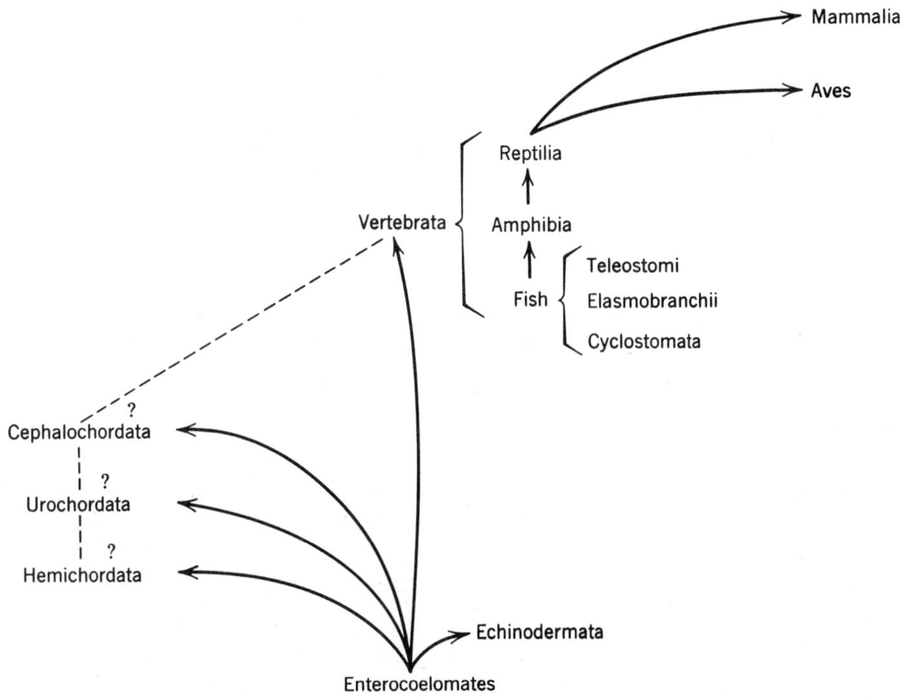

Figure 24.1

Phylum Echinodermata

General Characteristics

Radially symmetrical; some built in "five-part plan"; skeletal structures—usually interlocking calcareous plates—derived from mesoderm; water vascular system; bilaterally symmetrical larvae; organized organ systems.

Classification

Class Asteroidea—starfish.
Class Echinoidea—sea urchins and sand dollars.
Class Holothuroidea—sea cucumbers.
(The only class that will be considered is Asteroidea—the starfish).

Energy-Obtaining Systems

The digestive process of the starfish is very complicated. The mouth is located in the center of the ventral surface and is surrounded by a membranous area. The anus is located in the center of the aboral surface (side without the mouth or oral opening). The entire digestive system, stomach, intestine, and rectum is located between these two surfaces and does not branch into the arms of the starfish. The starfish feeds mostly on bivalve molluscs (clams, etc.). As the powerful muscles of the starfish pull the shells apart, the cardiac stomach is extruded through the mouth. The slightest opening between the shells permits the entrance of the powerful enzymes which are secreted by the digestive glands into the stomach. These digestive glands extend into the arms or rays. These enzymes convert the food material into a veritable broth, which is then "swallowed" and brought into the system where it is absorbed by the lining of the pyloric stomach and the digestive glands. The extruded cardiac stomach is then drawn back into the body.

Respiration in the starfish is accomplished by tiny fingerlike projections which emerge from the skin between the spines. These projections originate from the body cavity and contain coelomic fluid. Oxygen is absorbed from the water, enters the coelomic fluid, and is dispersed throughout the body. These dermal branchae or skin gills also release CO_2 into the surrounding water. The coelomic fluid is the only means of circulation, and is passive rather than involving a pumping mechanism.

Excretion is achieved by a rather unique system in the starfish. Ameboid cells collect wastes from the body and deposit these wastes into the coelomic fluid. Eventually the waste materials collect in the dermal branchae and the skin gill itself is excluded from the body.

Energy-Utilization Systems

The sexes of starfish are separate. A pair of either ovaries or testes lies in each arm; these open to the outside by small pores on the central disc between the arms.

The nervous system is very primitive, consisting of a nerve ring within the central disc and five major branches to each of the arms. The muscular system, however, is very highly developed. This would be expected in view of the method of obtaining food. The gross movements of the body involve the water vascular system, a unique adaptation that involves the use of a type of internal hydraulic pressure. On the aboral surface is a hard structure called the madreporite, which initiates the system. Specific amounts of sea water are taken into the madreporite and proceed into the five arms or rays of the starfish. On the ventral surface of each arm are a series of tube feet, each controlled by an ampulla. When water is forced out the tube feet, the tube foot becomes longer. When the tube foot is attached to an object, contraction will bring water into the ampullae and a suction is set up. Through the coordination of the tube feet, the animal may move in a slow awkward fashion.

1. Examine the preserved specimens of starfish, brittle star, and/or *Arbacia* (sea urchin).
2. If enough starfish are available, students may dissect. If not, several dissected specimens will be available for drawing.
   (a) Place the animal on a dissecting pan, aboral side up.
   (b) The two arms in between which is found the madreporite are called the bivium. Cut the tip of each of the arms of the trivium, or other three arms.
   (c) Cut along the sides of each of the arms, coming together at the central disc.
   (d) Place the animal in water to loosen the tissue and carefully remove the upper body wall, separating any internal organs which might adhere to the skin. Remove the body wall from the entire central disc, being careful to cut around the madreporite.
3. On Fig. 24.2, label and distinguish the following organ systems by filling in their characteristic textures:

   cardiac stomach—thick-walled structure just below the mouth.
   pyloric stomach—thin-walled structure following the cardiac stomach on the aboral side.
   digestive glands—extend into the five arms, and fill most of the space in the arm. Their secretions empty into the pyloric stomach.
   reproductive organs—feathery structures which proceed about one-third of the way up each arm.
   tube feet—found on the oral side.
   ampullae of the tube feet—found along the floor of the oral side.
   madreporite—hardened circular structure found just off center on the aboral side.
   dermal branchiae—found on the aboral side between the spines. Place the bivium (which should still be covered with the body wall) into some water and observe the pore openings.
   pedicellariae—pincer-like structures found around the spines and dermal branchiae. These have two jawlike parts and are thought to help keep the surface free of debris.
4. The digestive glands of the starfish secrete a powerful enzyme complement that is capable of liquefying animal matter. What are some of the ways in which the starfish stomach and digestive glands might be kept from digesting themselves?

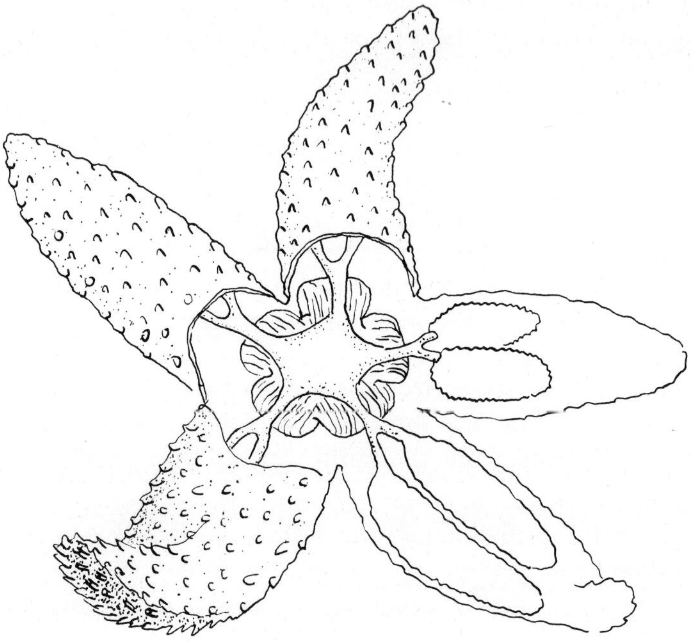

Figure 24.2

5. Why are the terms oral and aboral side used rather than dorsal and ventral?

Phylum Hemichordata:

General Characteristics

Wormlike; body divided into proboscis, collar, and trunk regions; short dorsal nerve cord in collar and a structure which may be a primitive notochord which is retained throughout life; definite pharyngeal gill slits.

1. Examine the preserved speciment of *Balanoglossus,* the acorn worm.
2. Locate those external elements associated with the general characteristics.
3. Observe the structure of the external elements. What is the habitat of this animal? How did you come to this conclusion?

Phylum Chordata

This phylum contains all animals that are characterized by paired gill clefts originating embryologically from the pharyngeal pouches, a notochord present at some time during the course of their development (a solid mass of tissue somewhat like cartilage, of mesodermal origin, which lies immediately ventral to the nerve cord, this is replaced by the vertebral column in vertebrates); and a dorsal nerve cord. Since the specimens that possess these characteristics are highly diverse, they will be considered under their subphyla or classes.

Subphylum Urochordata

General Characteristics

Resemblance to chordate found only in the larval form—only pharyngeal slits persist in the adult; larval form contains notochord and dorsal nerve cord.

1. If available, examine the preserved adult specimens of *Molgula* and/or *Corella.*
2. Examine a whole mount slide of a tunicate *(Molgula, Corella).*
3. In the space provided, draw the outline of the larva, labeling the following parts:
   nerve cord—found in the dorsal side of the body and can be seen most readily in the tail.
   notochord—just ventral to the nerve cord.

brain—thickening of the nerve cord at the anterior portion of the animal.
pharynx—ventral to the brain. Contains the pharyngeal gill slits.

Subphylum Cephalochorda

General Characteristics

Adult form shows three diagnostic features of chordates—gill slits, notochord and dorsal nerve cord; no well defined head or brain; muscle patterns similar to those found in fish.

1. If available, examine the preserved specimen of *Amphioxus* (lancelets).
2. Examine a whole mount under low power.
3. Draw the outline of the *Amphioxus,* in the space provided, labeling the following parts:
   nerve cord—dorsally located.
   notochord—ventral to the nerve cord.
   pharyngeal gill slits—anterior end of *Amphioxus.* Note that there are many gill slits; but they are paired.

Subphylum Vertebrata

General Characteristics

Backbone of bone or cartilage (as in some fish); brain of five fundamental divisions; cranial nerves.

Energy-Obtaining Systems

Because of the complexity of the systems and the student interest generated by organisms closely allied to themselves, the task of discussing the respiratory, digestive, and circulatory systems in vertebrate animals is not an easy one. The development of each of the systems will be generally traced through the vertebrate types.

DIGESTIVE SYSTEM

As the student has undoubtedly noticed, the general pattern of the digestive system throughout the invertebrate groups has been amazingly consistent. The pattern undergoes only slight variation in vertebrate forms. Food generally enters the mouth. The presence of teeth is directly related to the diet of the animal. The parasitic lamprey, which takes most of its food in liquid form, uses its teeth for holding rather than masticating. The food passes into the pharynx and then into the esophagus. From here the food usually goes into the stomach. In some reptilian forms and in all birds the food goes into a crop for mastication before entering the stomach or gizzard. The food is partially digested in the stomach by enzymatic secretions. The glands involved in these secretions are anatomically obvious as pyloric caeca (outpocketings of the stomach wall) in lower vertebrate forms, but become incorporated into the stomach wall in higher types. The partially digested food is released into the small intestine, where further digestion and absorption take place. The solid wastes proceed through the large intestine from which they are released to the environment with the aid of the rectal muscles. As forms increase in size and complexity, the length and interior surface area of the small intestine tends to increase. Why is this a selective advantage?

RESPIRATORY SYSTEM

This system undergoes much variation in the evolution of vertebrate forms. The adaptation in terms of habitat goes from aquatic environment to terrestrial. Concomitant with this, the system of respiration must undergo considerable change. The type of system found in primitive vertebrate forms involves a series of paired pharyngeal pouches. The pharynx, which is part of the digestive tract, and hence is lined by endoderm, begins to exhibit outpocketings which expand outward toward the ectoderm. The ectoderm, meanwhile, moves inward to meet the endoderm. As they meet, clefts are formed. These are the pharyngeal gill slits. Associated with each of the gill slits is a major artery. If the gills are to be retained as the major respiratory system, the artery develops into an elaborate capillary system capable of taking dissolved oxygen from the water as it passes over the gills. Water enters the mouth of the animal, passes over the gills, and passes back into the environment. As this passage across the gills occurs, the gas exchange is accomplished.

The development of pharyngeal gill slits also occurs in animals that are destined to have lungs. Although the gill slits may be functionally closed during the course of development, the aortic arches, which form in response to the pharyngeal pouches, and which in the aquatic organisms form the respiratory mechanisms, are modified to carry out some of the major aspects of circulation.

The lungs develop from mesodermal tissue. They are a spongy tissue with many tiny air sacs or alveoli. In lower forms the inflation of the lungs is accomplished by the expansion of the rib cage, or, as in the case of some amphibians, by the swallowing of air. Either process causes a negative pressure in the thoracic cavity, resulting in the inhalation of atmospheric air. The gas exchange then occurs in the alveoli. In higher forms the basic principle is the same. However, a diaphragm separates the abdominal and thoracic cavities and aids in the efficiency of the air intake by causing a greater negative pressure upon contraction than is possible with the rib cage alone.

CIRCULATORY SYSTEM

The major role of the circulatory system is the transport of nutrient material (fuel) and the oxygen necessary for aerobic respiration. Therefore, just as the respiratory system underwent modification in the transition from aquatic to terrestrial forms, the circulatory system has adjusted accordingly. Just as the digestive system has undergone relatively little change, the portion of the circulatory system involved in the transfer of food elements has likewise undergone little change. Generally, elemental nutrient material is picked up from the small intestine, where absorption is taking place. In vertebrate forms the liver is generally used as a storage area for these nutrients, primarily carbohydrates. The capillaries that collect nutrients from the small intestine join to form a major vein, the hepatic portal vein, which once again branches into capillaries as it enters the liver (fulfilling the definition of a portal vein as one that begins and ends in capillaries). Blood vessels take this nutrient material to other parts of the body as needed.

The circulatory system, in performing its function, must fulfill two major qualifications. First it must be circulatory; that is, it must come back to any given point after passing through the body. Second, it must be capable of carrying oxygen and nutrient material to *any* point of the body with relative efficiency. The fulfillment of these two prerequisites establishes a highly complex system. In its course around the body, all blood must come in contact with the lungs or gills for oxygenation; with the liver and digestive tract for nutrient material; and with the kidney for removal of metabolic waste material. The fish, with a two-chambered heart, sends a major blood vessel anterior to the gills. Because the blood reaches the gills after passing through the heart, the heart carries only deoxygenated blood. After circulation through the gills the oxygenated blood is circulated both anteriorly and posteriorly by a dorsal and ventral aorta. At the tissue level the blood vessels branch into capillaries, giving oxygen to the cells and receiving metabolic wastes and carbon dioxide from the cells. The blood vessels once again group and form a large vein, which joins the major vein of the body, the post cardinal or anterior cardinal, and returns to the heart. The amphibian heart has three chambers, and both oxygenated and deoxygenated blood passes through. Deoxygenated blood enters the right auricle from the sinus venosus. From there it proceeds into the ventricle and then into the pulmonary artery, which takes the blood to the lungs. After oxygenation and deposition of carbon dioxide, the blood returns to the heart, entering the left auricle and proceeding once again to the ventricle. This time the blood leaves the ventricle and enters the aorta. As the aorta arches dorsally, it forms several branches which bring oxygenated blood to head and forelimbs. The aorta then proceeds dorsally, bringing the oxygenated blood to posterior tissues. The exchange of gases is once again made on the cellular level through the thin capillary walls and the vessels rejoin to form the venous sytem. A single large vein, the postcaval vein, comes from the posterior part of the animal, and a single large vein, the pre-caval vein, comes from the anterior portion. These join at the sinus venosus and the blood once again enters the right auricle as deoxygenated blood. The mammalian system is not unlike the amphibian. The efficiency is greatly increased because there is a four-chambered heart and the oxygenated and deoxygenated blood are kept entirely separate (see Fig. 24.3).

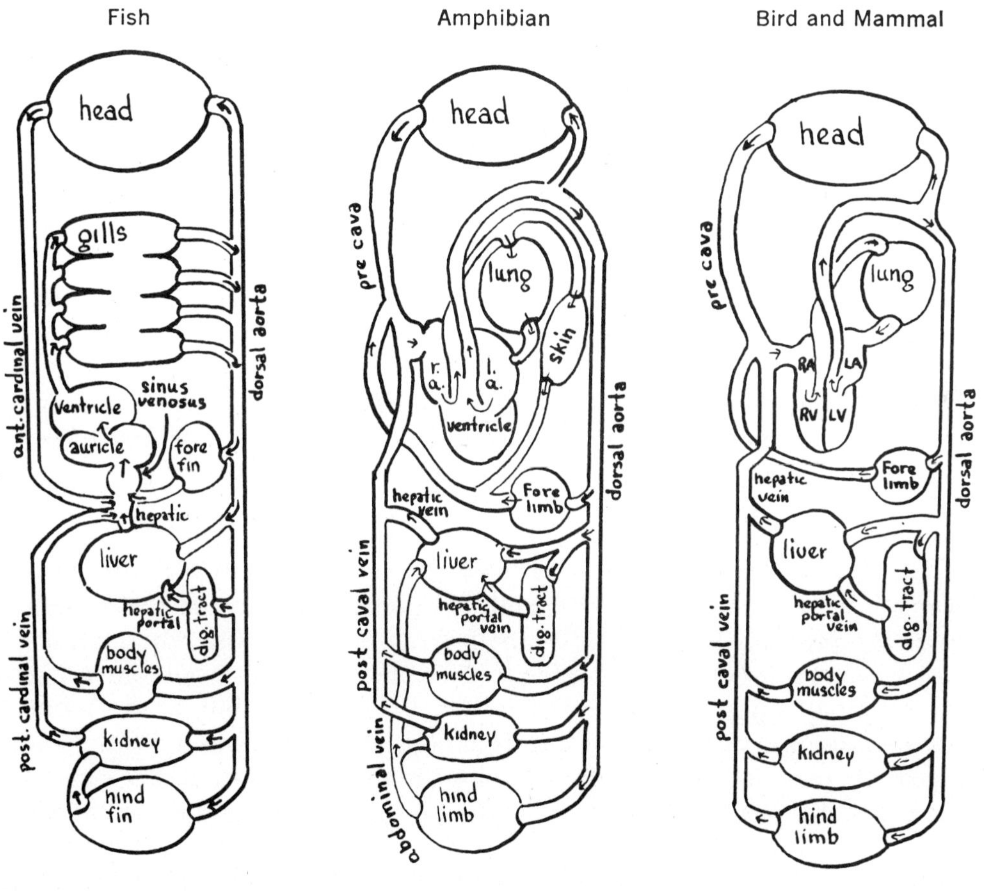

Figure 24.3

EXCRETORY SYSTEM

The development of the excretory system through the class Vertebrata is very complicated and will not be dealt with in any detail. The kidney extracts waste materials from the blood as it passes through. The structure of the kidney in primitive forms is very diffuse, owing in part to the primitive type of tubular structure within the organ. In the Reptilia, Aves, and Mammalia, the kidney is made up of a more compact and efficient type of tubule and thus becomes more compact. The collected waste, known as urine, descends down a tube or ureter into a muscular bladder. When the bladder becomes turgid, the muscles contract, dispelling the waste material through a urethra. In lower forms the urethra exists into the cloaca; in higher forms the urine is dispelled through organs associated with reproduction. The development of the urinary ducts and that of the reproductive ducts have been closely associated throughout the chordata. The means of expelling urine is dependent upon the type of reproductive duct provided by the organism.

Energy-Utilization Systems

REPRODUCTIVE SYSTEM

The reproductive system varies with the particular organism. In fish, where external fertilization is common, the eggs or sperm leave through the urogenital opening into the cloaca. The male usually lies over the eggs while dispelling the sperm, which increases the chances of fertilization. In amphibian forms the eggs may be fertilized externally or internally, depending upon the organism. In frogs, the male clasps the female in such a way as to squeeze the ripe eggs from the body. The eggs are then fertilized externally. Amphibian eggs must be laid in a water environment because there is no protective adaptation. Reptilian forms are totally adapted to terrestrial life by providing a shell for the developing embryo or bearing the young live. For either of these two methods, fertilization must occur internally. This pattern of internal fertilization is continued in both the bird and mammalian forms.

NERVOUS SYSTEM

The nervous system becomes more highly developed in the course of evolution in the vertebrates. The primary development is that of the brain. In fish, those areas of the brain which control reaction, coordination, and instinctual patterns are most highly developed. The cerebral hemispheres of the shark are pronounced, but do not control conscious action as in the mammal. As the functions of the different parts of the brain are relocated, the cerebral hemispheres develop in such a way that the once linear brain is forced into involutions. This reaches a pinnacle in the human brain, where almost all that is visible is the cerebral hemispheres. The general patterns of cranial nerves and those nerves associated with the dorsal nerve cord undergo relatively little change (see Fig. 24.4).

The muscular system is vertebrate forms becomes very highly complex as one proceeds through the classes. The fish have a relatively simple musculature consisting of a pattern

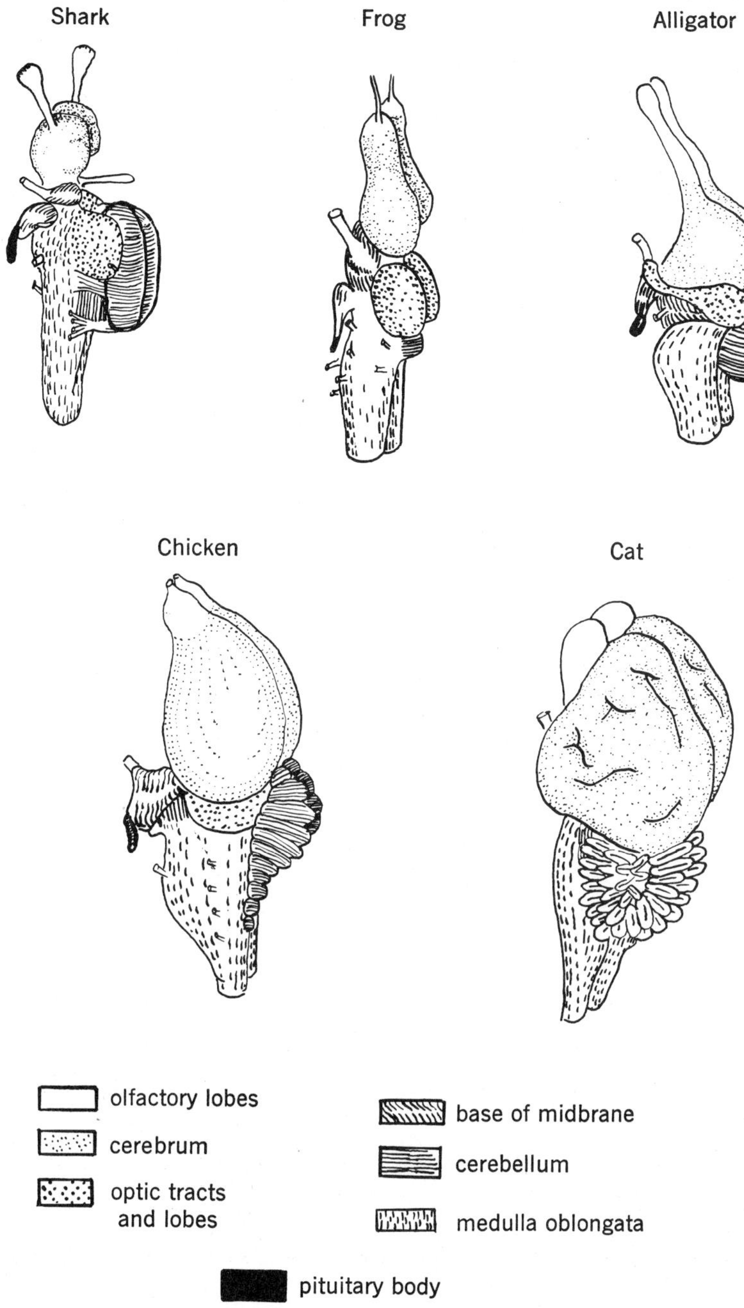

Figure 24.4

of V's which have their apex at the lateral body line. Anyone who has eaten fish, which is nothing more than eating the muscle tissue, is quite aware of this pattern. With the advent of limbs and the necessity of locomotion, the muscle system has become associated with the skeletal system. The skeletal or voluntary muscles of limbed animals are associated with the bones in such a way that contraction of a muscle brings a bone into a different position. Suffice it to say that the musculature of the human, which is involved in the coordination as well as the movement of the body is very complex.

Class Cyclostomata

General Characteristics

No jaws or paired limbs; primitively developed organ systems; typical vertebrate characteristics.

1. Examine the preserved specimen of the brook lamprey.
2. In the space provided, draw the lamprey, labeling the following parts:
   round mouth—hence the name (cyclostome)
   denticles or horny teeth—found inside the mouth and used for holding on to the host fish.
   tongue—in center of the mouth.
   lidless eyes
   urogenital sinus—posterior ventral side; the excretions of the urinary and the products of the genital systems exit from this opening.
   anal opening—located just anterior to the urogenital sinus.

3. How are the mouth parts adapted for parasitic existence?

Class Chondrichthyes

General Characteristics

Cartilaginous skeleton; gills; jaws; ventral mouth; scales.

1. Examine the preserved specimen of *Squalus* pup, the young dogfish shark.
2. If dissected specimens are available, fill in Fig. 24.5 with the texture of tissue and the following labels:

   teeth—modified scales found in the mouth.
   esophagus—leading from mouth to stomach.
   stomach—J-shaped structure used for storage and digestion of food.
   intestine—long tubular structure leading from the stomach to the colon or large intestine.
   spiral valve—within the intestine this spirally arranged structure delays the passage of food and increases surface area for absorption of nutrients.
   colon—the enlarged portion of the intestine anterior to the exit of the tube from the body.
   gill rays—found under the gill slits, these rays show marked vascularization (increased blood supply).
   heart—a pear-shaped organ found in the thoracic cavity.
   liver—large lobed structure found at the junction of the esophagus and the stomach. If the specimen has been injected to show the blood vessels, the liver may be blue, for the major blood vessels which contribute to its storage function are veins.
   kidneys—flat mass of tissue somewhat featherlike in appearance but not in texture which lie along the dorsal wall of the abdominal cavity.
   bladder—thin-walled bag found at the posterior region. Exits to the urogenital sinus.
   nasal openings—olfactory sense, sensitive to salt concentration of water. These do not communicate with the pharynx.
   pelvic fins—paired; found posteriorly.
   pectoral fins—found anteriorly.
   dorsal and caudal fins—found on the dorsal and tail region respectively.
   clasper fins—found on the male associated with the pelvic fins.

Class Amphibia

General Characteristics

Young stages aquatic; adult stage terrestrial; lungs; four limbs; eggs must be laid in water, since they have no protective shell.

1. Examine the preserved specimens of frog and newt (salamander).

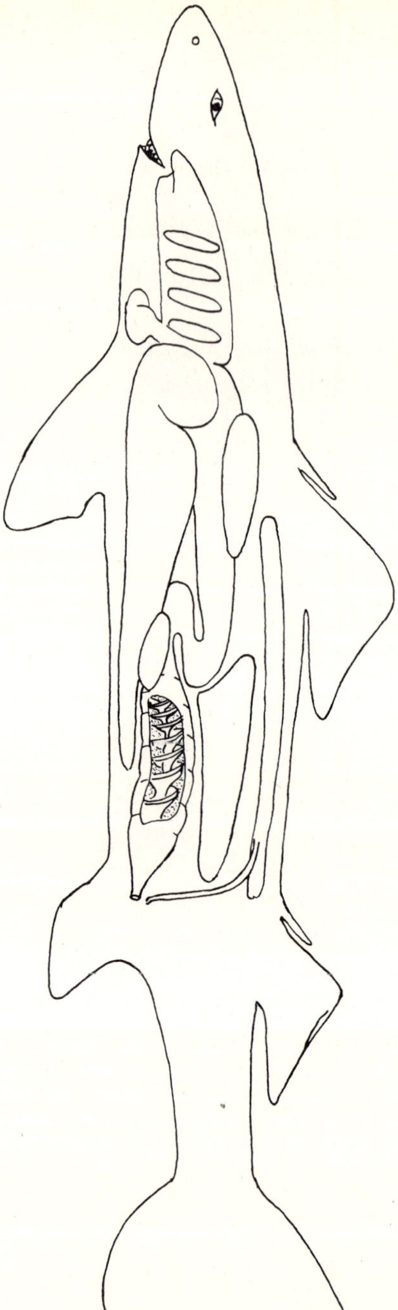

Figure 24.5

2. If enough frogs are available, students may dissect. If not, sufficient numbers of dissected specimens will be available for study. Note: Examine the circulatory system in demonstration frogs which are specially injected.
   (a) The center of the ventral side is marked by the linea alba or white line. Insert the scissors just to the side of the linea alba in the region of the hind legs, and carefully cut anteriorly to the lower jaw. Be careful to lift the skin as you cut so that none of the underlying organs are disturbed.
   (b) Cut around each leg, and along the jawline and open the skin outward, pinning it to the dissecting pan. If the muscle layer has not been cut, make the same incision and pin back with the skin.
3. Examine Figs. 24.6, 24.7, and 24.8, making sure that you can identify all the labeled internal structures.
4. How is the frog adapted for the land environment?

5. The heart of the frog has three pumping chambers. How does this increase the efficiency in terms of oxygen availability to the tissues?

6. The osteichthyes have lost the ability to generate a pelvic girdle. The amphibia standardly have a functional pelvic girdle. What does this indicate concerning the type of marine life that may have given rise to the amphibian stock?

7. Are the adult forms of amphibia completely free of the water environment? Why?

INTERNAL VISCERA

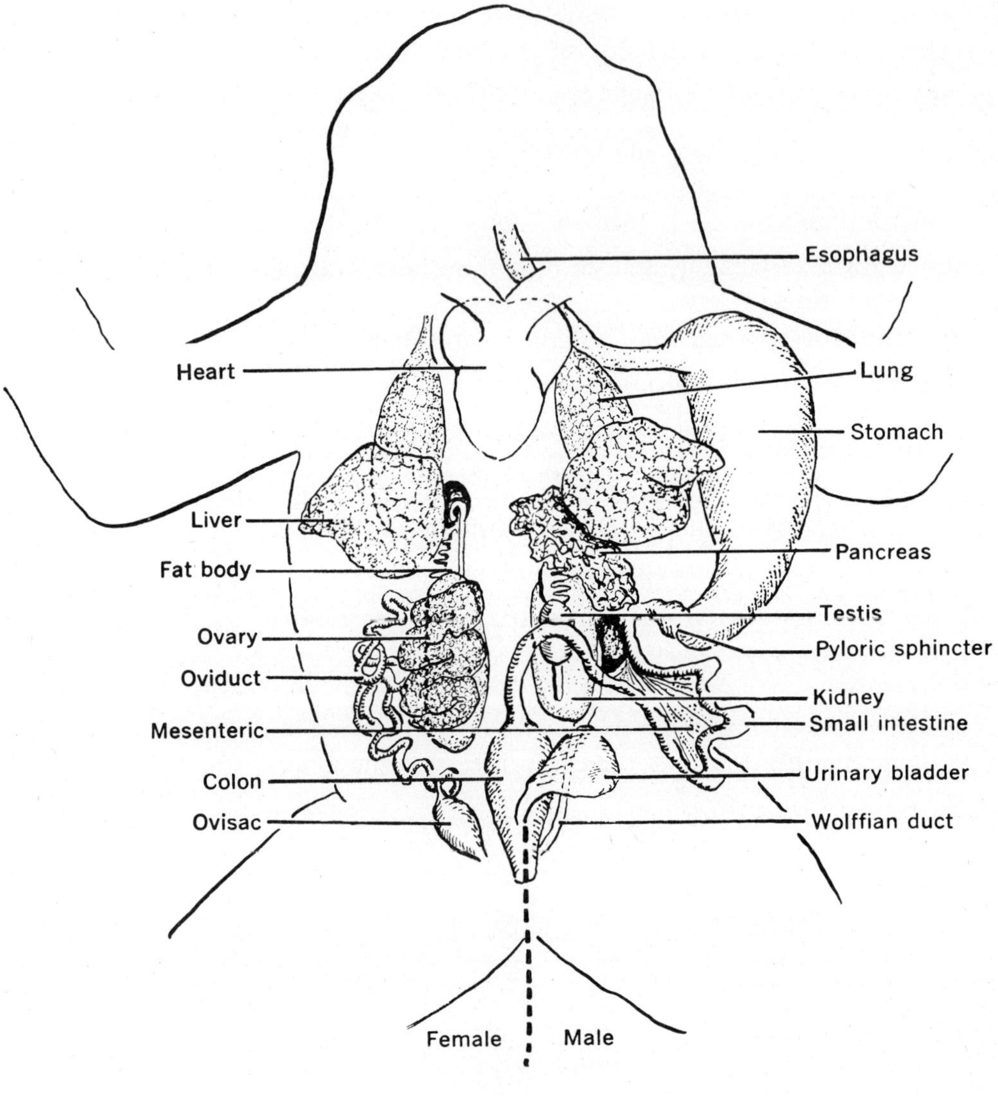

Figure 24.6

ARTERIES

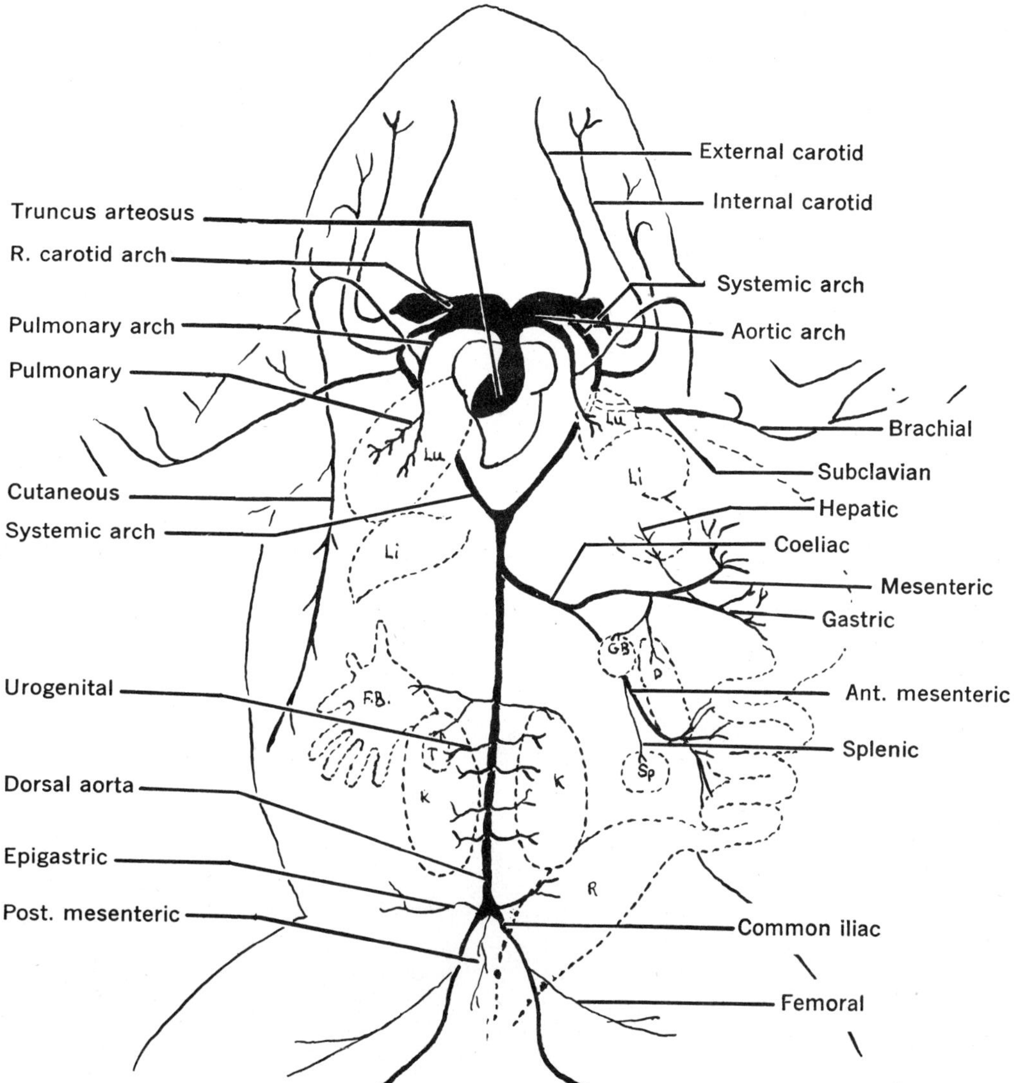

Figure 24.7

VEINS

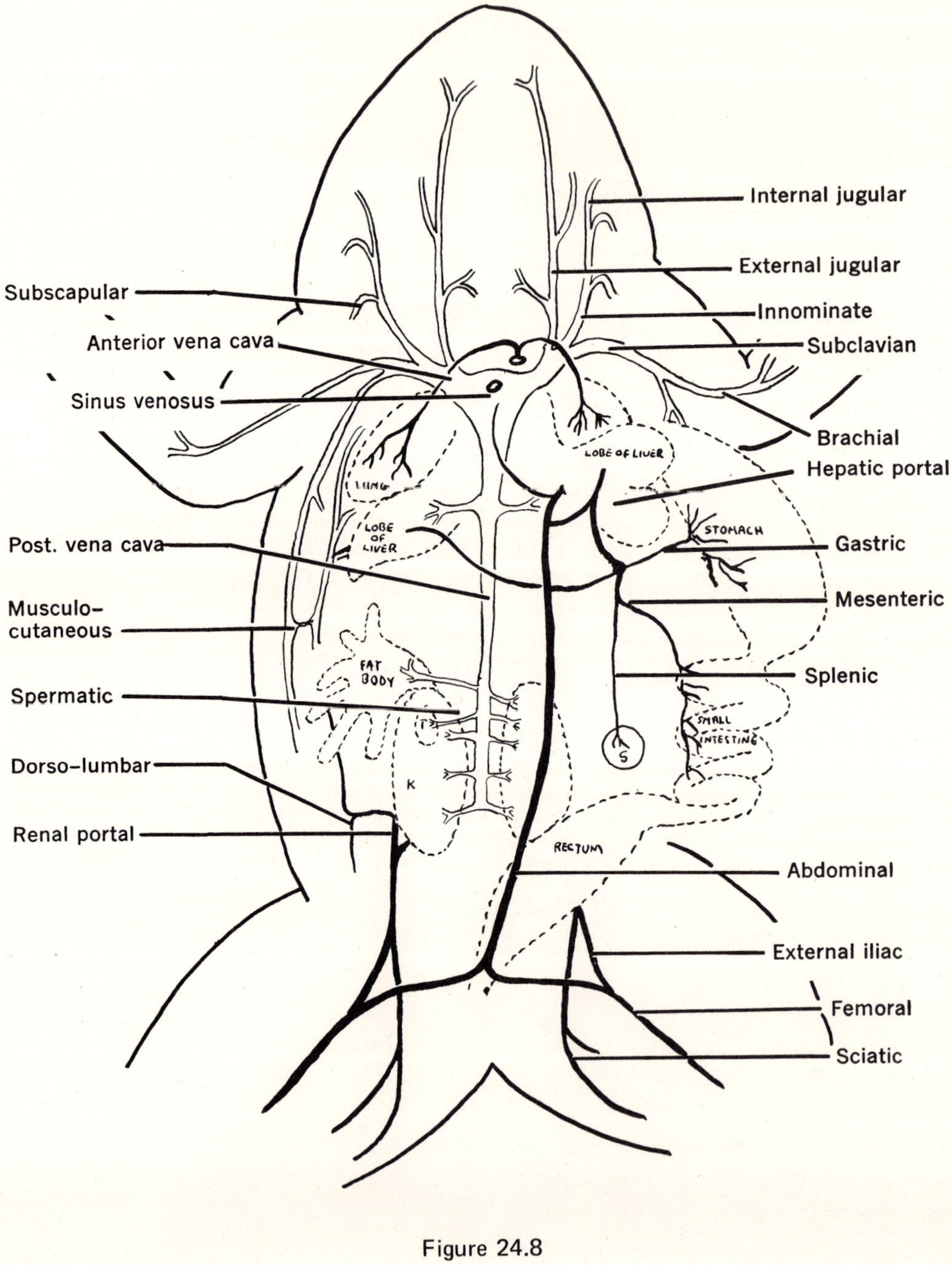

Figure 24.8

Class Reptilia

General Characteristics

Dry, scaly, impervious skin; production of shelled eggs capable of being laid without surrounding water. Sea snakes must leave water to reproduce.

1. Examine the preserved specimens of small turtle, small lizard, and small snake.
2. Note the differences between the Amphibian structure and the reptilian structure.

Class Aves

General Characteristics

Four-chambered heart; constant body temperature (homiothermal); feathers; anatomically closely related to reptiles.

1. Examine the preserved specimen of the chick.
2. Note the typical bird characteristics: the toothless, horny beak; the forelimb transformation into wings; the single posterior pair of limbs which are clawed and scaly, thus resembling the reptilian ancestor; and the body architecture which suits the flying animal.

Class Mammalia

General Characteristics

Four-chambered heart; constant body temperature; presence of true hair; young nourished by milk; muscular diaphragm separating thoracic and abdominal cavity; red blood cells without nuclei; young usually born alive.

1. Examine the preserved specimen of the pig embryo, and if available, the mouse and the bat.
2. Note obvious mammalian characteristics.

The following section deals at length with mammalian systems, their structure, and function.

Part VII

THE MAMMAL AS A PRODUCT OF EVOLUTION

25

Energy-Obtaining Systems

MATERIALS

CLASS STOCK

- chemicals
  - 1.0% starch solution
  - IKI solution
  - 0.2% NaOH
  - Benedict's solution
  - rennin tablets
  - albumen solution
  - pepsin solution
  - 0.2% HCl
  - pancreatin
  - 2.0% Na_2CO_3
  - 2.0% bile salts solution
  - 2.0% soap solution
  - phenol red solution
  - pyrogallol
  - conc. KOH
  - CO_2 in pressure tank
  - O_2 in pressure tank
- water baths: 37°C or incubator, boiling
- beef pancreas
- distilled water
- milk

STUDENT STATION

- test tubes
- wax pencil
- millimeter ruler
- Bunsen burner
- glass rod
- beakers
- glass tubes (straight and bent)
- 2-hole rubber stoppers
- rubber tubing

CLASS STOCK (cont'd)

salad oil
sheep heart
frog heart
defibrinated blood

INTRODUCTION

The energy-obtaining systems of the mammal are the result of complicated selective processes of evolution. The digestive system, for example, is structurally adapted to fulfill the needs of the mammalian diet in breaking down food material to the level where it can be used as a direct energy source in cellular metabolism. In principle, however, the mammalian digestive system is analogous to the food vacuole of the amoeba or the gastrovascular cavity of the coelenterates. In fact, many of the coenzymes involved in the breakdown of foodstuffs are the same in all metazoans.

The same is true for the aspects of the circulatory system directly involved in nutrient, oxygen, and waste transport, the respiratory system, and the excretory system. In principle they are the same in all metazoans; structurally, they are adapted to the mammalian needs.

The Breakdown of Foodstuffs

The general pattern of food breakdown is one of progressive degeneration. Carbohydrates are usually brought into the system in the form of polysaccharides. These polysaccharides are broken down to disaccharides and finally to the monosaccharide form that may be utilized directly by the cell. Proteins are reduced to polypeptides, then to dipeptides, and finally to the basic component, amino acids. Fats are first emulsified to afford an increased surface area for enzyme activity, and finally broken down into glycerol and fatty acids.

The degradation to the final usable form of each nutrient type must be completed by the time the food reaches the middle section of the small intestine, where most of the absorption into the body takes place. Therefore, all the digestive enzymes are found to be secreted by structures which precede this area.

Mouth
- polysaccharide ⟶ disaccharides

Stomach
- protein ⟶ polypeptides

Duodenum
- polysaccharide ⟶ disaccharides
- disaccharide ⟶ monosaccharides
- protein ⟶ polypeptides
- polypeptide ⟶ smaller polypeptides

polypeptide ⟶ dipeptides
fat ⟶ (emulsified)

Intestine
disaccharide ⟶ monosaccharides
dipeptide ⟶ amino acids
fat ⟶ glycerol and fatty acids

The following exercises are arranged in an order that will, hopefully, give maximum understanding. Some, however, take a longer time to execute. It is suggested that you read the entire exercise for understanding and start on those experiments that require 1½ to 2 hr incubation time.

ANATOMY OF THE DIGESTIVE SYSTEM

See Appendix, p.

ACTION OF DIGESTIVE ENZYMES

Carbohydrate Digestion

Amylase

Amylase is found in saliva and is responsible for the breakdown of plant starch (polymerized glucose), to the disaccharide maltose, a dipolymer of glucose. The activity of this enzyme on starch may be measured by the IKI test, which gives a blue-black color in the presence of starch, and Benedict's test, which gives a brick red color in the presence of reducing sugar, formed as the result of the enzyme action.

1. Mark a clean large test tube with a wax pencil at 3 cm and 10 cm from the bottom of the tube.
2. Collect 3 cm of your own saliva in the tube. Add water to the second mark. This is your enzyme preparation.
3. Label five clean test tubes 1 to 5.
4. Mark each tube 2 cm, 3 cm, and 4 cm from the bottom.
5. To each tube add 1% starch suspension up to the 2-cm mark.
6. Add to each tube as follows:
   (1) Water up to the 4-cm mark (control).
   (2) Enzyme preparation up to the 4-cm mark (full strength).
   (3) Enzyme preparation up to the 3-cm mark, water up to the 4-cm mark (½ strength).

(4) Enzyme preparation halfway to the 3-cm mark, water up to the 4-cm mark (¼ strength).

(5) Enzyme preparation (boiled 5 min and cooled) up to the 4-cm mark (effect of boiling on enzyme activity).

7. Incubate in a 37°C water bath or an incubator for 15 to 20 min (30 min if in a dry incubator.

Note: If incubating in a dry incubator, fill a beaker halfway with water adjusted to 37°C (tap) and place tubes in this.

8. Label another set of five tubes 1*a* through 5*a*.
9. Mark each tube 2 cm from the bottom.
10. At the end of the incubation time, pour each of the incubated samples into each of the tubes with the same number up to the mark. This will give you two series, each containing an equal amount of the test solutions.
11. Add 2 drops of IKI to each of the *a* series of tubes and record results in Chart 25.1.

CHART 25.1

	Control	Full	½	¼	Boiled
Starch test					
Benedict's test					

12. Add Benedict's solution up to the 3-cm mark of the incubated tubes. Place in boiling water for 5 min or carefully hold over a Bunsen burner until boiling. Record your results on Chart 25.1, indicating strength of reactions from ++ for intense to – for no reaction.
13. How did the experimental conditions simulate those of the body?

14. What does the effect of boiling on the activity of the enzyme show concerning the chemical nature of the enzyme amylase?

15. How does enzyme concentration affect activity?

16. Why does saliva usually not have the opportunity to work full strength on the starch that enters the body?

17. Since not all the starch is broken down by amylase in the mouth, (*a*) what is the selective advantage of an animal's having salivary amylase? (*b*) Where would you expect to find further breakdown of starch?

18. Why does diluting the saliva before the experiment bring the conditions closer to those in the body? (Hint: What is the ratio of water to amylase in saliva when one is actively eating?)

Pancreatin

Pancreatin is a conglomerate of digestive enzymes extracted from the pancreas. The pancreas is a gland that lies in the curve between the stomach and the duodenum. It acts as both an exocrine and an endocrine gland. As an endocrine gland it secretes insulin, a substance important in glucose metabolism. As an exocrine gland it secretes a host of digestive enzymes that enter the intestine through the pancreatic duct. Among these enzymes is pancreatic amylase.

Pancreatin solution may be either prepared by the instructor or extracted by the student from beef pancreas (sweetbreads).

EXTRACTION OF PANCREATIN

1. Cut the pancreas into ½-in. cubes.
2. Place two cubes in the base of a clean test tube.* Add water to a depth of about 1 cm.
3. Mash with a glass rod. Fill the tube with distilled water and pour off into a beaker.
4. Add 1 cm of water and mash again. Fill with water and pour into the same beaker. This is your enzyme preparation.

ACTIVITY OF PANCREATIN ON CARBOHYDRATE DIGESTION

1. Mark two clean test tubes 1 and 2.
2. Mark each 2 cm, 3 cm, and 5 cm from the bottom of the tube.
3. Add pancreatin solution up to the first mark in each tube.
4. Boil the pancreatin in the first tube for 30 sec over a Bunsen burner or place in a boiling water bath for 5 min. Add water to the first mark to restore the volume.
5. Add 0.2% NaOH to the second mark in each tube.
6. Add 1% starch solution to the 5-cm mark and shake.
7. Incubate for 15 min in a 37°C water bath, or for 30 min in a dry incubator.
8. Label two other tubes 1*a* and 2*a*. Mark each 2 cm from the bottom of the tube.
9. After incubation is complete, pour 2 cm of the material in each incubated tube into the *a* series tube.
10. Test tubes in *a* series for starch with IKI, and test those of the regular series for reducing sugar, adding Benedict's solution up to the 5-cm mark, and boiling.
11. Record the data on Chart 25.2

CHART 25.2

	1	2
Starch		
Benedict's		

12. What aspects of the experiment simulated conditions of the body?

*Or place 20 cubes with 20 ml. of water in a blender and run for 1 min. on high speed. Divide among students.

13. What does the effect of boiling the enzyme show about the chemical nature of pancreatic amylase?

14. Since there is amylase in the saliva, what is the advantage to the organism to have amylase secreted by the pancreas too?

15. What is the advantage of ensuring the complete breakdown of carbohydrates? Why are carbohydrates important to the body.

Protein Digestion

Protein digestion is begun in the stomach. First, rennin acts to change the consistency of milk, which contains a great deal of protein, but does not alter the bonds of the protein, structure. Pepsin, which enters the stomach in the inactive pepsinogen form, is activated by the acidic condition of the stomach and reacts on several specific sites of the protein molecules, breaking them into large polypeptides.

The material empties into the intestine. The trypsinogen secreted by the pancreas is activated by the basic condition of the duodenum. In turn, trypsin activates chymotrypsinogen to its active form, chymotrypsin. Each of these digestive enzymes acts at a different site on the protein molecules, thus breaking them into smaller polypeptides and dipeptides. Dipeptidases, secreted by the intestine and pancreas, are responsible for the final breakdown of dipeptides into amino acids.

Rennin

1. Mark two clean test tubes 1 and 2.
2. Mark each tube 1 cm and 6 cm from the bottom.
3. Place half a rennin tablet in each tube.
4. Fill with water to the first mark and shake to dissolve the rennin.
5. Boil the rennin solution in tube 2 for 1 min, and add water to restore initial volume.
6. Fill to the 6-cm mark with milk.
7. Incubate at 37°C for 15 min in a water bath, 30 min if incubation is dry.
8. Pour each tube into a beaker, noting consistency.
9. Fill in Chart 25.3 indicating whether the material in the tube showed curdling.

CHART 25.3

	Rennin	**Boiled Rennin**
Consistency		

10. What is the difference between coagulation and curdling?

11. What is the advantage of having a substance such as milk in a solid form in the stomach rather than in its natural liquid state?

Pepsin

1. Label five clean tubes 1 to 5.
2. Mark each tube 2 cm, 4 cm, and 6 cm from the bottom.
3. In each of the tubes, put an albumen solution up to the first mark. Try not to take the "sticky" albumen if the solution has been made from the egg.

4. Fill each of the tubes as follows:

	Second Mark	Third Mark	
(1)	water	water	control
(2)	pepsin solution	water	
(3)	pepsin solution	0.2% HCl	
(4)	pepsin solution	0.2% NaOH	
(5)	boiled pepsin solution (boiled one minute)	0.2% HCl	

5. Shake each tube.
6. Incubate in a bath or incubator for 1½ to 2 hr.
7. Apply the biuret test to each test tube.
8. Record the results on Chart 25.4, indicating the presence of protein or polypeptides as shown by the biuret test.

CHART 25.4

	1	2	3	4	5
Protein					
Polypeptide					

9. Which conditions of the experiment simulate those in the region of the body where pepsin is found? Is this substantiated by the results of the experiment? (Explain.)

10. What is the chemical nature of the enzyme pepsin? What has led you to this conclusion?

Pancreatin

1. Label four clean test tubes 1 to 4.
2. Mark each 2 cm, 4 cm, and 6 cm from the bottom.
3. In each of the tubes, put albumen solution up to the first mark. Try not to take the "sticky" albumen.
4. Fill the tubes as follows:

	Second Mark	Third Mark
(1)	water	water
(2)	pancreatin	0.2% HCl
(3)	pancreatin	0.2% NaOH
(4)	boiled pancreatin (boiled one minute)	0.2% NaOH

5. Shake and incubate tubes for 1½ hr at 37°C.
6. Test all tubes by the biuret test and indicate on Chart 25.5 the presence of protein and/or polypeptides.

CHART 25.5

	1	2	3	4
Protein				
Polypeptide				

7. From the results of the experiment, is the optimal pH for the activity of pancreatic peptidases above or below 7? Explain.

8. How does the optimal pH as shown by the experiment compare with natural conditions under which this group of enzymes operates?

9. What is the chemical nature of the enzymes responsible for the breakdown of protein? What question does this raise?

10. Under natural conditions would these enzymes be confronted with a protein of the complexity and size of albumen? Explain.

Fat Digestion

Because of the high degree of insolubility of fats in water, the body has devised indirect means of breaking down lipid material. Fat in a water environment will take a spherical form, yielding a maximum volume to a minimum of surface area. This is an adverse condition for chemical reactions because these reactions are dependent on the availability of a substrate on which to act. Bile salts act as emulsifying agents, reducing the surface tension of fat and causing the fat globule to break up into smaller particles which remain in suspended form. By increasing the number of spherical forms, with accompanying reduction in size, the surface area is increased.

Lipase from pancreatic juices then acts chemically upon the lipid to yield one glycerol and three fatty acid molecules, the components of a fat.

Emulsification

1. Label five clean test tubes 1 to 5.
2. Mark each 5 cm from the bottom.
3. To each add as follows to the mark:

   (1) distilled water
   (2) 2.0% Na_2CO_3
   (3) 2.0% albumen
   (4) 2.0% bile salts
   (5) 2.0% soap solution

4. Add 5 drops of salad oil to each tube.
5. Shake all tubes and time how long it takes for a definite layer of oil to be formed. Record the time on Chart 25.6.

CHART 25.6

	1	2	3	4	5
Time					

6. Which seems to have the finest particle size?

7. Which seems to have the coarsest particle size?

8. Is the size of the particles correlated with the stability of the emulsion? Explain.

Pancreatin

1. Label two clean test tubes 1 and 2.
2. Mark each tube at the 3-cm and 6-cm levels from the bottom of the tube.
3. Add pancreatin to each tube to the 3-cm mark.
4. Boil the pancreatin in tube 2 for 1 min, adding water to restore the original volume (up to the first mark).
5. To each tube add bile salt solution to the second mark.
6. Add 2 to 3 drops of phenol red indicator to each tube. This is a pH indicator which gives a red color under basic conditions and a yellow color in an acid medium. The solution after adding phenol red should have a slightly yellow appearance.
7. Add 0.2% NaOH drop by drop to one tube, shaking after each addition until the tube has a slight red color (2 to 3 drops should be enough).
8. To each tube add 5 drops of salad oil.
9. Shake and incubate for 1½ hr at 37°C.
10. After incubation examine each tube by holding it against a piece of white paper to note the color change. Record the data on Chart 25.7.

CHART 25.7

	1	2
pH		

11. What are the products of the action of lipase on fat?

12. What has caused a change in the pH in tube 1?

13. What is the chemical nature of the enzyme lipase in pancreatin? What has brought you to this conclusion?

14. Why were bile salts added to the tubes?

TRANSPORT OF NUTRIENTS, OXYGEN, AND WASTES

A highly organized circulatory system is peculiar to complex forms of animal life, both vertebrate and invertebrate. However, the principle involved is universal to all organisms. In unicellular forms, cytoplasmic streaming probably accomplishes the distribution of materials throughout the cell (organism), thus increasing the exposed surface area and potential for chemical reaction as well as providing all parts of the cell with necessary nutritional factors. The vertebrate body has an elaborate enclosed system which accomplishes much the same thing. The circulatory system, by virtue of evolutionary changes, is integrated with all the other systems of the body and aids in their functions by providing them with nutrient and catalytic materials and removing waste materials and products of their secretion. In addition to this relationship with other systems, in which the circulatory system acts primarily in its transport capacity as part of the energy-obtaining processes, certain processes, unrelated to its ability to transport materials, are carried out by this system. The processes are categorized as energy utilization and will be discussed in the next chapter.

ANATOMY OF THE CIRCULATORY SYSTEM

See Appendix, pp. 474-476

THE HEART

The most basic function of the circulatory system is the movement of blood in the body. The organ that provides the force to propel the blood through the vessels is the heart. The heart, located in the pericardial cavity of the chest or thoracic region, manifests a continuum of rhythmic contractions. Each of these contractions forces the blood further along the blood vessels. In addition to this, the heart handles both oxygenated and deoxygenated blood, and its ability to keep each type from contaminating the other is directly related to the degree of possible efficiency of the system.

Therefore, the structure of the heart in relation to its capacity for handling oxygenated and deoxygenated blood largely determines the possible efficiency of the circulatory system.

1. Obtain a sheep heart and examine carefully.
2. Correlate your findings with Fig. 25.1, identifying each part and determining how the heart operates.
3. Examine the heart of a frog. Correlate your findings with Fig. 25.2, identifying each part. Associate the functional with the structural aspects.
4. The sheep heart has a double beat, exactly like the human heart. Does the beat involve the contraction of the left side (both auricle and ventricle) and then the contraction of the right; or does this beat involve the anterior heart (both auricles) and the posterior heart (both ventricles)? Discuss the results on pressure in the blood vessels in both cases.

Sheep Heart

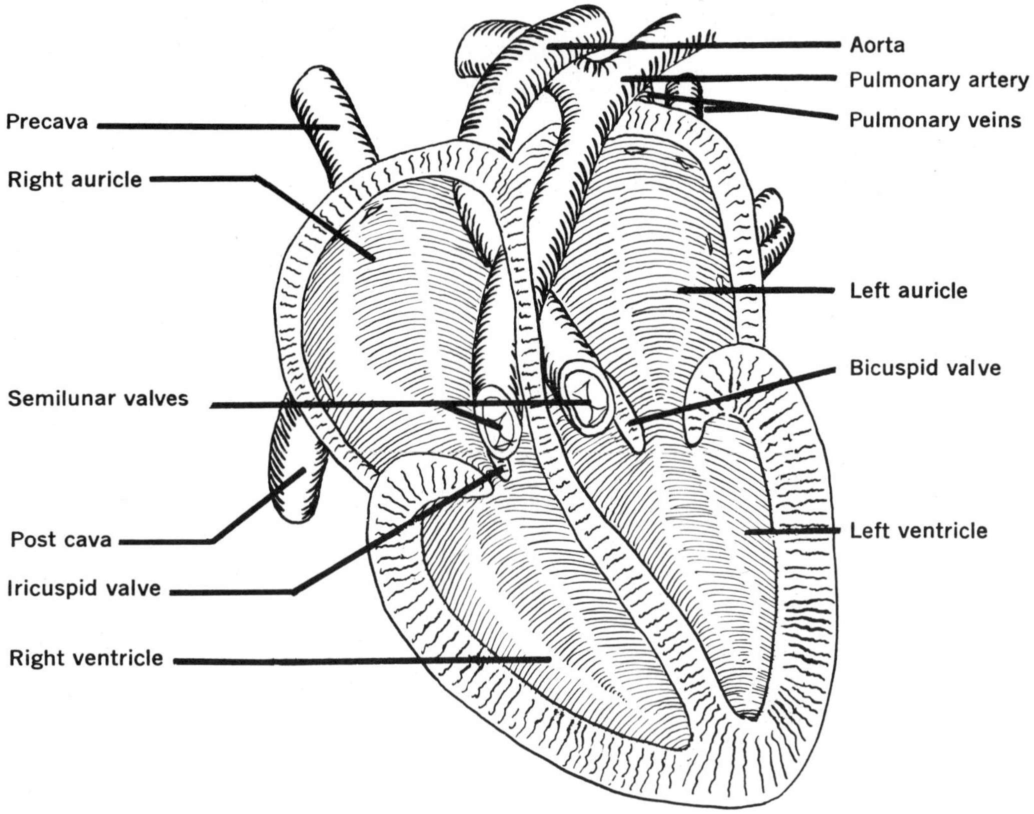

Figure 25.1

5. The right side of the heart receives venous blood from the body. What happens to this blood and how does it get to the left side of the heart?

6. In the frog heart, the right side also receives venous blood from the body. What happens to this blood and how does it get to the left side of the heart?

Frog Heart

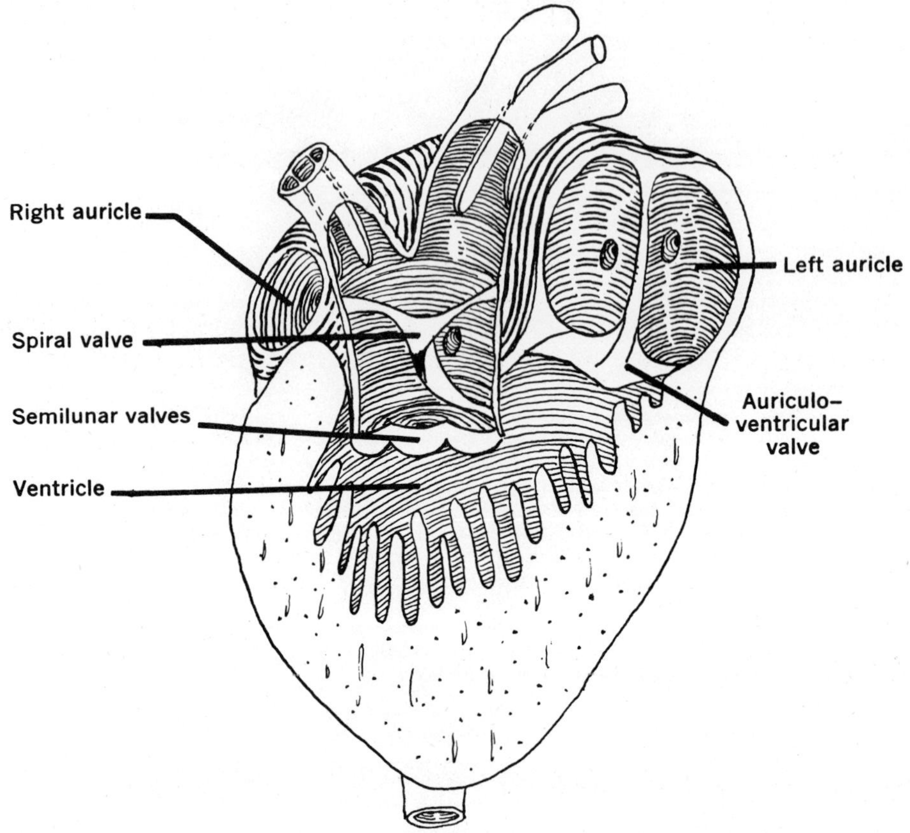

Figure 25.2

7. The frog heart has a triple beat. From working out the function of this heart, in which order do the contractions occur?

8. Compare the means of separation of oxygenated and deoxygenated blood in the frog and the sheep heart.

9. Compare the muscularity of the frog and the sheep heart. Refer to measurements taken in Chapter 18, and estimate the size the sheep heart would have to be if it had the degree of muscularity of the frog heart.

10. The sheep is totally dependent upon red blood cells receiving oxygen from the lungs to supply cellular respiration. What means does the frog have of obtaining oxygen other than the organized breathing system?

TRANSPORT OF GAS BY HEMOGLOBIN

Many enzymes or catalytic agents may attach themselves to more than one substance. This is true of hemoglobin. Hemoglobin combines with oxygen to form oxyhemoglobin, and serves to carry oxygen to the cells of the body in this form. It may also combine with carbon dioxide (although this is not the most important means of CO_2 transport) and carbon monoxide. The effect of these three gases upon hemoglobin will be examined.

1. Obtain three beakers containing defibrinated blood, and label them 1, 2, and 3. Using a glass tube, bubble the last part of several breaths (CO_2) into beaker 1 or, if available, use CO_2 from a pressure tank.
2. Continue until a color change is noted.
3. Record the color on Chart 25.8.

CHART 25.8

	O_2	CO_2	CO
Color			
Time of oxygenate			

4. Into beaker 2 bubble gas from the laboratory supply (city gas). If the speed of gas flow is difficult to regulate, set up the apparatus illustrated in Fig. 25.3. Regulate the flow of gas by adjusting the two screw clamps.

   Caution: This gas is dangerous. The apparatus should be set up in the back or at the side of the laboratory, preferably near an open window.

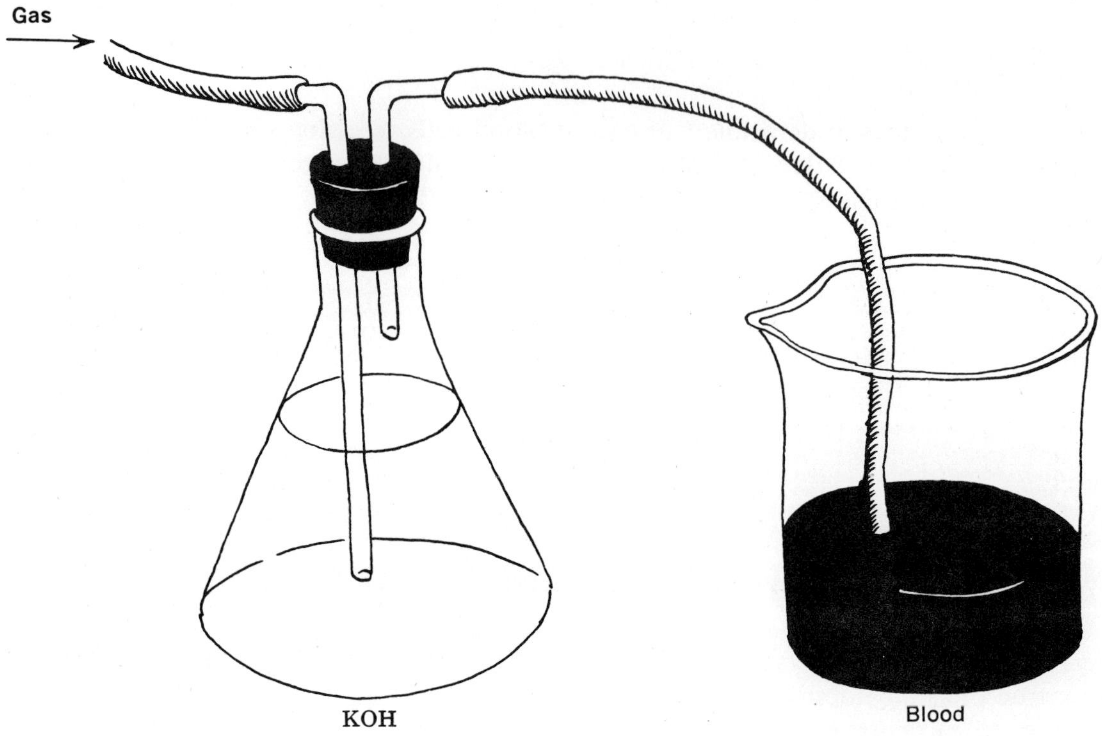

Figure 25.3

5. Continue until a color change is noted.
6. Record data on Chart 25.8.
7. Into beaker 3 bubble oxygen from a pressure tank or set up the apparatus illustrated. In the flask with potassium hydroxide the carbon dioxide is absorbed and only the oxygen will be bubbled through the blood.
8. Continue until a color change is noted.
9. Record on Chart 25.8.
10. After each reaction is complete and recorded, bubble oxygen through beakers 1 and 2 by inserting the glass tube of the oxygen apparatus into beaker 1 and proceeding as described in step 7 above.
11. Record the time it takes to obtain the color associated with oxygenated blood.

12. From your experimental results, what was the condition of the hemoglobin prior to the introduction of any of the gases?

13. Why is carbon monoxide considered toxic to the body?

14. In attending a person with carbon monoxide poisoning, would you use 100 percent oxygen, 95 percent oxygen and 5 percent carbon dioxide, or 75 percent oxygen and 25 percent carbon dioxide? Why?

15. What color is the blood when an artery is cut? A vein? What is the notable exception to this rule?

THE RESPIRATORY SYSTEM

Through the breathing process, the body is supplied with the oxygen necessary for aerobic respiration at the cellular level. Those organs directly involved in obtaining this oxygen from the environment are considered part of the respiratory system. The terminology is admittedly confusing, and although the respiratory system would be better termed "breathing" and the term "respiration" better reserved for the cellular aspects of the process, common usage is such that changes in terminology would undoubtedly lead to greater confusion.

The respiratory or breathing system in mammals is structurally adapted to the atmospheric environment. Since the internal environment of the body is liquid, the respiratory system bridges this transition from atmospheric to aquatic environment. The oxygen is dissolved in the liquid phase in the alveoli of the lungs, and from there it enters the circulatory system by combining with the hemoglobin molecule of the red blood cell. The selective transport of oxygen rather than other components of air is due to the differential diffusion resulting from the presence of the oxygen-binding compound hemoglobin. The oxygen dissolved in the liquid phase binds with the hemoglobin as the red blood cells pass through the minute capillaries of the alveoli. This oxygen then passes away from the lungs, and more oxygen may dissolve in the liquid. Since there is no factor for removing the other elements, they remain constant in their concentrations, both in the atmosphere and dissolved in the alveoli.

The respiratory or breathing system is also responsible for the elimination of carbon dioxide from the system. The process of this elimination is, to a certain extent, a reversal of the oxygen uptake process. However, the transport of carbon dioxide is somewhat more complicated and the description of this process should be read in your textbook.

THE ANATOMY OF THE RESPIRATORY SYSTEM

See Appendix, p. 471

MEANS OF DETERMINATION OF THE GAS CONTENT OF OF ATMOSPHERIC AND EXHALED AIR

One means of determining the kinds and relative quantities of gases that are used in the respiratory process is to calculate the relative percentages of the gases present in the atmosphere (i.e., the air that comes into the body during inhalation) and compare them with the gas content of exhaled air. The difference in composition then represents the turnover during the respiratory process on a cellular level.

Analysis for Oxygen

Pyrogallol is an organic compound capable of absorbing atmospheric oxygen and carbon dioxide, CO_2. If the oxygen and CO_2 are absorbed in a closed system, where only pyrogallol is permitted to flow freely, the space originally occupied by the oxygen and CO_2 will be replaced by pyrogallol. The principle is much like that of drinking a soda through a straw. When you begin, the air is drawn out of the straw. This creates a negative pressure (suction) and the liquid is brought up into the straw to replace the air. In application to the analysis for oxygen, oxygen and carbon dioxide are in essence removed from the system, as they become part of the pyrogallol molecule and revert to liquid condition. This creates a negative pressure in the system and, since the only compound free to enter is the pyrogallol, more pyrogallol will enter. This will, in turn, permit the absorption of more oxygen and CO_2, and the process will continue until all the oxygen and CO_2 are absorbed. A measurement of the volume of pyrogallol compared to the volume of the system gives the percentages of oxygen and CO_2 originally in the system.

Of course, for an analysis of respiration, the volume of oxygen must be distinguishable from the volume of CO_2. Because there are no simple means of analyzing oxygen alone, analysis of carbon dioxide may be done, and the results subtracted from those of the combined percentages, giving the figure for oxygen alone.

Analysis of Carbon Dioxide

As you have noted in the experiments dealing with cellular respiration, potassium hydroxide, KOH, absorbs CO_2. If the same operation is performed using KOH in place of pyrogallol, only the CO_2 is absorbed and its percentage may be determined.

Analysis for Nitrogen

Essentially, atmospheric or inhaled air is made up of nitrogen, oxygen, and carbon dioxide. The other inert gases are present in such minute quantities as to be negligible for our purposes. Since pyrogallol gives the volume in percent of both oxygen and CO_2, there remains only to subtract this from 100 percent to obtain the percent volume of nitrogen.

$$\%\ \text{Oxygen} = \frac{\text{Pyrogallol results} - \text{KOH results}}{\text{length of analyzer, mm}} \times 100$$

$$\%\ CO_2 = \frac{\text{KOH results}}{\text{length of analyzer, mm}} \times 100$$

$$\%\text{Nitrogen} = 100 - \%\ \text{oxygen} - \%\ CO_2$$

Assembling the Analyzer

Prepare a large test tube (25 X 250 mm) in the following manner:

1. Fit two bent glass tubes into a size 3 two-hole rubber stopper. One of the tubes should be long enough to extend within ½ in. of the base of the tube, the other should be within ¼ in. of the end of the stopper (see Fig. 25.4).
2. Attach rubber tubing to the end of each glass tube.
3. Clamp the rubber tubing so that the system, when assembled, may be closed.
4. Insert the stopper into the tube. This is your apparatus to analyze the gases.

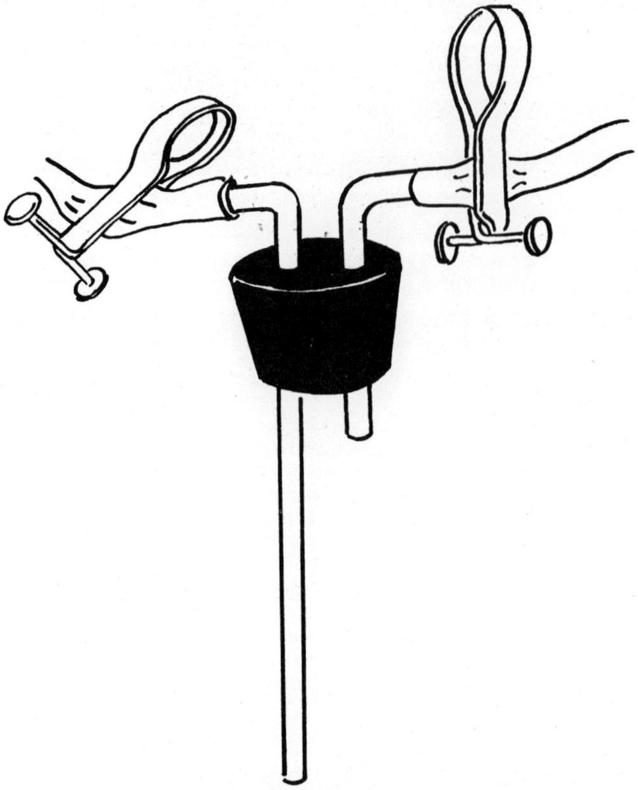

Figure 25.4

ANALYSIS FOR OXYGEN

1. Fill the apparatus with water. (Take off the top to fill.)
2. Replace the stopper.
3. Unclamp both rubber tubes and pour out the water. This is to replace the water with fresh atmospheric air.
4. Replace both clamps to seal the system.
5. Place the rubber tube of the long glass tube into a beaker containing pyrogallol.
6. When you are ready to begin, release both clamps and suck gently on the rubber tubing attached to the short glass tube to start the system (see Fig. 25.5).
7. Draw a small amount of the pyrogallol into the tube and reclamp immediately. It is essential that the system remain as closed as possible.
8. Gently rock the tube with the pyrogallol without removing the tube from the beaker. If this is too difficult and you find it necessary to remove the tube from the pyrogallol in the beaker in order to agitate the tube effectively, reclamp before removing.

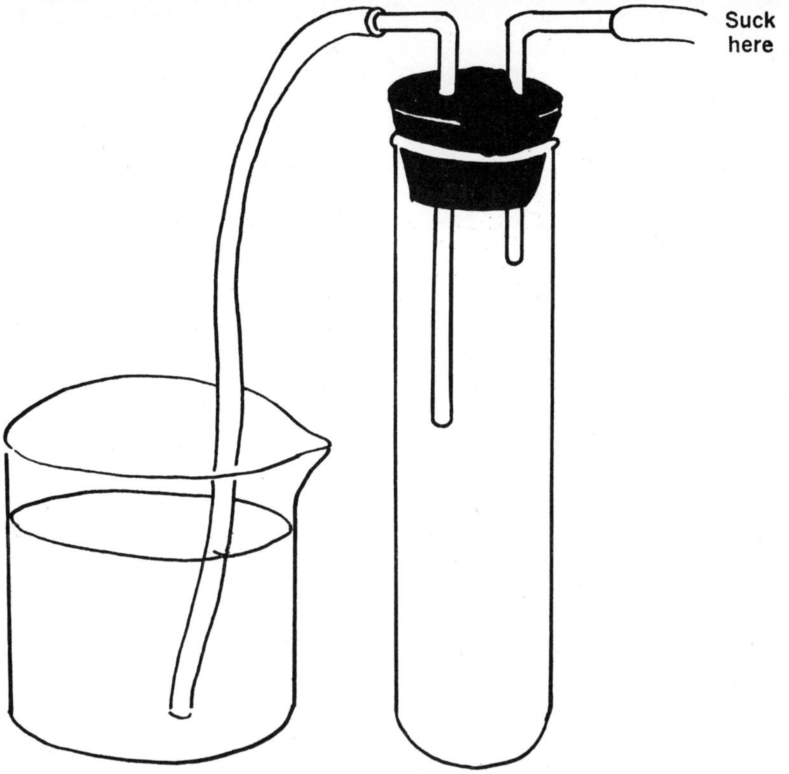

Figure 25.5

9. If the system does not operate spontaneously as it should (pyrogallol should flow drop by drop into the tube, but the success depends on the relative heights of the beaker and the tube and the degree of negative pressure established initially), continue to suck and rock gently after each small amount of liquid enters.
10. When the pressure in the system is such that no more pyrogallol enters the tube, the end point has been reached.
11. With a millimeter ruler, measure the length of the column of pyrogallol and the length of the whole tube up to the bottom of the stopper. Since the tube is curved at the bottom, and the ruler generally has 1 to 2 mm beyond the first mark, the errors cancel one another out, so that measurement should be made by resting both the tube and the edge of the ruler on the laboratory table.
12. Enter the data on Chart 25.9.
13. Empty pyrogallol back into beaker. It may be reused.

ANALYSIS FOR CARBON DIOXIDE

1. Wash the apparatus completely and reassemble.
2. Fill with water.

CHART 25.9

	mm Measurements			Percentage Composition		
	Pyrogallol (O_2 + CO_2)	KOH (CO_2)	Entire Tube	O_2	CO_2	N_2
Atmosphere						
Exhaled						

3. Evacuate the water by pouring out the short tube.
4. Close both clamps.
5. Place the rubber tube of the long glass tube into a beaker of concentrated KOH.
6. Open both clamps and suck gently on the short tube to start system. Replace mouthpiece clamp.
7. Rock tube to disperse liquid.
8. The system may have to be kept going by sucking since very little KOH will be needed to absorb the CO_2 in the atmosphere. The initial amount may have been sufficient.
9. When no more KOH will enter, the end point has been reached.
10. Measure the length of the KOH column.
11. Enter the data on Chart 25.9.
12. Permit KOH to run into beaker for reuse.
13. Clean the apparatus thoroughly.

CONTENT OF EXHALED AIR

1. Fill the apparatus with water.
2. Take a deep breath; exhale with lips parted around rubber tube and with the clamps closed. Exhale only the last part of each breath into the apparatus, opening both clamps only at that time.
3. When you come to the end of a breath, hold lips firmly on the tube until the clamps are replaced.
4. Repeat until all the water has been replaced by exhaled air.
5. Repeat the procedure for pyrogallol (O_2 and CO_2).
6. Enter data on Chart 25.9.
7. Clean apparatus and repeat the analysis of CO_2 (KOH).
8. Calculate the percentage of volume of the gases CO_2, O_2, and N_2 in atmospheric and exhaled air.

9. From the results of the experiment, which gas is utilized during the cellular respiratory process. How?

10. Which gas is released by cellular respiration and how is this shown in the experiments?

11. How did the carbon that is eliminated in the form of CO_2 originally enter the body?

12. Why did you use only the last part of the breath to fill the analyzer?

26

Energy-Utilization Systems – Maintenance of Self

MATERIALS

CLASS STOCK

- chemicals
  - 5.0% sucrose solution
  - salt solution
  - 1.0% vinegar solution
  - 0.05% quinine
  - citrate or oxalate
  - 0.85% saline
- prepared slides
  - human blood
  - thyroid tissue
  - pancreatic tissue
  - cross section of adrenal gland
- blood (oxalated)
- blood typing kit
- blood pressure cuff
- stethoscope
- 1% $CaCO_3$

STUDENT STATION

- millimeter ruler
- cotton applicators
- blindfold
- forceps
- beakers
- slides
- wax pencil
- toothpicks
- cover slips
- microscope
- microscope lamp
- lens paper

INTRODUCTION

Energy-utilization systems include those systems that permit the body to be aware of the environment, respond to the environment, and aid in creating homeostatic conditions within the body. It would be impossible to examine the wide variety of mammalian features that fulfill the aforementioned criteria, so only certain features amenable to student study will be discussed.

THE NERVOUS SYSTEM

The coordination of the body is controlled by a widespread circuit of nerve cells, called neurons, which have their origin or termini in the brain or spinal cord. In addition to coordination, the nervous system is directly responsible for awareness and physical reaction to surrounding environments.

Awareness and reaction to the external environment are common to all living things. Responses, either positive or negative, to light, chemicals, sound waves, and touch are present in the most primitive to the most advanced forms of life. Awareness will be the primary feature of the nervous system examined in this laboratory section.

Anatomy of the Nervous System

See demonstration.

Awareness Through Vision

The nerve associated with the visual sense is the second cranial nerve (II), the *optic.* The optic nerve carries the converted visual impulse to the brain, where interpretation and correlation takes place. The conversion of the visual impulses to electrical potential occurs in specialized cells of the retina. These specialized cells, the *rods* and *cones* are found dispersed in the retina—the cones more generally centrally located, the rods found primarily at the periphery. The point at which the optic nerve enters, however, is devoid of these light receptor cells. The absence of these cells produces a blind spot at the point of entrance. Generally we are so accustomed to the bioptic system, that we are unaware of the lack of vision from this area of the eye.

1. In Fig. 26.1 there is a circle and a cross.

Figure 26.1

2. Hold the manual about 20 in. from your face and cover your right eye with your hand. Have another student hold a ruler from the bridge of your nose, and measure the distance to the book.
3. Direct your sight on the circle. This will focus the circle on the retina at the point of sharpest vision, the *fovea.*
4. Slowly bring the book closer to your face. At a certain distance the cross will disappear from view. Move the book back and forth until you reach the point where the cross just disappears. Have another student measure the distance from your eye to the book at this point.
5. Record the distance in cm on Chart 26.1.
6. Slowly bring the book closer to your face. At a certain distance the cross will reappear. Have another student measure the distance and record on Chart 26.1.

CHART 26.1

	Distance
Cross disappears	
Cross reappears	

Measurement of the blind spot

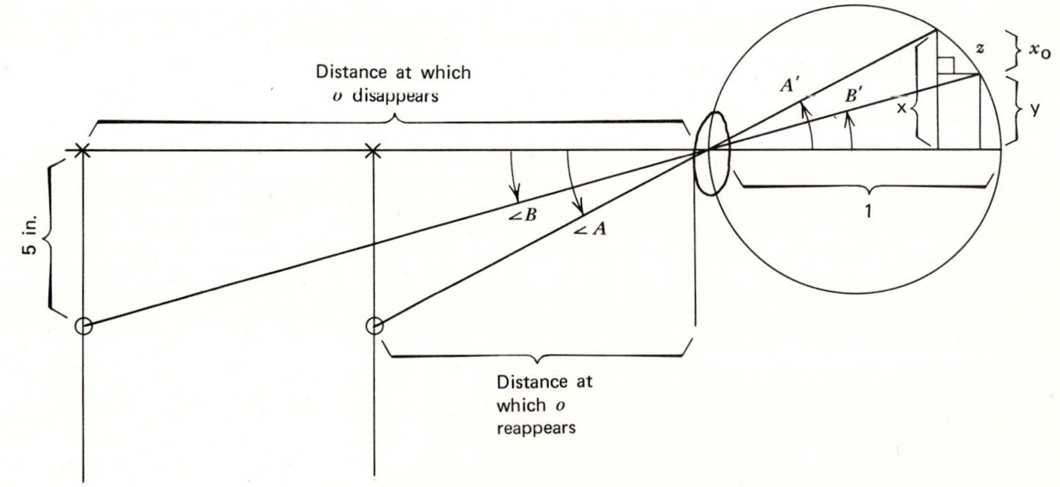

Figure 26.2

$$\tan = \frac{\text{opposite side}}{\text{adjacent side}}$$

$$\tan B = \frac{5}{\text{distance at which } o \text{ disappears}} = \frac{y}{1} \quad \text{Solve for } y$$

$$\tan A = \frac{5}{\text{distance at which } o \text{ reappears}} = \frac{x}{1} \quad \text{Solve for } x$$

$$x - y = \tan -45°* = \frac{z}{x_0} \cdot \frac{1}{2} = \frac{z}{2x_0} \qquad \tan 45° = 1 \qquad \text{solve for } z \qquad z = \text{approx. diameter of blindspot}$$

*Ignore (–) signs.

Chemical Responses: The Taste Sensation

There are two major cranial nerves that control the taste sensations on the tongue. The *facial* nerve receives impulses from the anterior portion of the tongue; the *glossopharyngeal* nerve receives impulses from the posterior portion of the tongue.

1. Obtain four cotton applicators.
2. Wash out your mouth with water so that no taste residue remains.
3. Dip one of the applicators in a 5% sucrose solution (table sugar).
4. Rest the applicator briefly on the tip, front center, back center, and side of your tongue.
5. Record on Chart 26.2 where the sweet taste is the most intense (+ +), less intense (+), and absent (–).
6. Wash out your mouth with water.
7. Repeat, using a fresh applicator for each solution and washing your mouth between each test.
   (a) Salt (NaCl)
   (b) Vinegar (acetic acid), 1%–sour
   (c) Quinine, 0.05%–bitter
8. Which senses are definitely controlled by the facial nerve?

CHART 26.2

Solution	Tip	Front Center	Back Center	Sides
Sweet				
Salt				
Sour				
Bitter				

9. Which senses are definitely controlled by the glossopharyngeal nerve?

10. From the experimental results, about which of the senses are you uncertain concerning their control?

CUTANEOUS SENSATIONS

Cutaneous sensations are received by the spinal nerves and conveyed to the conscious level. The receptors for these sensations are located beneath the epidermal layer of the skin. There are different types of nerve endings for various responses. The position and number of the different types of nerve endings determined the sensitivity or threshold of the sensations.

Touch Perception

The ability to discriminate between one- and two-point touches is helpful in arriving at the relative distance between the nerve endings that relay touch sensation. This exercise should be done in pairs.

1. Blindfold your partner.
2. Touch the parts of the body of your partner outlined on Chart 26.3 with the points of a pair of forceps.
3. Touch with the double points of the forceps spread apart and ask the subject if he feels one or two points touching. If he answers "two," reduce the distance and repeat. Continue reducing the distance until only one point is perceived.
4. Measure the distance between the points with a millimeter ruler and enter them on Chart 26.3.

5. Repeat for each part of the body and then exchange places with the subject.
6. Which areas of the face and arm have the highest number of touch receptors?

7. Of what advantage are the increased number of touch receptors on certain parts of the body?

CHART 26.3

Area	Subject 1	Subject 2
Back of hand		
Palm		
Finger tip		
Forearm		
Cheek		
Forehead		
Nose		
Neck		
Lip		

8. What other parts of the body would you deduce to have a large number of touch receptors and what advantage would it be to the individual?

ENERGY UTILIZATION FEATURES OF THE CIRCULATORY SYSTEM

Certain features of the circulatory system function in ways that characterize energy-utilization systems.

TYPES OF BLOOD CELLS

The blood contains several different types of cells. The red blood cells, or erythrocytes, function primarily to carry oxygen and carbon dioxide to and from the body cells. The white blood cells are defensive agents of the body. The role of specific leucocytes is not fully understood.

Erthrocytes

The erythrocyte is a disc-like cell, without a nucleus. It normally appears red because of the oxygenated hemoglobin.

Leucocytes (White Blood Cells)

Lymphocytes

The lymphocyte is the smallest of the white blood cells, only slightly larger than the erythrocyte. It has a distinct nucleus that fills almost the entire cell and stains dark blue with Wright's stain. There are no granules in the cytoplasm. The cytoplasm frames the nucleus, and stains a light blue.

Monocytes

The monocytes are the largest of the white blood cells, ranging from two to three times the size of red blood cells. The nucleus stains a muddy purple and is very indented, at times resembling a horseshoe. There is a large amount of nongranular cytoplasm which stains a light blue.

Neutrophils, Eosinophils, and Basophils

All have nuclei divided into areas or lobes which stain a dark blue-purple. The cytoplasm is granular, and the granules stain from red to lavender to dark blue. The cytoplasm stains a light blue. You are not expected to distinguish between these types.

1. Examine the prepared slides of human blood.
2. Locate and identify each type of cell described above.
3. In the space provided, draw each type of cell, being careful that distinguishing features of each type of cell are recognized.

PROTECTION BY MEANS OF THE BLOOD-CLOTTING REACTION

Among the important functions of the blood is to seal surfaces of the body that have been injured. When an injury occurs, air or the irregular cut surface causes the rupture of platelets. This sets in action the following series of reactions that result in the formation of fibrin:

$$\text{Platelets disintegrate} \longrightarrow \text{Thromboplastin}$$

$$\text{Prothrombin} \xrightarrow[\text{Ca}^{++}]{\text{Thromboplastin}} \text{Thrombin}$$

$$\text{Fibrinogen} \xrightarrow{\text{Thrombin}} \text{Fibrin}$$

The fibrin forms a network in which blood components become trapped, resulting in the formation of a clot.

Prepare three test tubes with 5ml blood in each the following way:

(1) Control—nothing done.
(2) Add 1ml $CaCO_3$
(3) Add 1ml $CaCO_3$ and whip with glass rod or applicator stick to remove fibrin as it forms.

1. Observe beaker 1. Why is the blood not clotting? [Hint: Ca^{++} + oxalate (or citrate) → calcium oxalate (or) calcium citrate].

2. Why did the blood in beaker 2 clot? (Hint: $CaCO_3 \rightarrow Ca^{++} + CO_3^{=}$)

3. Which blood components are in the clot?

4. Why are blood clots red?

5. Why do scabs usually become yellow before falling off?

BLOOD GROUPS

Each red cell contains the antigens that distinguish the so-called blood groups. If the red blood cells of an individual contain the antigen designated A, the person is said to be of blood group A. Red blood cells may contain antigen A, antigen B, both, or neither, and the blood types will then be A, B, AB, or O, respectively. The plasma of each individual contains antibody substances. These substances, when in the presence of the antigen, cause the red blood cells containing the antigen to clump or agglutinate. This clumping is typical of antigen-antibody reactions.

The plasma of blood group A contains anti-B, which is compatible with the A protein contained in the red blood cells of the system. This plasma will, however, cause the agglutination of any red blood cells containing protein B with which it might come in contact in the course of transfusions.

Blood Group	Antigen in RBC	Antibody in Plasma
A	A	anti-B
B	B	anti-A
AB	A and B	none
O	none	anti-A and anti-B

In order to determine your blood type, follow the directions below.

1. Prepare a clean slide by drawing a line down the center, marking the left side of the slide "anti-A" and the right side "anti-B."
2. Place a single drop of anti-A serum in the appropriate space, and one drop of anti-B in its appropriate space. Be careful not to mix the fluids or contaminate the droppers. If contamination occurs, wash the dropper thoroughly before replacing it in the bottle.
3. Wipe the tip of your third finger with alcohol and cotton.
4. Using a sterile needle or a packaged lancet, ask your neighbor to prick your finger. Doing it yourself usually causes difficulty.
5. Press out a drop of blood.
6. Quickly take up some blood on a toothpick and place into anti-A drop.
7. Using *another* toothpick take some blood and place in the anti-B drop.
8. Stir each with its own toothpick, being careful not to mix the drops.
9. Place the slide on white paper and observe clotting. If you cannot determine which are agglutinating, cover each drop with a cover slip and observe under the microscope.
10. Fill in Chart 26.4 with the reaction you would expect for the blood types given.

CHART 26.4

Blood Type	Agglutination	
	Anti-A	Anti-B
A		
B		
AB		
O		
Your blood		

11. Why is type AB considered the universal recipient and type O considered the universal donor?

THE ENDOCRINE SYSTEM

The endocrine system is composed of all glands that produce and secrete an extracellular chemical compound into the body without a structural duct. The secretions of these ductless glands are called hormones. The scope of hormonal control is equaled only by the bodily control exerted by the nervous system. In many cases, the effect brought about by hormonal secretion is identical to the response to a nerve impulse. In these circumstances, the hormone tends to maintain the response initiated by the nerve impulse.

Many hormones, however, influence activities which are solely under endocrine control. The removal of the gland or the disruption of its activity by disease deprives the body of specific functions. The extent to which the body is impaired is correlated with the importance of the hormone in maintaining a steady state.

The endocrine system and the nervous system contribute to the maintenance of a steady or homeostatic state in the body. Any organism, in order to survive, must be able to exist in a variety of environments. In Chapter 18, the importance of variation to the species was emphasized. Part of this variation, which ensures survival on a species level, is the degree of adaptability on the individual level. The nervous system enables the individual to respond quickly to stimuli, for example, to jump from the path of an on-coming car. Parts of the endocrine system provide the means for raising the glucose level in the blood, thereby giving the muscles the added energy required for stamina under stress. The efficiency of these factors may determine the survival of the species.

The endocrine system is extremely broad in scope, and hence it is impossible to cover every aspect. We shall examine several glands microscopically and shall observe the effect of pituitary hormones in detail.

ANATOMY OF THE ENDOCRINE GLANDS

Gross Anatomy

See Appendix, p. 474.

Microscope Examinations of Several Endocrine Glands

The Thyroid Gland

The thyroid gland is located in the neck, adjacent to the trachea, near the voice box or larynx. This gland secretes thyroxin in response to the thyrotropic hormone of the pituitary gland. Thyroxin is active in determining the rate of energy exchange of chemical reactions.

1. Obtain a prepared slide of thyroid tissue.
2. Observe under low and high power.
3. Draw a follicle and some surrounding interfollicular areas in the space provided, labeling the following parts:
   follicle—ovate bubblelike vesicle, which makes up the thyroid.
   secretory epithelium—the periphery of the follicle is made up of cuboidal and columnar epithelial cells. These cells produce and secrete thyroxin.
   colloid—in the center of the follicle is the secretory product of colloid.
   fat cells—the large globular cells in the interfollicular tissue. Their function is unknown.

Pancreas

The pancreas acts as both an exocrine and an endocrine gland. The significance of the exocrine portion of its function has already been covered. The secretions that aid in digestion are collected and progress down the pancreatic duct to the intestine. The cells that secrete pancreatin surround the ducts.

Many masses of cells are found in the so-called acinar area of the pancreas. These masses are characterized by their lack of ducts and the increased vascular tissue (blood vessels which appear as fragile bubblelike structures in cross section) in comparison with surrounding tissues. These are called the islets of Langerhans and produce insulin. The hormone insulin enhances the conversion of glucose to its storage from glycogen.

1. Obtain a slide of pancreatic tissue.
2. Draw a section showing the islets of Langerhans in the space provided.
3. Label the following structures:

   pancreatic acinus—pyramidal epithelial cells that surround a duct. Usually only one layer thick around the duct.

   islets of Langerhans—knots of tissue of smaller and less dense cells than the cells of the acinus; they give a fainter stain.

   blood vessels—cross sections of blood vessels tend to appear as bubbles or holes in the tissue. Note the increased number of blood vessels in the regions of the pancreas containing the islets.

4. Why do you expect a greater number of blood vessels in the region of the pancreas containing the islets of Langerhans?

The Adrenal Glands

The adrenal gland is actually two distinct glands combined structurally in one. The medulla is ectodermal in origin, originating and functioning very much like the sympathetic nervous system of the autonomic nervous system. The cortex is mesodermal in origin and functions primarily in the regulation of metabolism.

1. Obtain a slide of a cross section of the adrenal gland.
2. Draw a section of the adrenal gland showing the relation of the cortex and the medulla.
3. Label the following parts:

   capsule—outermost layer of cells; made up of connective-type tissue giving the appearance of long stringlike cells.

cortex—loosely packed chain cells. The area closest to the capsule is called the zona glomerulosa. The cells of the zona fasciculata are generally larger and contain oil droplets. These are actively involved in secretion. The next layer of cells, the zona reticularis is thought to be made up of degenerating cells.

X zone—The X zone separates the medulla and the cortex. The size of the X zone depends on the maturity of the animal at the time the slide was prepared. The zone may range from a large area of dense cells in the immature animal to a narrow band of degenerate cells which have a stringlike appearance in older animals.

medulla—the cells of the medulla are in groups of ten to twenty. The cells are slightly less dense than those of the cortex.

HOMEOSTASIS—MAINTENANCE OF STEADY—STATE CONDITIONS

The survival potential of a species is directly related to the degree of adaptability of that species. This adaptive potential is essential to survival because it provides a wider range of phenotypes to cope with possible environmental conditions that might confront the species. Under adverse conditions, only those organisms capable of existing in the environment will survive, whereas others will perish.

The principle of adaptation operates on an individual as well as on a population level. Each system is capable of activity over a range of conditions. Upsetting the balance of the body sets adjustments in motion on every level of activity. These adjustments involve the counterbalancing of systems to retain a state of normal function. The tendency to re-balance and adjust systems to achieve this normal state is termed homeostasis. The four systems in the human that are most directly involved in homeostatic process are the nervous system, the endocrine system, the circulatory system, and the respiratory system.

The compensatory mechanisms operative in these systems range in their activity from minor to major adjustments. Minor adjustments may be regarded as the day-to-day regulatory processes that coordinate the body functions. Major adjustments are the gross changes in which all systems are involved in counterbalancing one another to achieve a steady state. The conditions of adjustment that will be studied in the laboratory require gross changes in the balance of systems as well as the natural changes which occur to maintain the steady state.

RETURN TO HOMEOSTASIS UNDER ADVERSE CONDITIONS

The breathing rate and heart rate are interdependent and codependent upon the metabolic rates of the cells of the body. The following exercises demonstrate the adjustments of the breathing and heart rate under conditions of stress.

The exercises should be done in groups of four students, if possible, consisting of two smokers and two nonsmokers. One smoker and one nonsmoker should be the subjects for the odd-numbered tests, while the other record pulse and breathing rate. The recorders should be subjects for the even-numbered tests and the first subjects record breathing and pulse rate. This provides each student with the opportunity to get sufficient rest. Since this exercise requires endurance, a student subject to dizzy spells should not be a subject, and should act only as a recorder.

Recording Breathing Rate

1 inspiration + 1 expiration = 1 breath

Recording Pulse

Place the tips of four fingers on the inner part of the wrist, resting the thumb on the back of the wrist. Do *not* use your thumb to try to get the pulse, since it is insufficiently sensitive and the pulse of the blood vessel of the thumb may be confused with that of the subject. Press lightly with the four finger tips until a pulse is noted. Record the number of double beats per minute by measuring for 2 min and dividing by 2. A stethoscope may be used on the subject's chest or back if one is available.

1. *(a)* Count the number of breaths per minute of the smoker and the nonsmoker.
   (b) Count the number of heartbeats per minute of the smoker and the nonsmoker.
   (c) Record on Chart 26.5.
2. *(a)* The nonsmoker will take very deep, forced breaths for 1 min, after which the subject will try to breathe as involuntarily as possible.
   (b) Directly after the 1 min of forced breathing, the breathing rate should be recorded.
   (c) At the same time, record the pulse rate of the subject.
   (d) Test both breathing rate and pulse rate every 2 min, until normal rates are achieved.
   (e) Record the length of time required for the normal pattern to be established.
   (f) Repeat the experiment with the smoker subject.
   (g) Record all data on Chart 26.5.

3. *(a)* The nonsmoker will hold a paper bag over his nose and mouth in such a way that the same air is breathed over and over again. Continue this for 2 min, breathing as involuntarily as possible.
   (b) Record the breathing rate and the pulse rate during the second minute.
   (c) Retest the pulse and breathing rate every 2 min until normalcy is restored.
   (d) Record the length of time required for the normal pattern to be established.
   (e) Repeat the experiment with the smoker subject.
   (f) Record all data on Chart 26.5.
4. *(a)* The nonsmoker will make 20 to 30 deep knee bends and then run quickly in place or up and down a flight of stairs for 1 min.
   (b) Record the breathing rate and pulse rate for 1 min after the strenuous exercise.
   (c) Test every 2 min, recording the length of time until the normal breathing and pulse rate return.
   (d) Repeat with the smoker subject.

CHART 26.5

			Return to Normal	
	Breath/Min	Pulse/Min	Pulse	Breathing
Normal condition				
Forced breathing				
Rebreathing same air				
Strenuous exercise				

5. Describe the mechanisms that control breathing and pulse rate.

6. What is meant by the statement "breathing and pulse rate are interdependent and codependent upon the metabolic rate"?

7. Explain the results obtained by forced breathing in view of the mechanisms that control breathing and heart rate.

8. What are the events responsible for the return of normal heart and breathing rates?

9. Explain the results obtained from rebreathing the same air in view of the mechanisms that control breathing and heart rate.

10. Explain the results of the breathing rates after exercise according to oxygen use and control of breathing and heart rate.

11. Are the differences in rates and recovery of the smoker as compared with the non-smoker significant?

12. What conditions of the smoker's lungs tend to increase the recovery time?

13. Under adverse conditions requiring stamina, which has the greater chance of survival, the smoker or the nonsmoker? Why?

27

Energy-Utilization Systems - Reproduction, the Maintenance of Species

MATERIALS

CLASS STOCK

- prepared slides
  - mammalian ovary
  - mammalian testis
  - cross section of blastula
  - cross section of yolk plug
  - cross section of neural groove
  - whole mount of 48 hr chick
- preserved specimens
  - frog embryos
  - human fetuses
- nonovulating female frog
- male frog
- Holtfretter's solution
- unfertilized frog egg
- Ringer's solution

STUDENT STATION

- probe
- slides
- dissecting microscope or hand lens
- forceps
- microscope
- microscope lamp
- lens paper

INTRODUCTION

The maintenance of species through reproduction is an essential feature of all organisms. Accordingly, some of the energy derived from nutritional sources is directed toward the operation of those factors involved in reproduction.

Gross Anatomy of the Reproductive System

See Appendix, p. 500.

The Gonads—Ovary and Testis

The gonads serve a dual function, being part of the reproductive system of an organism and producing the hormones that help regulate the system. Complete regulation of the system is accomplished by the interaction of the sex hormones, produced by the ovary or testis, and the gonadotropins, produced by the pituitary gland.

1. Obtain a slide of mammalian ovary.
2. In the space provided, draw a section of the ovary which includes the periphery and a Graafian follicle.
3. Label the following parts:
   Graafian follicle—large bubble-like area which may be seen bulging from the ovary. Because it is seen in cross section, the area may seem to have cells only at the periphery.
   follicular cells—the cells at the periphery of the Graafian follicle and surrounding the ova.
   ovum—the large cell that may show mitotic figures. The ovum is surrounded by follicular cells.
   interstitial tissue—general filler tissue found in the ovary between developing Graafian follicles.
   primary follicle—immature follicle that consists of a large oocyte surrounded by small nutritive cells. These are found mixed in the interstitial tissue at the periphery of the ovary.

4. Which cells are responsible for hormonal secretions?

5. Obtain a prepared slide of cross section of mammalian testis.
6. In the space provided, draw a section of the testis which includes the periphery, lobule, and septula.
7. Label the following parts:
   tunica albuginea—outer covering of the testis.
   lobules—areas of the testis resembling compartments which contain the coiled seminiferous tubules.
   septula—connective tissue which separates the lobules.
   seminiferous tubules—seen in cross section, these tubules are circular. The development of the spermatozoa occurs within the tube and the mature sperm are released into the lumen.
   lumen—the central core of a seminiferous tubule.

8. Which cells of the gonads are responsible for hormonal secretion?

PITUITARY-GONAD ENDOCRINE RELATIONSHIP

This exercise may be done by demonstration. In the spring, the female frog ovulates. The secretion of the follicle-stimulating hormone (FSH) by the pituitary is initiated by the changes in bodily conditions that occur after hibernation. The male frog, on the other hand, is capable of producing viable sperm throughout the year.

The laboratory frogs have been kept at 10°C to slow metabolic processes and inhibit the secretion of FSH. The instructor will demonstrate that the frog has not ovulated by pressing downward on the abdomen so as to strip any ripe eggs.

A male frog will be killed by severing the brain and spinal cord at the level of the tympanic membrane. The top of the head will then be cut off. By inserting the blade of a sharp small-bladed scissors into the spinal cord of the remnant of the head, and cutting laterally, and by bringing down the flap formed by this cut, the pituitary, a small pink ovoid body encased in adipose tissue, should be made visible. The pituitary will be removed and placed in a dish containing 3 ml Ringer's solution. The pituitary is then mashed and pushed in and out of a 5-ml syringe until completely macerated. The syringe is then filled and the solution is injected into the abdomen of the female with a #23 needle. The female will be kept in an upside-down position for 30 sec to prevent leakage, and will be left at room temperature for 24 hr.

To illustrate the effect of the hormone extract, the frog that was injected the day before will be stripped of eggs. The eggs will be placed in a small dish containing enough Ringer's solution to moisten. The jelly will be clipped to separate the eggs and expose a greater surface area.

The testes from the male will be removed. These are bright yellow ovoid bodies on the dorsal wall of the abdomen. The testes will be macerated in 8 to 10 ml of Holtfretter's solution and a few drops placed on the stripped eggs. These will be set aside for the laboratory exercise on Development.

1. What may you conclude about the FSH hormone of the female and the gonadotropins of the male?

2. Would this have worked if a female pituitary was used and the season was autumn? Spring?

3. What function does the enlarged thumb of the male frog have? (Hint: The mating procedure in frogs is very similar to the instructor's method of stripping eggs.)

4. Describe the hormonal events that lead to ovulation, beginning with the secretion of FSH from the pituitary and comparing this initiation of the process with the injection of pituitary extract.

5. Of what survival value to the species is the maturing of all the eggs in one season?

DEVELOPMENT

Perhaps one of the most significant results of evolution is the establishment of developmental patterns followed by an individual of a particular species. The pattern is highly specific for each organism, and involves a consistent repetition of the stages each organism in that species has undergone to reach adulthood. The pattern of development is the result of many species adaptations to environmental conditions.

The frog is associated with an aquatic environment. Many amphibians, however, are relatively terrestrial. Even these must return to the water in order to ensure the development of the young. The entire developmental pattern—the small amount of nutrient material provided by the egg, the protective coverings of the egg that easily dehydrate, and the early age at which the embryo becomes free-living—depends on an aquatic environment. The mating time of the amphibian also depends on environmental conditions; temperature, for example, plays an important role in development. Other organisms have requirements just as rigid as these for the development of the young and the perpetuation of the species.

In the course of this laboratory, you will examine the development of several organisms in an attempt to determine the ideal factors for their development.

FROG DEVELOPMENT

The class will be supplied with a vial of preserved frog embryos of various stages. These will be poured into a finger bowl and available for your use. Find the stages necessary for your study and lift them from the finger bowl with a spoon. After you are finished return them so that others may use them.

1. Obtain a fresh unfertilized egg if there is a frog still ovulating from the previous week, or take a preserved unfertilized egg.
2. Examine with a hand lens.
3. Draw the unfertilized egg in the space provided, labeling the following:

animal pole—darkly pigmented side of the egg.
vegetal pole—lightly or nonpigmented side of the egg.
yolk—nonpigmented substance of the egg.
gelatinous membranes—surrounding the egg.

4. Obtain a two-cell stage from the preserved specimens and examine with a hand lens.
5. Draw the two-cell stage in the space provided.
6. In which direction is the first cleavage furrow?

7. Does the cleavage divide the egg into two equal parts?

8. Obtain a four-cell stage from the preserved specimens and examine with a hand lens.
9. In which direction is the second cleavage furrow?

10. Does it divide the egg into four equal parts?

11. Obtain an eight-cell stage from among the preserved specimens and examine with a hand lens.

12. In which direction is this cleavage furrow?

13. Does this divide the egg into eight equal cells?

14. Obtain a blastula (thirty-two cells) and examine with a hand lens.
15. Draw the blastula in the space provided, indicating its size in relation to that of the unfertilized egg.
16. How much growth has occurred between the time of fertilization and the blastula?

17. Examine a slide of the cross section of a blastula.
18. Draw the cross section of blastula in the space provided, labeling the following parts:
    macromeres—large nonpigmented cells that fill the area of the vegetal pole. Presumptive endoderm.
    micromeres—the smaller pigmented cells that cover the animal pole. Presumptive ectoderm.
    blastocoel—cavity seen as a hole in cross section. This is the result of the differential rate of division that has occurred.

19. Obtain a yolk plug stage. This stage is characterized by being primarily covered with the presumptive ectoderm material and is therefore pigmented throughout. Through a differential rate of divison, a plug of nonpigmented large yolk cells appears on one side of the embryo. Examine this stage with a hand lens.

20. Examine a cross section of yolk plug stage.

21. Draw the cross section of yolk plug stage in the space provided, labeling the following parts:

    yolk plug—plug of nonpigmented cells that protrudes through the opening in the presumptive ectoderm.

    blastopore—the hole that is plugged by the yolk plug. Will become the anus of the organism.

    gastrocoel or archenteron—cavity within the embryo. This is the beginning of the digestive tract.

    blastocoel—may be seen as a small cavity that has been pushed into the area of the yolk cells as the ectoderm has developed and migrated.

    yolk endoderm—nonpigmented cells on the now ventral side of the developing embryo. These are large and yolk-filled.

22. The yolk plug has been formed by the division of the micromeres at a more rapid pace than that of the macromeres. The micromeres hence overgrow the macromeres and envelop them. The cells that now contain large quantities of yolk will eventually become part of the digestive system. How are they now taking part in the role they will eventually play in the adult organism?

23. How does the size of the yolk plug stage compare to that of the unfertilized egg?

24. What major event that has been discussed in earlier exercises has occurred in the yolk plug stage which permits the reference to ectoderm, mesoderm, and endoderm rather than presumptive ectoderm, presumptive endoderm, etc.?

25. Obtain an embryo that is in the process of neurulation. This stage involves the formation of the primitive brain and spinal cord. During this stage, the dorsal side of the embryo develops two ridges or crests which come together to form the neural canal. Examine with a hand lens.

26. What is the most obvious feature of the size of the embryo during neurulation?

27. Examine a cross section of late neural groove stage.

28. Draw the late neural groove stage in the space provided, labeling the following parts:

neural groove—depression between the two ridges on the dorsal side of the embryo.

neural crests—the ridges which are forming and which will enclose the neural groove, thus making it the neural canal.

mesoderm—delineated tissue directly under the ectoderm. This tissue will split in the later neurula to form the body cavity or coelom. (Refer to the formation of coelom.)

notochord—small mass of tissue directly under the forming neural canal. The neural groove and crests are made up of ectoderm. The notochord, however, is mesoderm.

archenteron—the cavity within the embryo. This will soon become an active digestive tract.

endoderm—the lining of the archenteron.

ectoderm—the outer covering of the embryo.

29. If there are embryos from the endocrinology experiments of the past week, examine them. These embryos should date from the last day of the week (the youngest) to the first day of the week (the oldest) and should range from neurula to the young tadpole.
30. Poke the embryos of different stages lightly with the end of a probe. At what age does muscle reaction to external stimulation occur?

31. At what age do the tadpoles become free-living? What does this indicate about the yolk in the cells?

32. What stages of the frog's existence depend on a water environment and what aspects depend on a land environment?

33. In what ways has the development of the young been ensured by the particular developmental patterns of the frog?

CHICK DEVELOPMENT

1. Examine the unfertilized chicken egg by breaking the shell and permitting the "egg" to fall into a finger bowl.
2. Draw the egg in the space provided, labeling the following parts:

   yolk—bright yellow material, used as food for the developing embryo.
   blastodisc—small whitish round disc on the upper surface of the yolk. This contains the single nucleus of the female, since this egg is unfertilized. It is from this blastodisc alone that the embryo would develop.
   yolk membrane—the membrane which keeps the yolk from spreading throughout the egg.

light albumen—watery albumen found at the egg periphery.
dense albumen—heavier egg "white" found closely associated with the yolk.
chalaza—twists of albumen on the lateral sides of the egg.

3. The shell of the chick egg develops around the egg in the last third of the fallopian tube. Where must fertilization of the chick egg therefore occur?

4. The yolk is obviously for nourishment of the developing embryo. What role can you deduce for the albumen?

When a fertilized chick egg undergoes division, the division is initiated by the cytoplasmic material of the blastodisc. The cells pinch inward and cut through part of the yolk. The yolk density, however, is so great that cleavage can not split the yolk into two halves. The cleavage patterns for the chick are not the same as those seen in the frog, and the blastodisc does not separate from the yolk by a cell membrane for several divisions.

In the frog you have observed how the overdevelopment of the micromeres eventually caused the envelopment of the yolk. This is not possible in the chick. The chick develops on a flat surface rather than as a sphere (see Fig. 27.1).

5. Examine a whole mount of 48-hr chick. No drawing is necessary, but note the following developmental features:

heart—two bulbous tubes found under the head flexure. Notice the major blood vessels that leave the body and go outward. In the developing organism where do they terminate and what is their probable function?

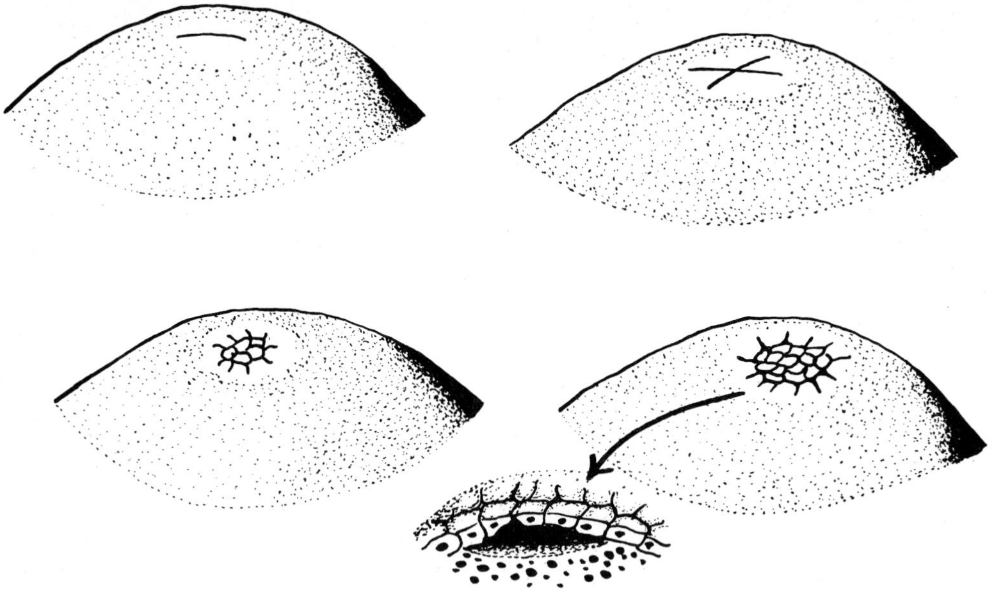

Figure 27.1

head flexure—the head is forced into this position by the development of the enlarged areas of the brain. Notice the irregular "bumps" on the dorsal side of the head. These are the enlargements of the various sections of the brain, which are, at this point, laid out linearly. As further development occurs, the brain will enfold and become more compact.

eye—is in the process of developing. The inner lens may also be seen.

otic vesicle—may be seen at about the level of the heart. It is a small button-like structure on the lower head region and will eventually develop into the ear.

neural tube—extends from the brain to the mid-dorsal posterior region of the embryo. In the chick, this also develops from ectoderm.

somites—small square segments found near the neural tube. These will develop primarily into muscle.

6. The chick will remain in its shell for 21 days before becoming free-living, How does this compare with the development of the frog?

7. How is the chick egg adapted for terrestrial environment?

HUMAN DEVELOPMENT

The adaptations of organisms indicate that there is more than one way of coping with a problem presented by the environment. The birds and some reptilian forms have developed the shelled egg to permit complete release from an aquatic environment. Other reptilian forms and mammals have adapted an internal growth of the young to accomplish the same ends, freedom from the aquatic environment. This method of internal growth poses certain problems in the study of mammalian embryos. The egg, fertilized or not, is not readily available and there are moral objections to the interruption of human development for student study. A sufficient number of developing organisms have been spontaneously interrupted, however, to supply practically all schools with preserved specimens of human fetuses.

1. Examine the human fetuses available on demonstration. Each should be labeled with the approximate age of the fetus.
2. At what age (from fertilization) does the fetus become recognizable as a human?

3. At what age is its sex discernible?

4. A harelip and cleft palate are usually associated deformities found in some children. They involve an opening in the mid-ventral line of the upper lip and the hard palate. From the developing fetuses, can you see what would cause such a deformity?

5. The fetuses of four months' gestation appear quite complete. What further development must occur before birth?

6. The human fetus develops within the uterus of the mother. An elaborate placenta develops around the developing fetus and nourishes it by means of blood vessels in the umbilical cord. To what blood vessels in the chick are these umbilical vessels analogous?

7. The size of the human egg is microscopic. Compare the two means of coping with a terrestrial environment, the shelled egg and the placental pattern, in terms of the amount of yolk, time of gestation (mouse, which is a placental animal, takes 21 days), free-living ability at the end of the gestation period, and protection mechanisms?

Appendix 1

DISSECTION OF THE FETAL PIG

Place the fetal pig on the dissecting pan. Before beginning the actual dissection, identify the following external features of the pig:

head
neck
trunk—from which extend two pairs of appendages and the tail.
thorax—anterior portion of the trunk encased by the ribs.
abdomen—posterior portion of the trunk.
eyes
ears
nostrils
mouth—open the mouth and determine if the pig is born with teeth or if they are cut later, as in the human.
anus—posterior region of the trunk, found just below the tail.
umbilical cord—structure containing blood vessels for the removal of wastes from the developing pig and the allantoic duct.

Once these identifications have been made, proceed with the actual dissection. Place the pig, ventral side up on a dissecting pan. Using cord, tie a knot around the forelimb on the right side. Run the string beneath the pan and tie to the left limb in such a way that the limbs are spread apart. Repeat this procedure with the hind limbs. With a scissors, make an incision from anterior to the umbilical cord to the neck, along the mid-ventral line (see Fig. A1.1). Lift the skin as you make the incisions, being careful not to cut too deeply. Make lateral cuts under each of the forelimbs, at right angles to the mid-ventral incision. Next, starting at the umbilical cord once again, cut carefully around the cord posteriorly to the hind limb, making the incision about ¼ in. to the side of the mid-ventral line. Once again, return to the umbilical cord, cutting around the other side

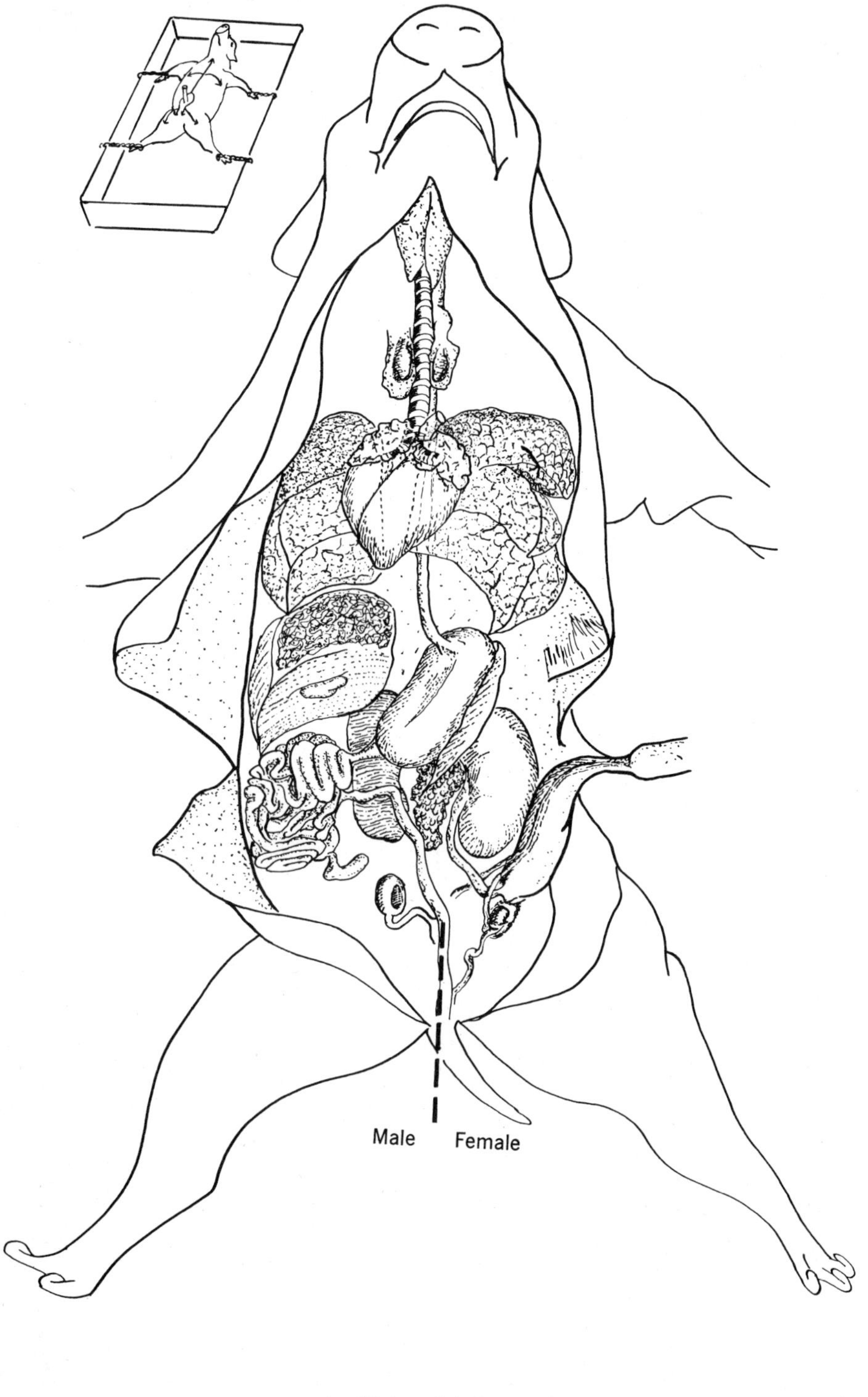

Figure A1.1

and posteriorly, keeping about ¼ in. to the other side of the mid-ventral line. This will leave a mid-ventral flap of skin about ½ in. wide posterior to the umbilical cord. Make lateral cuts around the anterior portion of the hind limbs. The cut skin should now be lifted outward, forming two flaps. These flaps may be pinned back with dissecting pins. The sternum, or breastbone, which is cartilaginous at this stage in development, may be seen on the mid-ventral line of the thoracic cavity. Cut to one side of the sternum, forcing the ribs apart so that the organs in the thoracic cavity may be examined.

THE DIGESTIVE SYSTEM

Cut some of the muscle tissue of the neck region, being extremely careful to dissect around large arteries and veins. Expose the trachea, a whitish tube made up of cartilaginous rings.

Open the mouth of the pig by cutting with the angle of the jaw. Run a blunt-edged probe along the tongue of the pig and back into the pharynx or throat, pressing lightly on the probe. The probe will go down the tube behind the trachea. This is the esophagus. The esophagus passes dorsally to the heart and is very difficult to distinguish in more posterior regions. At the anteriormost portion of the abdominal cavity, the esophagus passes into the stomach.

The stomach is a horn-shaped organ that lies cradled by the lobes of the liver. There are three major lobes of the liver, two centrally located, which conceal the stomach and may be dissected away. Notice that on the underside of the right lobe is a greenish sac. Do not disturb this in cutting away part of the liver. The stomach is made up of two parts: a large anterior portion which is called the cardiac region, and the smaller or narrower posterior portion referred to as the pyloric region. This pyloric region adjoins the duodenum, the first region of the small intestine. Between the loop formed by the curvature of the stomach with the duodenum is the pancreas. This gland is characterized by an irregular spongy tissue. The small green sac on the underside of the liver that was noted before is the gall bladder. The gall bladder connects to the duodenum through a duct that brings bile salts into the small intestine. This duct branches to join the pancreas, which secretes pancreatic juices into the intestine. The small intestine is a continuous tube structure which is held tightly together by a type of connective tissue, the mesentery. The terminal portion of the intestine, which lies flat against the dorsal wall in the posterior region, is the large intestine or colon. At the junction between the small intestine and the colon is a blind pocket, the vermiform appendix. The anus is the terminal opening which is preceded by a set of rectal muscles that control the egestion of wastes from the final portion of the gastrointestinal tract, the rectum.

1. How does the pancreas aid in digestion?

2. How do the products of the liver which are collected in the gall bladder aid in digestion?

THE RESPIRATORY SYSTEM

Once again force open the mouth of the pig. Run a blunt probe along the tongue, pulling slightly forward when you reach the pharynx. This should bring the probe into the trachea. The trachea has alread been identified as the whitish tube made up of cartilaginous rings. Note the enlargement at the anterior portion of the trachea. This is the larynx or voice box (referred to sometimes as the Adam's apple). Notice that just before its entrance into the spongy lung tissue, the trachea bifurcates (divides) into two bronchi. The lungs are a light pink spongy tissue made up of three lobes on the right side and two on the left.

1. What is the function of the lungs in the fetal pig?

2. Would you expect the color of the lungs of the fetal pig to be the same as those of the newborn? Why?

3. What consequences does your answer to 2 have in terms of the circulatory system of the fetal pig?

THE CIRCULATORY SYSTEM

Figures A1.2 and A1.3 are representative of the typical or average circulatory system found in the fetal pig. Minor variations in the exact position of the arteries and veins may exist in your specimen. Refer judiciously to the illustrations provided, and follow each artery or vein to the organ to which blood is brought or from which blood is taken to make sure you are examining the proper one.

The heart is found encased in a membranous pericardium. Directly anterior to the heart and sometimes partially covering it is the thymus gland. This gland has been associated with the antibody formation of the animal. Since the antibodies are usually found in the protein portion of the blood, the thymus may be considered as auxiliary of the circulatory system. Carefully remove the pericardium, being careful not to damage any of the structures of the heart in the process. The major part of the heart is made up of the ventricles. These large muscular regions form the posterior portion of the heart. The anterior chambers, the auricles, sometimes appear as spongy flaps lying over the heart. Locate the right auricle and ventricle and the left auricle and ventricle. In the adult form the right side of the heart handles only deoxygenated blood, the left side of the heart only oxygenater blood.

Nonoxygenated blood enters the heart from the precaval and postcaval veins, which enter on the dorsal side of the right auricle. Find these points of entrance by lifting the right auricle. The blood then passes into the right ventricle and out the pulmonary arteries to the lungs. After oxygenation in the lung (referring, of course, to the adult circulation), the blood returns to the heart by means of the pulmonary vein, which enters the left auricle. The blood passes into the left ventricle and then into the aorta. The aorta arches dorsally (the aortic arch), giving off two major branches as it arches. The first of these branches, the brachiocephalic trunk, subdivides almost immediately to form the two common carotid arteries, which go to the head region on either side of the trachea, and the right subclavian artery, which goes to the right forelimb. The second branch of the aorta is the left subclavian artery, which goes to the left forelimb. Notice that shortly after the branching of the left subclavian there is a small branch that connects the aorta to the pulmonary artery. This is the ductus arteriosus, a structure that permits the by-passing of the lungs by bringing the pulmonary blood into the aorta. This structure collapses owing to a pressure sequence which automatically occurs when the lungs are filled with atmospheric air for the first time and which may be aided by contraction of the muscles of the ductus. The aorta passes behind the heart, and is then called the dorsal aorta. After passing through the diaphragm, the aorta gives rise to two single (unpaired) arteries. The coeliac axis supplies the stomach, liver, and spleen. The spleen is a smooth fingerlike projection found anatomically associated with the stomach. Its function is the destruction of red blood cells. The next branch, the anterior mesenteric artery, supplies oxygenated blood to the small intestine and pancreas. Paired arteries are given off next by the dorsal aorta; one pair, the renal arteries, goes to the kidneys; the next, the spermatic or ovarian arteries (depending upon the sex of the pig) supplies the gonads with oxygenated blood. A single artery, the posterior mesenteric, goes to the colon and rectum. Finally, the aorta gives rise to the external iliac arteries, which goes into the legs. The remainder of the dorsal aorta passes into the tail as the caudal artery.

The return of blood to the heart is governed by a similar system. Blood from the hind limbs passes into the internal iliac vein and the external iliac vein which joins to form the common iliac vein. The common iliac veins from each of the hind limbs join to form the postcava, the major vein of the posterior region of the body. Renal veins and spermatic or ovarian veins join the postcava, coming from their specified organs to do so. A single mesenteric vein brings waste products from the metabolic action of the intestine, stomach, and spleen cells. This mesenteric vein enters the liver and joins the umbilical veins in the fetal circulatory setup. In the adult, the wastes are brought to the postcava by way of the liver. The postcava enters the heart at the right auricle. The anterior region of the body is drained by the precava. The external and internal jugular veins carry deoxygenated blood from the head region. These join the right and left subclavian veins, which take deoxygenated blood from the forelimbs, to form the precava. The precava enters the heart at the dorsal side of the right auricle.

1. The circulatory system is so called because the blood may be carried to any part of the body and returned to a given point. Show that this is so by following a red blood cell from the external jugular vein to the kidney, where waste products are removed. Give all the arteries and veins as well as the parts if the heart that the red blood cell must pass through on its journey.

2. How does fetal circulation differ from adult circulation? Does this difference reflect the different needs of the organism in both stages? Why?

ARTERIES

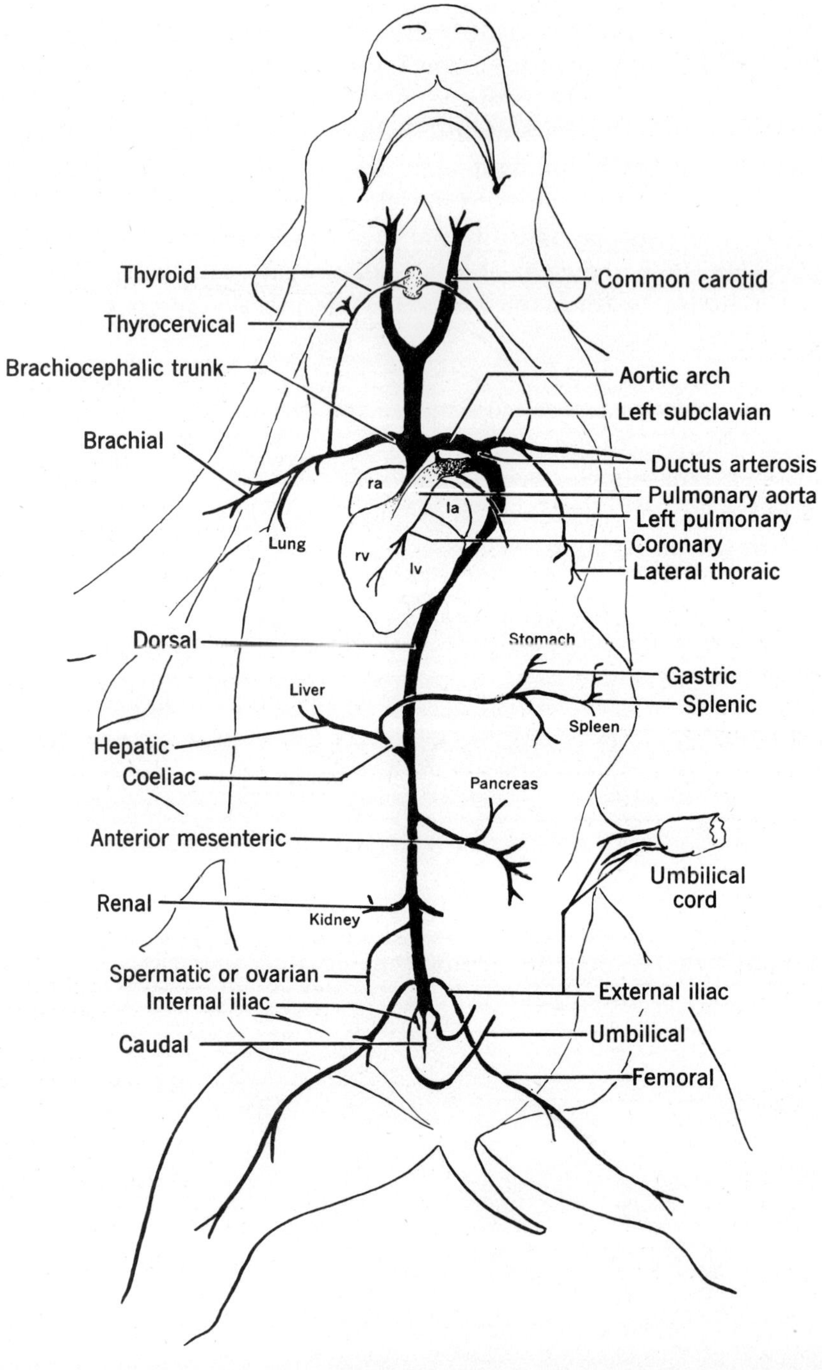

Figure A1.2

VEINS

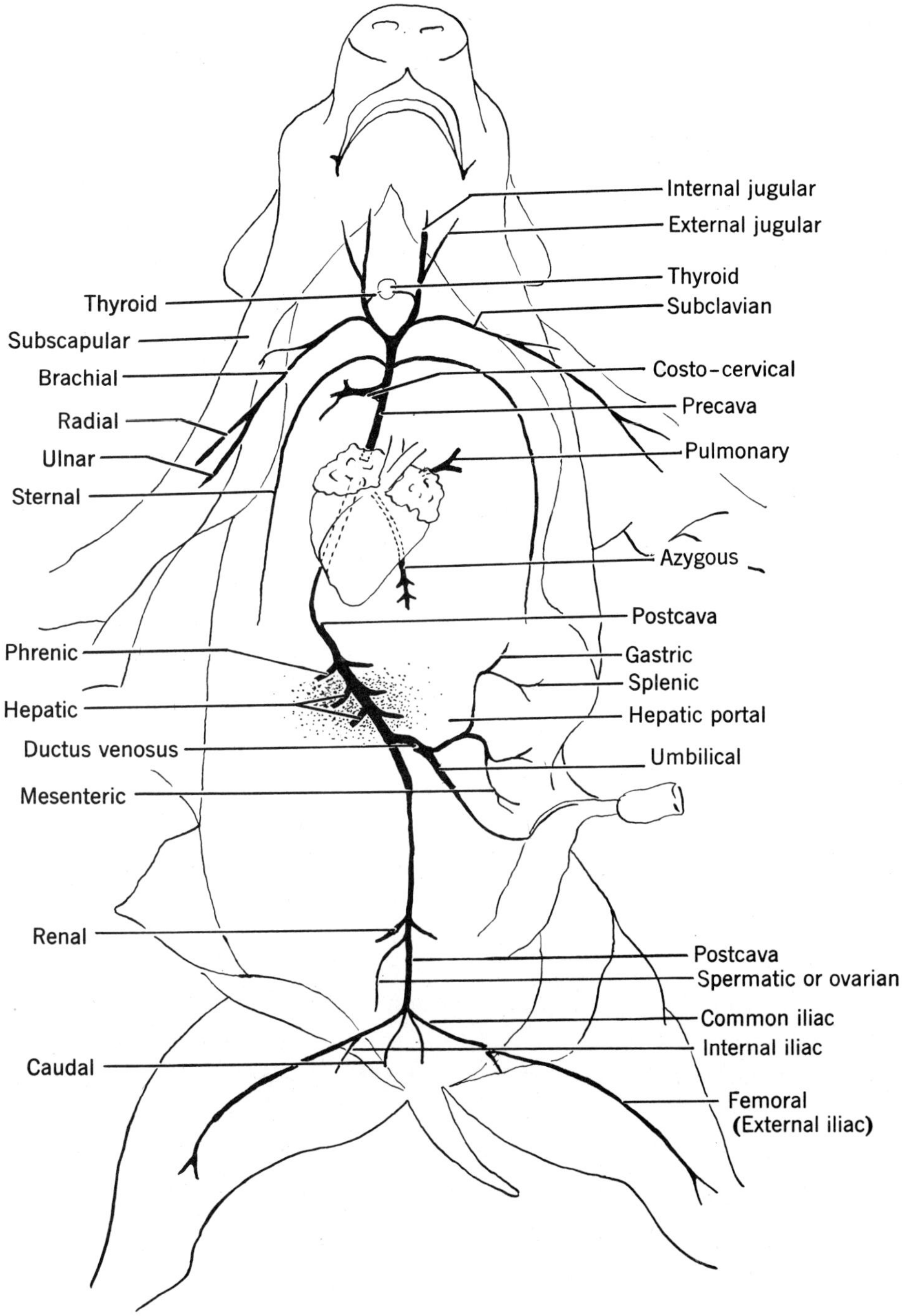

Figure A1.3

THE ENDOCRINE SYSTEM

A small red ovoid body lies on the ventral side of the trachea. This is the thyroid gland. Notice that there is no duct.

The pancreas has already been seen in connection with its function in digestion.

The adrenal glands may be seen closely applied to the anterior portion of the kidney, which is composed of two large bean-shaped structures on the dorsal wall of the abdominal cavity. From the gross anatomy it is possible to distinguish between the cortex and the medulla.

1. How does the pancreas act as an exocrine gland? An endocrine gland?

2. What is the function of the thyroid gland?

THE REPRODUCTIVE SYSTEM (UROGENITAL SYSTEM)

Determine the sex of your fetal pig by examining the external genitalia. The female urogenital sinus is found directly anterior to the anus. In the male, the penis lies in a sheath in the abdominal wall, and the opening is found directly posterior to the umbilical cord.

Carefully cut through the public symphysis, the bony pelvic structure which protects the urogenial system.

The Female

Small oval-shaped ovaries lie closely applied to the oviducts or Fallopian tubes. These structures are directly posterior to the kidneys. The Fallopian tubes are flattened and coiled. Careful examination of the tube near the ovary will reveal the tube opening or ostium tuba. The oviducts continue posteriorly, enlarging to form the horns of the uterus, which finally come together to form the common uterus. A posterior constriction of the common uterus shows the position of the cervix or neck of the uterus. Posterior to this point the tube is the vagina. The vagina joins with the urethra from the bladder to enter the urogenital sinus. Trace the urethra to the point where it enters the bladder. The bladder is a balloonlike muscular sac which receives liquid waste from the kidney. The urine is conducted from the kidney by the ureter.

The Male

On either side of the posterior region of the public symphysis the inner portion of the scrotal sac should now be exposed. If the fetal pig is at an early stage of development, the testes may not be descended. In this case they will be found in the abdominal cavity or in the inguinal canal which leads from the scrotal sac. The testes are somewhat oval, and are dark in color. They are surrounded by the epididymis, a white coiled tube that stores the sperm in the adult. The epididymis continues and straightens, becoming the vas deferens and passing into the body cavity. The vas deferens loops over the ureters coming from the kidney, passes dorsally of the bladder and enters the urethra. In this way both sperm and the wastes from the kidneys are released from the body by the same organ, the penis.

WRITING A RESEARCH REPORT

Biology, similar to other natural sciences, does not exist to be taught; instead, it should answer certain questions. The answers are communicated to others working in the same or similar areas and thus contributes to the body of subject matter on which further experimentation and theorical work can be carried out. Teaching science is a secondary consequence.

Perhaps there is no better means to understand scientific work than by performing it and writing the work as a scientific report. Scientific writing is generally impersonal and unimaginative because it should precisely convey: (1) *how* a study was made (methods), so that others can perform the same work; (2) conclusions in such a way that they can be drawn by anyone examining the data presented. In both of these endeavors, personal feelings are distractful. The following pattern is provided for use in writing a report on individual or group research projects.

Title Page

The title page should be a full page containing only the title of the report and the name(s) of the author(s) centered on the page.

Summary

The summary should consist of a full page containing the condensation of the main features of the report, one or two paragraphs in length, centered on the page.

Body of the Report

Title and authors names repeated.

Introduction—A statement of the problem explained in terms of some broad biological principle; how other investigators have contributed to the premise of your experiments. References must be given.

Materials and Methods—A description of the procedures used, both those adapted from other investigators and those that have been developed in the course of the work. Sources of materials (supply houses) and full identification of strains or varieties of organisms must be given. This section should be written as though someone elsewhere wanted to repeat your work, and you wish to guide him by your experience.

Results—The descriptions of what you have found preferably expressed in quantitative terms in the form of tables and graphs. Where quantitative expression is not possible, use simple, clear drawings or photographs. The written part should refer to all illustrations and point out salient features of the data. The more condensed and concise the results, the more powerful is the data. Remember that important points can be lost admist a mountain of charted figures and a tidal wavc of curves.

Discussion—An exposition of what the results mean with reference to the problem presented in the introduction. This discussion should entail some statements regarding what others have found and whether you are in agreement with their conclusions. Mention should sometimes be made about the value of your procedures, especially when they are lacking in accuracy. Finally, some broad interpretation can be given, with ways that the interpretation can be experimentally checked.

Literature Citations—Only those references used in the body of the text should be cited. The format for literature citations vary. One common usage follows.

Papers:

Jones, C. 1969. Isozymes of Lactic Acid Dehydrogenase of the Squid. *Physiol. Jour.* 44(6): 22-47.

Books:

Smith, A. Methods of Protein Isolation, Carefree and Footloose, Inc. New York, 1968. pp. 234-267.

Papers in Symposia:

Peters, N. 1963. The Action of Insulin on Fetal Dogs. *In:* Physiology of Mammals, C. S. Samuels, *Ed.* Denver Press. 104-122.

General Comments

Reports must be typed doubled-spaced, having margins of 1 in., except at the left, where 1½ in. are required.

Illustrations should be done in ink, preferably India ink.

Expressions To Be Avoided	*Use Instead*
I did the . . .	The procedure employed was . . .
This project . . .	This investigation . . . This study . . .
This procedure was tried . . .	This procedure was attempted . . .
I wonder if . . .	The question can now be asked . . .
I found out . . .	It was demonstrated . . . It was ascertained . . .
A good way to . . .	A successful method for . . .

Appendix 2

PREPARATIONS AND SUPPLIES

Chemical Preparations

Acetocarmine stain

Add carmine to 45% acetic acid (boiling) until no more dissolves. When cool, filter the solution.

Benedict's reagent

800 ml distilled water
173 g Na_3 citrate
100 g Na_2CO_3
Warm to dissolve, cool and filter.
100 ml distilled water
17.3 g $CuSO_4$
Add to first solution, stirring constantly.
Add distilled water to total volume of 1 liter.

Brodie solution

46 g NaCl
10 g Na cholate
0.3 g Evans blue
Add water to 1 liter.

Diché reagent

97.25 ml glacial acetic acid
2.75 ml concentrated H_2SO_4
Mix together; then add
1 g diphenylamine
Mix, then add
100 ml water

Frog Ringer's solution
7.0 g NaCl
0.15 g KCl
0.15 g $CaCl_2$ (anhydrous)
0.1 g $NaHCO_3$
distilled water to 1 liter.

Giesma stain
10% aqueous solution of Giesma stain in distilled water.

Gram stain–Jensen's version
solution 1. Dissolve 0.5 g methyl violet in 100 ml distilled water.
solution 2. Dissolve 2 g potassium iodide and 1 g iodine (ground together in a mortar to a fine powder) in not more than 20 ml distilled water. The iodine will not dissolve in a weaker solution of potassium iodide. Make up to 100 ml when dissolved.
solution 3. Basic fuchsin 0.1% (as above) or 1% aqueous safranin
95% ethyl alcohol

Holtfretter's solution
3.5 g NaCl
0.05 g KCl
0.1 g $CaCl_2$ (anhydrous)
0.2 g $NaHCO_3$ *
distilled water to 1 liter

IKI
20 g potassium iodide
1 liter distilled water
Stir to dissolve
4 g iodine crystals
Allow time for dissolving
Keep in dark-colored stoppered bottle.

Janus green solution
Janus green 0.01 g
0.25 M sucrose (100 ml)

Nessler's reagent
50 g potassium iodide in distilled water
Add saturated solution of mercuric chloride until slight ppt remains.
400 ml 50% NaOH (which has been clarified by sedimentation)
Dilute to 1 liter with distilled water.
Allow to settle for 1 week before using.
Decant the clear solution and keep in a well-stoppered bottle.

*Sterilize, dry, and add to remainder of ingredients after autoclaving to prevent precipitation (ppt).

Neutral Red
1% aqueous solution for stain

Pyrogallol
160 g KOH
130 g distilled water
Cool.
Add 10 g resublimed pyrogallic acid and stir.

Trommsdorf's solution
To 4 g of starch in water, add a boiling 20% aqueous solution of $ZnCl_2$ slowly. Continue heating until the starch dissolves as much as possible and the solution appears almost clear.
Dilute with water and add 2 g of either zinc or potassium iodide.
Dilute to 1 liter, filter, and store in dark, well-stoppered bottles.

MEDIA PREPARATION

Chlorella medium
1.25 g KNO_3
1.25 g KH_2PO_4
1.0 g $MgSO_4 \cdot 7H_2O$
1 liter distilled water
1.0 ml trace solution
Place in vessels and autoclave.
Trace solution
0.3 g $CaCl_2 \cdot 2H_2O$
5.0 g EDTA
1.0 g H_3BO_3
0.15 g $CuSO_4$
2.20 g $ZnSO_4 \cdot 7H_2O$
0.5 g $MnCl_2 \cdot 4H_2O$
0.5 g $FeSO_4 \cdot 7H_2O$
0.15 g $CoCl_2 \cdot 6H_2O$
0.10 g $(NH_4)_6Mo_7O_{24} \cdot 4H_2O$
100 ml distilled water
Add KOH to pH 7.
(The trace stock may be prepared and kept rather than making repeated preparations.)

Defined bacteria medium (for limited growth)
13.6 g KH_2PO_4
2.0 g $(NH_4)_2SO_4$
0.2 g $MgSO_4 \cdot 7H_2O$
2.0 g Maltose

1.0 ml trace solution (see *Chlorella* medium above)
1 liter distilled water
Bring to pH 7.4 by addition of 10 N KOH, then autoclave.

Drosophila medium
18 g powered agar
875 ml cold water
Dissolve by boiling.
Add 125 ml unsulfured molasses.
Heat again to boiling.
To 125 ml dry corn meal (fine grain) add 50 ml brewer's yeast and mix with 250 ml water. Break up lumps with fingers.
Add cornmeal and yeast mixture to boiling agar and cook for 20 min, stirring frequently.
Cool for 5 min and add 5 ml proprionic acid as a mold inhibitor.
Pour into sterilized bottles to depth of ¾ in. or pour into vials to form slants, and plug.
Store in refrigerator for not more than 2 weeks.
Just before use sterilize strips of white paper 1 × 3 in. in double thickness. Add packaged dry yeast to water to prepare a light yeast suspension. Add one or two drops of yeast suspension to food bottles or vials before use. Stab sterile white paper strip into medium along the edge of bottle. Etherize flies and drop onto this paper.

EMB agar
10 g peptone
10 g lactose
5 g sucrose
2 g K_2HPO_4
0.4 g eosin Y
0.06 g methylene blue
15 g agar
1 liter distilled water
Dissolve by boiling. Autoclave and pour plates (makes approximately 27 to 30 plates).

Nitrite formation medium
1.0 g $NaNO_3$
1.0 g K_2HPO_4
0.3 g $MgSO_4$
1.0 g Na_2CO_3
0.5 g NaCl
0.4 g $FeSO_4$
Add distilled water to 1 liter.
This medium does not need to be sterilized.

Nutrient agar
3.0 g beef extract
5.0 g peptone
15 g agar

1 liter distilled water
Dissolved by boiling, stirring constantly. Autoclave and pour plates. Makes approximately 30 plates.

Nutrient broth
3.0 g beef extract
5.0 g peptone
1 liter distilled water
Dissolve by heating. Pour into bottles or flasks and autoclave.

Peptone broth
4% aqueous solution peptone
Pour into bottles, flasks, or tubes and autoclave.

Potassium hydroxide–Maleic acid buffer
(a) 11.61 g maleic acid in 1000 ml H_2O
(b) 5.61 g KOH in 1000 ml H_2O
50 ml of (a) + 55 ml of (b) and add H_2O up to 200 ml.

Potato dextrose agar
39 g potato dextrose agar
1 liter distilled water
Autoclave, cool, pour plates or slants.

Starch agar
10 g tryptone
10 g yeast extract
5 g K_2HPO_4
3 g soluble starch
Add starch to small amount of water. Heat, stirring, until starch goes into solution. Add water to 1 liter and bring to boil. Add remainder of ingredients.

Streptomycin agar
5 g peptone
3 g beef extract
15 g agar
1 liter distilled water
Boil to dissolve, autoclave, and permit to cool to 60°C.
Add 20 mg streptomycin.
Pour plates.

Soil extract medium (algae and ferns)
To 500 ml distilled water add 100 ml alkaline soil and a pinch of $CaCO_3$. Boil for 20 min.
Filter while hot and then either autoclave or store in refrigerator. Soil extract should be yellow.
Good soil will filter easily, since it contains mostly large particles. Fine particles in the soil often inhibit filtration and indicate that medium is not too rich.

Yeast medium

- 10 g dextrose
- 5 g yeast extract
- 5 g peptone
- 1 g malt extract
- 1 liter distilled water

Place in vessels and autoclave.

Note: The media for the study of inorganic ion requirements and vitamin B_{12} requirements are found in Chapter 7.

Normalities of Concentrated Acids

Acid	Concentration (% by wt)	Molarity	Normality
Acetic	99.5%	17.5	–
Hydrochloric	35.4%	11.4	12.1
Sulfuric	99.0%	18.3	27.3
Nitric	70.0%	15.6	15.5

Preparation of Base Solutions

Base	1N	Concentrated
NaOH	40 g/liter	approx. 130 g/100 ml
KOH	56 g/liter	approx. 125 g/100 ml

% Solutions

1% = 1 g/100 ml

Stock Lists

Reagents

acetic acid
acetocarmine
acetone
actinomycin D
agar
alanine
albumen
amino acids (20 standard forms)
ammonium chloride

ammonium molybdinate
ammonium nitrate
amylase
antisera A
antisera B
arginine

beef extract
Benedict's reagent
benzene

benzoadenine
bile salts
boric acid
brewer's yeast

caffeine
calcium carbonate
calcium chloride
calcium nitrate
calcium sulfate

catechol
citric acid
cobalt chloride
colchicine
copper chloride
copper sulfate
corn meal
crystal violet
cycloheximide
cysteine

dextrose
diphenylamine

EDTA
eosin Y
ethyl alcohol
ethyl ether
Evan's blue

ferric chloride
ferric cyanide
ferric sulfate
formaldehyde

gibberellic acid
Giesma stain
glucose (dextrose)
glucose–1–phosphate
glutamic acid
glycine

histidine
hydrochloric acid
hydroxylysine
hydroxproline

indole acetic acid
iodine crystals
isolysine

Janus green dye

lactose
lanolin
leucine
lysine
lysozyme

magnesium chloride
magnesium nitrate
magnesium sulfate
maleic acid
malt extract
manganese chloride
mercuric chloride
methionine
methyl alcohol
methyl cellulose
methylene blue
mitomycin C
molasses, unsulfured

neutral red
ninhydrin spray

pancreatin
pepsin
peptone
petroleum ether
phenol red (water soluble)
phenylalanine
phosphoric acid
potassium bicarbonate
potassium chloride
potassium hydroxide
potassium nitrate
potassium phosphate
potato dextrose agar
proline

pyrogallic acid

quinine

rennin tablets

salad oil
serine
sodium bicarbonate
sodium carbonate
sodium citrate
sodium chloride
sodium cholate
sodium fluoride
sodium hydroxide
sodium molybdinate
sodium phosphate
sodium sulfate
starch
steapsin
streptomycin (Mann Res. Lab.)
sucrose
Sudan IV
sulfuric acid

threonine
trypsin
tryptone
tryptophane
tryosine

Vitamin B_{12} (cyanocobalamin)

xylol

yeast extract

zinc chloride
zinc iodide
zinc sulfate

Prepared Slides

Plant

Allium root tip
dicot leaf xs
dicot stem xs
fern prothallia showing male and female sex organs wm
fern rhizome xs
gymnosperm stem xs
Ligustrum leaf xs
Lilium anther xs
Lilium embryo sac xs
Lilium fertilization
monocot leaf xs
moss antheridium ls
moss archegone, ls
Pinus leaf xs
Tilia 2 to 3 yr old xs

Animal

adrenal cortex and medulla xs
Amphioxis wm
chick 48-hr wm
early starfish development
Fasciola wm
frog blastula xs
frog late neural groove xs
frog neural tube (2.60 mm)
frog yolk plug xs
horse chestnut stem xs
human blood smear
mammalian ovary xs
mammalian testis xs
pancreatic tissue
Planaria wm
sickle cell anemia Cat #52492 MacAllister Sci Corp.
spermatogenesis-chicken
striated muscle
Taenia wm
thyroid sx
tunicate *(Molgula* or *Corella)*
whitefish blastula

Models

human systems

Skeletons

cat
frog
human

Preserved Specimens

clam[a]
crayfish[a]
earthworm[a]
fetal pig

[a]May be done as demonstration instead of student dissection.

frog
grasshopper[a]
preserved embryos-one, two, four, eight celled stages blastula, yolk plug, late neural groove (frog) states (Carolina Biological Supply)
Ward DC3 abridged collection or representative specimens

Live Material

Bacterial Strains

B. megatherium
B. subtilis (American Type Culture Collection)
E. coli (American Type Culture Collection)
E. coli mutabula (ATCC; Bacteriology Dept., Indiana University) or
E. coli lac^- (Carolina Biological Supply)

Viruses

T-4 (Turtox)

Protistans

Amoeba
Chlamydomonas moeweusii
Chlorella (Algal Culture Collection)
dinoflagellates
Euglena gracilis
Mixed Desmid culture (Turtox)
Neurospora
Nostoc
Oedogonium (Algal Culture Collection)
Paramecium
Rhizopus nigricans (Turtox or Carolina Biological)
Schizosaccharomyces (American Type Culture Collection)
Spirogyra
Tetrahymena

Plant Material

bean seedlings
corn for 9:3:3:1 ratio
fern spores (collect locally or Turtox or Carolina)
Juniper branches
lettuce seeds (Cat #51510 MacAllister Sci Corp)

[a]May be done as demonstration instead of student dissection.

moss sporophyte-*Polytrichum* (collected in spring)
radish seeds
wheat or barley seeds

Animal Material

Drosophila
Frog, male and female
Planaria (Turtox or Carolina)

Addresses for Special Material

American Type Culture Collection
2112 M Street NW
Washington, D.C.

Algal Culture Collection
Botany Department
Indiana University
Bloomington, Ind.

Carolina Biological Supply Co.
Burlington, N.C.
and
Powell Laboratories Division
Gladstone, Ore.

MacAlister Scientific Corporation
Educational Division
243 Broadway
Cambridge, Mass.

Turtox Biological Supply
General Biological Supply House
8200 South Hoyne Ave.
Chicago 20, Ill.

Ward's Natural Science Establishment, Inc.
P.O. Box 1712
Rochester 3, N.Y.

Appendix 3

PREPARATION OF LABORATORIES

Since the topics in this laboratory manual have not been given equal emphasis, some chapters are longer than others. Some laboratory exercises may be completed in one week, others may take several weeks. The particular laboratory exercises that take more or less periods depends on the emphasis of the particular course. Hence, organization beyond the scope of this manual is necessary.

Chapter 1 Measurement, Data Organization, and Analysis

No preparation necessary.

Chapter 2 Levels of Biological Organization

No preparation necessary.

Chapter 3 Chemical Background for Biology

Prepare necessary reagents.

Egg yolk solution: transfer the yolk of a broken egg back and forth in the shells, permitting the albumen to fall into a beaker. Place the yolk in another beaker. Add 300 ml distilled water to the yolk and stir.

Yeast culture: Add 200 ml distilled water to five packages of yeast. Stir until dissolved; boil.

Chapter 4 Protein Chemistry and Enzymes

Prepare necessary solutions (see Appendix 2).

Chapter 5 Intermediary Metabolism

Prepare necessary solutions (see Appendix 2).

Wheat or barley seedlings. Soak seeds overnight, drain and place in trays lined with wet paper toweling for 2 days before use. Each day's classes will require fresh seedlings (sprouts should be no longer than 0.3-1 inch). These sprouts however, may be reused several times daily if no KOH is spilled on them. Seedlings must be kept moist.

Yeast culture. Add 200 ml distilled water to one package of dry yeast. Stir until dissolved.

Chapter 6 Photosynthesis

Simple continuous light source. A single- or double-tube fluorescent strip may be obtained at a local hardware store. Place on a table with two supports 14-18 in. high. This will give a light intensity of approximately 200 ft-c.

Pasteur pipettes. These are used because they uniformly give 1/24 ml per drop, are inexpensive, and may be easily used with a rubber bulb. Disposable Pasteur pipettes come in 7- and 9-in. sizes; either is suitable.

Growth of Chlorella stock. Inoculate *Chlorella* medium (Appendix 2) which has been spiked with 2% glucose to encourage heavy growth of *Chlorella* from the stock. Grow in continuous light for about 5 days. *Chlorella* from this stock is the inoculum for class cultures.

Chlorella grown under light and dark conditions. Prepare a sufficient number of 25-ml Erlenmeyer flasks with straight *Chlorella* medium (Appendix 2) so that each laboratory day will have two flasks (ten flasks if laboratories run all week) and autoclave. Wash the stock cells by centrifuging in a clinical centrifuge for 1 min and resuspending in straight *Chlorella* medium twice. Inoculate each flask using sterile procedures in such a way that a final concentration of approximately 5×10^4 cells/ml is achieved. The student counting chamber may be used for a check. Place half the flasks in continuous dark, the other half above the continuous light source and grow for 1 week. One 25-ml flask will last for 1 day's laboratory, after which contamination may invalidate its use. All flasks may be made simultaneously and refrigerated or frozen after 1 week of growth and then de-frosted in time for class inspection.

Growth of yeast stock. Inoculate yeast medium (Appendix 2) with stock culture of *Schizosaccharomyces* and incubate at 37° C (if possible) for 3 days or until a heavy growth is noted.

Yeast grown under light and dark conditions. Prepare Erlenmeyer flasks with yeast medium and autoclave, using the same number of flasks as for the *Chlorella.* Wash the stock cells by centrifuging in a clinical centrifuge 1 min and resuspending in straight *Chlorella* medium twice. Inoculate each flask in such a way that a final concentration of approximately 5×10^4 is achieved, using the student counter as a check. One 25-ml flask will last for 1 day's laboratories. Both the yeast and the *Chlorella* cultures must be kept free of contaminants, for bacteria will overrun the cultures and making counting difficult.

Chlorella grown in different qualities of light. Prepare a sufficient number of 25-ml Erlenmeyer flasks to provide each laboratory day with a series (25 for a full week's laboratories) by filling with *Chlorella* medium and autoclaving. Wash the stock cells by centrifuging in a clinical centrifuge for 1 min and resuspending in straight *Chlorella* medium twice. Inoculate each flask to a total of 1×10^3 cells/ml, checking with the student counter. Submit the tubes to the light treatment in one of the following ways. (Note: Each uses DuPont cellophane, which may be purchased at variety stores.)

1. Cellophane may be wrapped around the flasks.
2. Filters may be made by placing layers of cellophane between two 2 × 3-in. slides. Filters would rest on the rack above the light source with the flasks placed over them. This latter method is preferred because *Chlorella* tends to settle to the bottom and this method permits uniform exposure of cells, since light is not absorbed when passing through the liquid. Flasks should be shaken once a day to ensure sufficient CO_2 dissolution.

The following number of layers for each color should be used in order to give uniform intensity: red, 1 layer; orange, 5 layers; yellow, 19 layers; green, 1 1/3 layers (use a strip of cellophane one-third the width of plate or tube); blue, 1 1/2 layers, violet, 2 1/2 layers.

Chapter 7 Nutrition

Ammonification. Two days before each laboratory, inoculate 4% peptone broth, which has been autoclaved in 25 ml Ehrlenmeyer flasks, with 3 to 4 pinches of soil. Incubate at room temperature. A single flask may be used for the entire day's laboratories.

Nitrite formation. For each of five weeks before the laboratory day, inoculate nitrite formation medium (see Appendix 2) made up in Ehrlenmeyer flasks with 0.2 g of neutral or alkaline soil. Incubate at room temperature. Each flask will last for a laboratory day; and each laboratory day should have five flasks, one from each week's inoculation (25 flasks for a five-day laboratory schedule).

Starch agar. Inoculate one plate with *E. coli,* one with *B subtilis.*

EMB agar. Inoculate one plate with *E. coli,* one with *B subtilis.*

The vitamin requirement. (Optional). The media may be prepared by the instructor. Instructions for the medium are in Chapter 7.

Chapter 8 Cellular Organization

Prepare necessary solutions.

Chapter 9 Reproduction of Cellular Components

Seedling preparation. Germinate radish seedlings in the manner described for wheat or barley seedlings p. 489. After 20-24 hr growth, transfer 100 seedlings to a petri plate and flood the plate with 2 ml of 10% formaldehyde. Transfer another 100 seedlings to a second petri plate with 2 pieces of circular filter paper and add 2 ml of 0.03% colchicine. After 20 hours, flood the plate with 2 ml of 10% formaldehyde. Transfer another 100 seedlings to a third plate with filter paper and add colchicine, 0.03%. After 20 hr, wash the seedlings under running water for 5 min, and place in another plate with mitomycin C, 50 ug/ml. Transfer another 100 seedlings to a petri plate with filter paper, and permit to grow for 20 hr. Flood the plate with formaldehyde. One series of plates will serve for all laboratories.

Chapter 10 The Theory of Particulate Heredity

Preparation of *Drosophila* medium and vials or bottles (see Appendix 2).

Collecting virgin ♀ ♀. Two to three weeks before the mating day (5 weeks before laboratory day) start several fresh bottles of both the white-eyed stock and the red-eyed stock. When hatching begins, etherize flies and examine for young females. Non-fertile females may be distinguished by their light color, which sometimes has a greenish tint. The intestine may be seen through the transluscent body wall when the eggs have not formed yet. Place no more than five virgins in a vial of fresh medium in case errors are made. Since females are incapable of mating until 10 hr after hatching, the vials may be cleared after the morning collection by killing any flies that have not come out of the bottle for etherization and considering all the females in the afternoon collection virgins. Male collections should also be made at this time, and, of course, kept in separate vials.

Mating of Drosophilia. Matings may be done directly in vials, in which case at least 2 ♀♀ (virgin) and 2 ♂ ♂ should be placed on the dry paper within the vial and left for 4 days at room temperature. The flies (P_1) should then be removed and discarded, and the vials left at room temperature for 10 more days (total of 2 weeks) for offspring to emerge in sufficient numbers to give significant counts. A single set of vials may be used for one class day if the flies are not overetherized. Because of this danger, 100 vials per class week should be available.

Another acceptable method, which might prove more expedient when dealing with large numbers of classes, is to prepare the medium in pint bottles or baby food jars, mating four ♀♀ (virgin) and four ♂ ♂ of appropriate types. After the parents are removed, the slab of medium containing the larvae may be removed, divided into ½-in. cubes, and placed in sterile vials. These vials should then be sprayed or inoculated with yeast solution by addings one or two drops as in the preparation of the initial bottles. Fresh sterilized paper should be placed in each to provide a dry place for pupa cases to form. One hundred vials per class week should be available to offset the possibility of overetherization.

Construction of a simple etherizer. A small plastic funnel, the end of which has been covered with three layers of cotton gauze and taped, may be inserted into a baby food jar, the bottom of which has been lined with absorbent cotton. The bottom of the funnel should not come directly in contact with the cotton. A few drops of ether placed in the cotton is sufficient to etherize flies for one class period.

Chapter 11 Genetic Regulation of the Cell

Seedling preparation. Germinate radish seedlings in the manner described for wheat or barley seedlings p. 489. After two-days growth, transfer about 100 seedlings to a petri plate and flood the plate with 4 ml of 10% formaldehyde. Transfer another 100 seedlings to a second petri plate with 2 pieces of filter paper and leave for 20 hours. Add formaldehyde after 20 hr. Place another 100 seedlings in a third plate with filter paper. Add 4 ml of 100 μg/ml of cyclohexamide. After 20 hr, flood with formaldehyde. Add another 100 seedlings to a fourth plate with filter paper and add 4 ml of 50 μg/ml actinomyocin D. After 20 hr flood with formaldehyde.

The translated coded message.

One set of the following messages are needed for each class. Make copies and cut along dotted lines. Distribute messages on the tables of the classroom.

- -

CCCCCAAAAAUAAUUCGGAACCGGAAUAUUCCACCUCUCCGGAAAAAAAUA

- -

CCCCCAAAAAUAAUUCGGCUACGGAAUAUUCCACCUCUCCGGAUAGGUAAA

- -

CCCCCAAAAAUAAUUCGGUACCGGAAUAUUCCACCUCUCCGGGGUAAAAUA

- -

CCCCCAAAAAUAAUUCGGUUUCGGAAUAUUCCACCUCUCCGGGGUGGUGGU

- -

CCCCCAAAAAUAAUUCGGUGACGGAAUAUUCCACCUCUCCGGGGUGACAAA

- -

CCCCCAAAAAUAAUUCGGGGUCGGAAUAUUCCACCUCUCCGGGACGACGGU

- -

CCCCCAAAAAUAAUUCGGGACCGGAAUAUUCCACCUCUCCGGGACAAAAUA

- -

CCCCCAAAAAUAAUUCGGGCGCGGAAUAUUCCACCUCUCCGGGACAUAGAC

CCCCCAAAAAUAAUUCGGCGGCGGAAUAUUCCACCUCUCCGGAUAGACGAC

UAAAUUUACAACCUACCUGCGCGGAUAAUUCCACCA

CCCCCAAAAAUAAUUCGGAAACGGAAUAUUCCACCUCUCCGGAAAAAAAAA

CCCCCAAAAAUAAUUCGGAAUCGGAAUAUUCCACCUCUCCGGAAAAAAGGU

CCCCCAAAAAUAAUUCGGAUACGGAAUAUUCCACCUCUCCGGAAAGGUAAA

CCCCCAAAAAUAAUUCGGUAACGGAAUAUUCCACCUCUCCGGGGUAAAAAA

CCCCCAAAAAUAAUUCGGAUUCGGAAUAUUCCACCUCUCCGGAAAGGUGGU

CCCCCAAAAAUAAUUCGGUUACGGAAUAUUCCACCUCUCCGGGGUGGUAAA

CCCCCAAAAAUAAUUCGGCUCCGGAAUAUUCCACCUCUCCGGAUAGGUAUA

CCCCCAAAAAUAAUUCGGCCUCGGAAUAUUCCACCUCUCCGGAUAAUAGGU

CCCCCAAAAAUAAUUCGGCCCCGGAAUAUUCCACCUCUCCGGAUAAUAAUA

CCCCCAAAAAUAAUUCGGCCACGGAAUAUUCCACCUCUCCGGAUAAUAAAA

CCCCCAAAAAUAAUUCGGCACCGGAAUAUUCCACCUCUCCGGAUAAAAAUA

CCCCCAAAAAUAAUUCGGACCCGGAAUAUUCCACCUCUCCGGAAAAUAAUA

CCCCCAAAAAUAAUUCGGAUCCGGAAUAUUCCACCUCUCCGGAAAGGUAUA

CCCCCAAAAAUAAUUCGGCAUCGGAAUAUUCCACCUCUCCGGAUAAAAGGU

CACAAAACCAUUCGGAUUAUCCAUGCGCACAUUUACCGGUUACCUCUACGG

AUAAUUCCACCACGGUAAUUUUGAUUUUACUUUCCUAUC

Place each of the following statements on a 3″ x 5″ card.

PLACE *P* BELOW *CCC*

PLACE *R* BELOW *CUA*

PLACE *D* BELOW *UAA*

PLACE *O* BELOW *CCU*

PLACE *B* BELOW *AAU*

PLACE *F* BELOW *UUA*

PLACE *S* BELOW *UAC*

PLACE *I* BELOW *UUU*

PLACE *L* BELOW *CCA*

PLACE *M* BELOW *CAC*

PLACE *V* BELOW *UGA*

PLACE *Z* BELOW *CAU*

PLACE *Y* BELOW *GCG*

PLACE *C* BELOW *AUA*

PLACE *A* BELOW *AAA*

PLACE *K* BELOW *ACC*

PLACE *G* BELOW *GAC*

PLACE *T* BELOW *AAC*

PLACE *W* BELOW *CUC*

PLACE *N* BELOW *AUC*

PLACE *E* BELOW *AUU*

PLACE *U* BELOW *GGU*

PLACE – BELOW *CGG*

If there are more than 22 students in a class make extra copies of some of the above statement so that the total number of cards equals the number of students. If there are less than 22 students in a class, give some students more than one card.

Chapter 12 Growth

Growth of yeast stock. Inoculate 25-ml Erlenmeyer flask which has been autoclaved with yeast medium (see Appendix 2) with 1 ml of *Schizosaccharomyces* stock culture. Grow at 37° C or room temperature on a shaker, if possible, or shaking the flask at every opportunity, until a heavy growth is achieved (2 to 3 days). Use only sterile technique when handling the yeast, since bacterial contamination will distort the results.

Yeast cultures. Using sterile technique, inoculate 30 25-ml Erlenmeyer flasks autoclaved with yeast medium with enough of the stock culture grown above to give about 3×10^5 cells/ml. Check with the student counter. Plug with cotton or loosely fitting cap. Incubate on a shaker, if possible; if not, shake the flasks as often as possible. One set of Erlenmeyers may be used for the entire day's laboratories. Place five of the flasks directly in the refrigerator or freezer for 0 time counts. Incubate the remaining tubes, removing five flasks at 2 hours' incubation and refrigerating or freezing them; five flasks at 4 hours'; five at 6 hours'; five at 8 hours' and the remaining five at 12 hours'. The five flasks of different times may be defrosted in time for the first laboratory of the day and used for that entire day.

Pea seedlings. For each of 10 days before the laboratory day, soak pea seeds overnight, drain, and place in Petri plates lined with wet filter paper. The paper must be kept wet throughout the growth period. Remove moldy seedlings and/or replace filter paper.

Chapter 13 Differentiation

Prepare necessary solutions (see Appendix 2).

Methylene blue standards. Methylene blue solutions made with water have a distinct blue tint, whereas methylene blue solutions in bacterial suspensions take on a greenish color. Because comparison between a blue color and the green coloration of the student tubes may be difficult, it is suggested that the standards be made up in a 2×10^8 *E. coli* suspension with $Fe(CN)_3$ added to make a final concentration of 0.05M.

Fern prothallia. Fern spores may be grown either in liquid cultures or on agar plates. The general time pattern is as follows:

4-7 days: germination
8-10 days: filamentous growth stage
12-16 days: prothallial stage

Four sets of fern cultures must be prepared for each laboratory day: I. grown 10 days in red light; II. grown 10 days in blue light; III. grown 6 days in white light; 6 days in red light; IV. grown 6 days in white light, 6 days in blue light.

Stock culture of E. coli: Inoculate a 500-ml Ehrlenmeyer flask that has been autoclaved with defined bacterial medium for limited growth (see Appendix 2) with a loopful of *E. coli* from a stock slant, using sterile technique. Incubate at 37° for 2 to 3 days (3 days if room temperature is used). Growth will reach a plateau at about 2×10^8 cells/ml. Make sure that the *E. coli* used is adaptive for lactose, since some strains always produce β-galactosidase.

Culture for fern spores.

1. *Liquid culture.* Using *Chlorella* medium or soil extract medium at pH 6.0 to 6.5, inoculate with sufficient numbers of fern spores for each student to have three. 25-ml Ehrlenmeyer flasks may be used and autoclaved with media. Inoculate two flasks 12 days before class, and onc 10 days before class.
2. *Agar culture.* Using *Chlorella* medium or soil extract medium adjusted to pH 6.0 to 6.5 with 15g agar per liter added, prepare petri plates. Inoculate with enough spores for each student to have three. Inoculations should be made 12 and 10 days before class.

Chapter 14 Sources of Variation

Drosophila crosses. Five weeks before the laboratory collect virgin ♀♀ and wild type ♂ bw y cy♀. Place 3 females and 2 males in fresh vials about 4 weeks before the laboratory. Remove the parents after one week. Two weeks before the laboratory, collect virgin F_1 females, and bw y cy males. Place 3 females and 2 males in fresh vials. Remove the parents after one week. Keep all flies at room temperature, or in an incubator at 35°C.

Five weeks before the laboratory, collect y w virgin females, and wild type males. Four weeks before the laboratory, place 3 females and 2 males in each vial. Remove the parents after one week. Two weeks before the laboratory collect F_1 virgin females, and y w males. Place 3 females and 2 males into fresh vials. Remove the parents after one week.

Chapter 15 Adaptation

Selection of E. coli for streptomycin resistance. Stocks of *E. coli* are prepared by adding inoculum to Erlenmeyer flasks containing one-fifth their capacity of nutrient broth.

Culture for 24 hr, preferably at 37° C, after which time cell numbers will be about 6 × 10^8 cells/ml.

Prepare 1½ liters nutrient broth containing 30 mg streptomycin to give the antibiotic a final concentration of 20 μg/ml. Pour 100 ml medium into 14 125-ml Erlenmeyer flasks, plug, and autoclave. After cooling, add bacteria from 24-hr stock cultures as follows:

cells transferred	aliquot of stock
10^{10}	16.6 ml of stock
10^9	1.7 ml of stock
10^8	0.17 ml of stock
10^7	10 ml of above flask
10^6	10 ml of above flask
10^5	10 ml of above flask
10^4	10 ml of above flask

Culture flasks for at least 2 days. To inoculate for F series, follow above procedure using the 10^{10} flask as stock and permit growth to continue again for at least 2 days. Transfer about 10 ml aliquots to sterile test tubes and have a series of P and F in one wire test tube rack. Up to ten sets can be made and left in the laboratory for a week without any deleterious effects.

Chapter 16 The Origin of Life

Prepare necessary solutions.

Chapter 17 Ecology

Winogradsky column. Instructions in body of chapter. Black mud must be used. The quality of mud may be ascertained by its H_2S odor.

Chapter 18 Monera and Viruses

Preparation of E. coli culture. See p. 496

Plating of *E. coli* and *B. megatherium.* Prepare nutrient agar plates and streak from slant cultures.

Preparation of phage suspension. See instructions in Turtox package of T-4 phage and *E. coli.*

Preparation of plating bacteria. Same as above.

Preparation of B. subtilis cultures.

1. Old culture: Inoculate a 25-ml flask autoclaved with nutrient broth with a loopful of *B. subtilis* culture using the sterile technique. Incubate for 1 week or more at room temperature to obtain heavy growth. One flask should be available for each

laboratory, but because active growth is not an important factor in spore formation, all flasks may be inoculated at the same time and kept until needed.

2. Fresh cultures: Inoculate a 25-ml flask autoclaved with nutrient broth with a loopful of *B. subtilis* culture using the sterile technique. Incubate for 1 day at room temperature or 5 hr in a 37°C incubator. One flask should be available for each laboratory.

Chapter 19 Protista

Mating types in Rhizopus. One series of plates may be used for the laboratories for the entire week. Prepare 20 petri plates with potato dextrose agar. Arbitrarily designate plates 2 to 4 as + or – strains. Inoculate the plates by placing small clumps of agar from the stock plate on each side of the potato dextrose plate. Mark the plates with the mating numbers (i.e., 1 X 2, 1 X 3, etc.) above the inocula and incubate at room temperature for 1 week.

Oedogonium. Prepared slides or living material of a sexual strain may be used. (The address for live material may be found in Appendix 2.)

Preparation of Chlamydomonas for mating reaction.

1. Grow heterothallic strains separately on soil extract agar plates for 1 week. Keep in bright light.
2. Wash cells from agar plates with distilled water and place in separate flasks.
3. Place in bright light for several hours.
4. Place in dark until 2 hr before class, leaving no more than 2 days.
5. Two hours before class, place in light.
6. When drops are added together, clumping will occur within a few minutes.

Flasks of + and – strains will be needed for each class.

Preparation of Neurospora life cycle stages.

1. Fourteen days before the laboratory, inoculate plates with both mating types.
2. Ten days before the laboratory, inoculate plates with both mating types.
3. Five days before the laboratory, inoculate plates with a single mating type (eight + or –).

Chapter 20 Land Invasion by Plants

Fern prothallia preparation. Using media outlined on p. 496, Appendix 3, inoculate media 14 days and 10 days before laboratory. Place culture flasks or plates in white light for the period of time.

Moss prothallia preparation.

Moss Cultivation. Dust moss spores into petri plates containing 25 ml of Bristol's solution and expose to light. Moss spores will germinate within 3 to 7 days and well-developed protonema (and "buds" if gibberellic acid at 10^{-4} M solution has been used) will form with 2 to 3 weeks.

Bristol's solution –
Prepare the following stock solution:

$NaNO_3$	10 g in 400 ml H_2O
$CaCl_2$	1.0 g in 400 ml H_2O
$MgSO_4 \cdot 7H_2O$	3.0 g in 400 ml H_2O
K_2HPO_4	3.0 g in 400 ml H_2O
KH_2PO_4	7.0 g in 400 ml H_2O
NaCl	1.0 g in 400 ml H_2O

Add 10 ml of each stock to 940 ml H_2O, add 1 drop of 1% $FeCl_3$ and 1 ml of trace element solution (not really necessary).

Chapter 21 The Seed Plants

Bean seedlings. Germinate and grow bean seedlings in small pots.

Chapter 22 Physiological Adaptations in Plants

Radish seedlings. Germinate radish seedlings as described for wheat or barley seeds, p. 489. Use five-day-old seedlings.

Prepare necessary solutions.

Chapter 23 Trends in Animal Evolution

No special preparation.

Chapter 24 Animal Evolution—The Deuterostomate Line

No special preparation.

Chapter 25 Energy-Obtaining Systems

Prepare necessary reagents.

Chapter 26 Energy-Utilization Systems—Maintenance of Self

Demonstrations. These are left to the discretion of the instructor.

Ovulation. Prepare an ovulating female frog for the first laboratory by injecting with pituitary extract the day before.

Chapter 27 Energy-Utilization Systems—Reproduction, The Maintenance of Species

Drosophila. Larval stages may be seen in vials 7 to 10 days old, and pupal stages in vials 10 to 13 days old. Therefore, only two sets of vials need be made in order to covertthe laboratory week. Ten days before class day, mate two ♀♀ (nonvirgins may be used) and two ♂♂ in vials or baby food jars, removing parents after 5 days. Repeat this procedure 10 days and 7 days before the middle-of-the-week classes.

PREPARATION SCHEDULE

The following schedule is designed as a general guide for laboratory preparations. It does not take into consideration any vacations, omitted laboratory exercises, or exercises that may extend beyond a week's time. It does, however, provide a general pattern for preparations that must be done at varying times prior to the actual execution of the exercise by the students.

WEEK	EXERCISE	EXERCISE TO BE PREPARED	PREPARATION
1	1	1	No special preparation.
		20	Collect sporophyte of moss *Polytrichum.*
2	2	2	No special preparation.
		7	Inoculate nitrite formation media.
3	3	3	Prepare necessary reagents.
		7	Inoculate nitrite formation media.
4	4	4	Prepare necessary reagents.
		5	Prepare wheat or barley seedlings.
		7	Inoculate nitrite formation media.
		6	Start yeast and *Chlorella* stocks.
5	5	5	Prepare wheat or barley seedlings.
		6	Innoculate *Chlorella* and yeast flasks and grow in light and dark.
		7	Innoculate nitrite formation media.
		10	Collect virgin femals for *Drosophilia* cross.
6	6	6	No immediate preparation.
		7	Inoculate peptone medium.
			Inoculate nitrite formation media.

WEEK	EXERCISE	EXERCISE TO BE PREPARED	PREPARATION
		10	Mate flies for crosses.
7	7	7	Prepare necessary reagents.
		10	Remove parents from *Drosophila.* crosses.
8	8	8	No special preparation.
		9	Prepare radish seedlings.
		10	Collect F_1 virgin females and make back cross.
9	9	9	No immediate preparation.
		10	Remove parents from back cross.
		11	Prepare radish seedlings.
		14	Collect virgin females for *Drosophila* crosses.
10	10	10	No immediate preparation.
		11	Prepare radish seedling plates.
		14	Make *Drosophila* crosses.
11	11	11	No immediate preparation.
		13	Prepare for fern prothallia.
		14	Remove parents of crosses.
12	12	12	Prepare necessary materials.
		13	Prepare fern prothallia.
		14	Collect virgin females and make appropriate crosses.
13	13	13	Prepare necessary materials.
		14	Remove parents from *Drosophila* crosses.
14	14	14	No immediate preparation.
		15	Prepare Streptomycin resistance flasks.
		16	Prepare Winogradsky column.
15	15	15	Prepare necessary solutions.
16	16	16	Prepare necessary solutions.
17	17	17	Prepare necessary solutionsl
		18	Prepare old *B. subtilis* culture.
		19	Prepare *Rhizopus* mating.
		19	Prepare *Neurospora* life cycle.
18	18	18	Prepare necessary materials.
		19	Prepare *Neurospora* life cycle.
		20	Prepare fern prothallia.
19	19	19	Prepare necessary materials.
		20	Prepare fern prothallia.

All other exercises require typical preparatory work, 3 to 4 days before the laboratory.

A ONE-SEMESTER PROGRAM

This laboratory manual may be used for a one-semester, experimentally oriented program to prepare students for individual or group research in the second semester. For such a program, the following exercises are recommended for use during the first semester:

Chapter 1 Measurement, Data Organization, and Analysis
Chapter 2 Levels of Biological Organization
Chapter 3 Chemical Background for Biology
Chapter 4 Protein Chemistry and Enzymes
Chapter 5 Intermediary Metabolism
Chapter 6 Photosynthesis
Chapter 7 Nutrition
Chapter 8 Cellular Organization
Section on organelles: chloroplasts and mitochondria
Chapter 9 Reproduction of Cellular Components
Chapter 10 The Theory of Particulate Heredity
Chapter 11 Genetic Regulation of the Cell
Chapter 12 Growth
Section on growth curve of yeast
Chapter 13 Differentiation
Section on induction of differentiation
Chapter 14 Sources of Variation
Section on gene mutation
Chapter 15 Adaptation
Section on selection
Chapter 16 The Origin of Life
Chapter 17 Ecology
Section on succession
Chapter 18 Monera and Viruses
Section on mass effect (may be done in conjunction with growth curve of Chapter 12).
Section on gram stain
Section on viruses
Chapter 19 Protista
Section on mating type concept
Section on isogamous reproduction
Chapter 22 Physiological Adaptations in Plants
Section on plant hormones

The above suggested exercises should give the student a broad experience in experimental approaches, and should prepare him for any independent research project on respiration, enzyme studies, or hormone studies.